Green and Smart Technologies for Smart Cities

Green and Smart Technologies for Smart Cities

Edited by
Pradeep Tomar and Gurjit Kaur

CRC Press is an imprint of the
Taylor & Francis Group, an **informa** business

CRC Press
Taylor & Francis Group
52 Vanderbilt Avenue,
New York, NY 10017

© 2020 by Taylor & Francis Group, LLC
CRC Press is an imprint of Taylor & Francis Group, an Informa business

No claim to original U.S. Government works

Printed on acid-free paper

International Standard Book Number-13: 978-1-138-31809-0 (Hardback)

This book contains information obtained from authentic and highly regarded sources. Reasonable efforts have been made to publish reliable data and information, but the author and publisher cannot assume responsibility for the validity of all materials or the consequences of their use. The authors and publishers have attempted to trace the copyright holders of all material reproduced in this publication and apologize to copyright holders if permission to publish in this form has not been obtained. If any copyright material has not been acknowledged please write and let us know so we may rectify in any future reprint.

Except as permitted under U.S. Copyright Law, no part of this book may be reprinted, reproduced, transmitted, or utilized in any form by any electronic, mechanical, or other means, now known or hereafter invented, including photocopying, microfilming, and recording, or in any information storage or retrieval system, without written permission from the publishers.

For permission to photocopy or use material electronically from this work, please access www.copyright.com (http://www.copyright.com/) or contact the Copyright Clearance Center, Inc. (CCC), 222 Rosewood Drive, Danvers, MA 01923, 978-750-8400. CCC is a not-for-profit organization that provides licenses and registration for a variety of users. For organizations that have been granted a photocopy license by the CCC, a separate system of payment has been arranged.

Trademark Notice: Product or corporate names may be trademarks or registered trademarks, and are used only for identification and explanation without intent to infringe.

Visit the Taylor & Francis Web site at
http://www.taylorandfrancis.com

and the CRC Press Web site at
http://www.crcpress.com

Contents

Preface vii
Editors xi
Contributors xv
Abbreviations xvii

1. **Green Smart Cities** 1
   Akanksha Srivastava, Mani Shekhar Gupta and Gurjit Kaur

2. **Green Smart Town Planning** 19
   Yaman Parasher, Prabhjot Singh and Gurjit Kaur

3. **Green Smart Buildings for Smart Cities** 43
   Dushyant Singh Chauhan and Gurjit Kaur

4. **Green Smart Environment for Smart Cities** 75
   Yaman Parasher, Gurjit Kaur and Pradeep Tomar

5. **Green Healthcare for Smart Cities** 91
   Prabhjot Singh, Varun Dixit and Jaspreet Kaur

6. **Green Smart Education System** 131
   Aditya Pratap Singh and Pradeep Tomar

7. **Green Smart Agriculture System** 147
   Garima Singh and Gurjit Kaur

8. **Green Smart Security System** 165
   Yaman Parasher, Prabhjot Singh and Gurjit Kaur

9. **Green Smart Transport Systems** 185
   Arsh Javed Rehman, Shweta Yadav and Pradeep Tomar

10. **Green Smart Energy Management System** 203
    Garima Singh and Gurjit Kaur

11. **Green Smart Waste Management System** 223
    Dimpal Tomar, Pradeep Tomar and Gurjit Kaur

12 Green Smart Water and Sanitation System 239
    Priya Singh, Garima Singh and Gurjit Kaur

13 Innovation Opportunities through Internet of Things (IoT)
    for Smart Cities 261
    Rajalakshmi Krishnamurthi, Anand Nayyar and Arun Solanki

14 Application of a Smart City Concept to the Leading
    Destination: Evidence from Istanbul – A Case Study 293
    Aysegul Acar, Eda Kocabas, M. Fevzi Esen and Fatih Canitez

15 Green Roof Garden Concept for Smart Cities – A Case Study 321
    Carlos Alberto Ochoa and Aida Yarira Reyes Escalante

16 School Bus Routing Problems and Solutions for Smart
    Cities – A Case Study 343
    Carlos Alberto Ochoa and Aida Yarira Reyes Escalante

Index 359

Preface

With the recent surge in population rate and global trend in urbanization around the world, there has been massive pressure for the development of smart and sustainable cities that can ensure a prosperous and healthy environment for the residents. Therefore, the concept of green smart cities is a promising solution which is useful in dealing with the hazardous effects of the prevailing conventional systems.

This could be clearly seen from the fact that modern cities are currently responsible for about 70% of greenhouse gas emissions and 60% of the energy that is consumed worldwide because of the variety of energy inefficient buildings and transportation systems that prevail inside a city network. As a result, carbon dioxide emission was reported to be around 37 gigatonnes in 2018, which was approximately over 50% of what it was in 1990. All these hazardous effects primarily accounted for 40% of the energy usage of a building, which generally results in the emission of 36% CO_2 and 40% particulate matter.

It is due to the same reason that every year about 12.6 million people die (which is also roughly equal to one-fourth of the total deaths in the world) because of the pollution caused by the harmful gases and particulate matter that are generated from the modern day city. Apart from this, other factors such as water and soil pollution, climate change and ultraviolet radiation also contribute to the increase in illnesses and other injuries to many people around the world.

With the current scenario, modern smart cities of the present world need to enact sustainable strategies and action plans to counteract enormous environmental and social strains at different scales related to rampant pollution and energy consumption. Also, the growing carbon footprint from burning fossil fuels and other vital resources in industries and transportation also play a considerable role in affecting the health of millions of city residents. It is, therefore, essential to effectively include comprehensive carbon-focused planning practices that emphasize the concept of energy conservation, environment and climate change impact assessment, and so on, for effective resilience planning.

The green smart city concept in general addresses these issues and promotes green and socially inclusive urban development which usually takes into consideration the systematic and smart integration of modern day smart information and communication technologies (ICTs) infrastructure, with smart cities planning architecture, transportation systems, security, healthcare, agriculture, education, environment, waste and energy-related management systems. Therefore, the prime goal of this book is to deliver a comprehensive and consolidated overview of the green technologies for future smart cities that will improve the quality of life their residents by offering a lasting opportunity for social, cultural and economic growth within a dynamic, safe, stimulating and healthy environment.

Researchers, scientists, and academicians from around the world contributed their chapters and gave realistic solutions to reduce greenhouse gas emissions and carbon footprint levels. In general, collectively the book explores and highlights the technologies related to green smart cities: planning, grids, transportation systems, buildings, education systems, healthcare systems, waste recycling and disposal methods and energy efficient measures which can reduce carbon emissions while designing each part of a complete green smart city architecture. Through this book, readers will also learn how IoT solutions and green technologies can disrupt business models and bring new approaches to the urban environment. They will understand how the book can meet the needs of the tech-savvy customer and prepare for the changing face of society. With the same vision in mind, most of the countries around the world have already started implementing some of these self-sustainable solutions into the mainstream development of their respective cities. This can only bring fruitful results if all these technologies and solutions are adapted in a comprehensive or holistic way, rather than addressing each one of them individually.

Green and Smart Technologies for Smart Cities is therefore designed to provide a general overview of green science and technologies in different application services, that play an essential role in ensuring the energy efficient and sustainable development of smart cities of the future. With extensive analysis of the smart city landscape and in-depth focus on the drivers, hurdles and opportunities through green smart technologies, this book will lead the way with its innovative and interactive agenda. It will also act as a broad reference book for the identification of sustainable plans and programs at the urban level for green smart cities of the future. This book differs in a fundamental way from the other standard books because it also includes the energy efficient green technologies to design smart cities and describe green technologies or solutions which will impact smart cities of the future.

Therefore, with the concept of smart green technologies in mind for future smart cities, the book is organized into three major sections which are outlined below.

Chapters 1–3 provide the essential background on the understanding of smart cities, in general, along with the planning procedures for smart green towns. Chapter 1, "Green Smart Cities," is an introduction to urban centers that represents an integrated, habitable and sustainable ecosystem for the residents of a particular region. It recognizes the interconnected ICT technologies that serve a variety of interrelated services required for the development of a city. Chapter 2, "Green Smart Town Planning," defines the green smart solutions that need to be incorporated into urban development planning procedures and explains the vitality behind the segregation of useful sectional areas distributed in the whole city region. In addition to this, the chapter also provides a comprehensive analysis of the green urban space landscape, highlighting the relevant, vital techniques, approaches, challenges and benefits of the whole ecosystem and its social and cultural services. Chapter 3, "Green Smart Buildings for Smart Cities," gives a brief overview of the passive design technologies that are adapted as an alternative to the existing conventional design methodologies required for building structures. These topics are covered in recognition with building standards, codes and certifications that are based on international standards for green buildings.

Chapters 4–9 deal with the systems related to the green technologies-based ecosystem that forms the essential part of modern-day smart cities. Chapter 4, "Green Smart Environment for Smart Cities," gives an introduction to the green technologies associated with each of the five fundamental spheres of the environment. Chapter 5, "Green Healthcare for Smart Cities," discusses the concept of green building designs of hospitals, green hospital energy management systems, and green medical devices, along with detailed description of green smart digital healthcare systems in modern-day smart cities. Chapter 6, "Green Smart Education System," covers the essential role of more efficient and resourceful learning environments and the development of existing technology-enhanced learning approaches by incorporating new green-based technologies and unique criteria for learning. Chapter 7, "Green Smart Agriculture System," provides a comprehensive study that encompasses the features and technologies required, together with the benefits and challenges for making green smart agriculture feasible. Here, starting with the introduction, it lists various essential features and design methodologies that define the modern-day green technology-based agricultural solution.

A comprehensive review related to green technologies-based security solutions is discussed in Chapter 8, "Green Smart Security System." The growing pollution and carbon footprint from conventional transit systems is ripe for improvement. Chapter 9, "Green Smart Transport System," provides an effective solution for green smart transportation and also describes various projects and initiatives for green and integrated transportation.

The next three Chapters, 10–12, cover the management aspects related to the conservation of energy, waste and water. These three are the main vital factors needed to be managed properly in a sustainable way, for the survival of humans on earth. Chapter 10, "Green Smart Energy Management System," introduces the concept of smart grid architectures that ensure proper flow and distribution throughout the entire smart city network. The features and technologies described in this chapter also provide sustainable measures through which energy wastage can be compensated. The essential features and technologies related to green smart waste management like smart bins and integrated IoT-based waste management system are outlined in Chapter 11, "Green Smart Waste Management System." Chapter 12, "Green Smart Water and Sanitation System," elaborates essential features like smart toilets, smart waste transportation, smart water loss management, smart collection, purification, and so on, related to a green smart water and sanitation system. This chapter also discusses the technologies necessary for all these essential features. Some of the technologies mentioned in the chapter include smart sensors and pipe networks, an intelligent sensing unit for water quality and flow monitoring, and so on. In Chapter 13, "Innovation Opportunities through Internet of Things (IoT) for Smart Cities," the first part is dedicated to the study of technological and service aspects of the smart city while the second part focuses on the principle components for enabling a smart city through IoT.

Chapters 14–16 are a discussion of the case studies related to smart tourism, integrated green roof, and school bus routing problems prevailing in modern day smart cities. Chapter 14, "Application of a Smart City Concept to the Leading Destination: Evidence from Istanbul – A Case Study," discusses detailed research on how Istanbul manages and applies sustainable smart city concepts as a leading tourism destination attracting nearly 11 million international visitors a year. Chapter 15, "Green Roof Garden Concept for Smart Cities," discusses various roof garden concepts and propositions to design roof gardens using Cultural Algorithm (CA) for Mexico. Chapter 16, "School Bus Routing Problems and Solutions for Smart Cities – A Case Study," begins with a discussion on the issues related to the school bus routing problem. Here a school bus algorithm was proposed for Mexico using ubiquitous computing. A new Particle Swarm Optimization (PSO) algorithm was implemented to solve the problem of routing school buses and define the best routes.

The chapters are all contributed by experienced specialists from around the world, providing a unique compilation of topics relating to the role, significance and impact of green smart technologies on future smart cities. It is believed that the reader of the book will surely benefit from the comprehensive and consolidated methods and technologies introduced in each chapter regarding green smart technology. This book will therefore expect to target a vast, broad spectrum of readers at different hierarchical levels, ranging from undergraduate/graduate students to researchers, and so on. We firmly believe the book will provide readers with valuable insights and understanding that will eventually help to incorporate all the described systems into practical use.

Dr. Pradeep Tomar
Dr. Gurjit Kaur

Editors

Dr. Pradeep Tomar is Assistant Professor in the School of Information and Communications Technology, Gautam Buddha University, India (since 2009). Dr. Tomar earned his PhD from Maharishi Dayanand University (MDU), India. Before joining Gautam Buddha University, he worked as a Software Engineer in a multinational company, Noida, and was a lecturer at MDU and Kurukshetra University, Kurukshetra, Haryana. Dr. Tomar has excellent teaching, research and software development experience, as well as vast administrative experience at university level in various coordinator posts, that is, examination coordinator, university admission coordinator and timetable coordinator, Proctor and hostel warden. He is also a member of the Computer Society of India (CSI), the Indian Society for Technical Education (ISTE), the Indian Science Congress Association (ISCA), the International Association of Computer Science and Information Technology (IACSIT) and the International Association of Engineers (IAENG). He qualified for the National Eligibility Test (NET) for Lecturership in Computer Applications in 2003, as a Microsoft Certified Professional (MCP) in 2008, a SUN Certified Java Programmer (SCJP) for the Java platform, standard edition 5.0 in 2008 and qualified for the IBM Certified Database Associate – DB2 9 Fundamentals in 2010.

Dr. Tomar was awarded the Bharat Jyoti Award by the India International Friendship Society in the field of technology in 2012 and the Bharat Vikas Award by the Institute of Self Reliance, in 2017. He has been awarded the Best Computer Faculty Award by the Government of Pondicherry and the ASDF society. His biography is published in *Who's Who Reference Asia, Volume II*. Dr. Tomar has been awarded the distinguished Research Award from the Institute for Global Business Research for his work in "A Web-Based Stock Selection Decision Support System for Investment Portfolio Management" in 2018. He delivered expert talks at FDP (faculty development program) workshops, national and international conferences. He has organized three conferences: one national conference with the COMMUNE group and two international conferences, in which one international ICIAICT 2012 was organized by CSI (Computer Society of India), Noida Chapter, and the second international conference 2012 EPPICTM was organized in collaboration with MTMI, USA, University of Maryland Eastern Shore, USA and Frostburg

State University, USA, at the School of Information and Communications Technology, Gautam Buddha University, India.

Apart from teaching, he runs a programming club for ICT students and he guides various research scholars in the areas of software engineering, reusability of codes, soft computing techniques, Big Data and IoT (Internet of things). His major current research interest is in Component-based Software Engineering. He is working as co-investigator in a sponsored research project in high throughput design, synthesis and validation of TALENs for targeted genome engineering, funded by the Department of Biotechnology, Ministry of Science and Technology Government of India. Dr. Tomar has edited two books at national level: *Teaching of Mathematics* and *Communication and Information Technology*, and has authored and edited two at international level: *Examining Cloud Computing Technologies through the Internet of Things (IoT)* and *Handbook of Research on Big Data and the IoT*. He has also contributed to more than 100 papers/articles in national/international journals and conferences. He served as a member of the editorial board and reviewer for various journals and national/international conferences.

Dr. Gurjit Kaur is an Associate Professor in the Department of Electronics & Communication Engineering at Delhi Technological University (DTU), Delhi, India. She has been a topper throughout her academic education. As a testimonial, she was awarded by the chief minister S. Prakash Singh Badal for being the topper of the Punjab state. After that, she was awarded a Gold Medal by former President of India Dr. A. P. J. Abdul Kalam for being the overall topper of Punjab Technical University (PTU), Jalandhar, in her B. Tech program. She also received an Honour by the Guru Harkrishan Education Society for being topper among all the colleges and all the disciplines of PTU. She then proceeded to PEC University of Technology, Chandigarh, to complete her M. Tech in 2003 and also earned her PhD from Panjab University, Chandigarh, in 2010, with distinction.

She has spent over 16 years of her academic career in research and teaching in the field of Electronics and Communication in well-reputed institutes such as Punjab Engineering College, Punjab University, Jaypee Institute of Information and Technology, Gautam Buddha University, and Delhi Technological University, Delhi. During this academic tenure, she has guided four PhD students and more than 50 M. Tech students. Her research interests include Optical CDMA, Wireless Communication systems, high-speed interconnect and IoT. She has also authored three books at the international and national level. Her two books, *Handbook of Research on Big Data and the IoT* and *Examining Cloud Computing Technologies through the Internet of Things*, were published by IGI Global, International Publisher of Progressive Information Science and Technology Research in 2018 and 2017, respectively. She also authored a national level book entitled *Optical Communication*, which was published by Galgotia Publications in 2005. She has presented her research work as short courses/tutorials in many national and international conferences.

In recognition of her outstanding contribution as an active researcher, Dr. Kaur also received the Bharat Vikas Award by the Institute of Self Reliance at the National Seminar on Diversity of Cultural and Social Environment held in Bhubaneswar, Odisha, India, and the Prof. Indira Parikh 50 Women in Education Leaders Award at the World Education Congress, Mumbai, India, in 2017. Her career accomplishments over the past 16 years include over 90 peer-reviewed publications, including 55 journal publications, 35 conference publications, and over 10 invited talks/seminars. Due to her pioneering research, her name is also listed in *Who's Who in Science and Engineering Directory*, USA, and her biography was also published by International Biographic Centre as 2000 Outstanding Scientists.

Dr. Kaur also worked as a convener for two international conferences: International ICIAICT 2012, which was organized by CSI, Noida Chapter and International conference EPPICTM 2012, which was held in collaboration with MTMI, USA, University of Maryland Eastern Shore, USA and Frostburg State University, USA, at the School of Information and Communications Technology, Gautam Buddha University, India. To date, she has delivered many expert talks on the emerging trends in optical communication for all domain applications at various universities and served as a reviewer of multiple journals.

Contributors

Aysegul Acar
Karabuk University, Turkey
Safranbolu/Karabuk

Fatih Canitez
Istanbul Technical University
Management Engineering
Turkey

Dushyant Singh Chauhan
Department of Electronics and
 Communication Engineering
Delhi Technological University
Delhi, India

Varun Dixit
Research and Development
VMWare Inc.
Palo Alto, California, US

Aida Yarira Reyes Escalante
Department of Administrative
 Sciences
Institute of Social Sciences and
 Administration
Juarez City University
Juarez City, Mexico

M. Fevzi Esen
Istanbul Medeniyet University
Istanbul, Turkey

Mani Shekhar Gupta
Department of Electronics and
 Communication Engineering
National Institute of Technology
Himachal Pradesh, India

Gurjit Kaur
Department of Electronics and
 Communication Engineering
Delhi Technological University
Delhi, India

Jaspreet Kaur
Technical Research
Golden Sparrow LLC
Union City, California, US

Eda Kocabas
Istanbul Medeniyet University
Istanbul, Turkey

Rajalakshmi Krishnamurthi
Department of Computer Science
 and Engineering
Jaypee Institute of Information
 Technology
Uttar Pradesh, India

Anand Nayyar
Graduate School
Duy Tan University
Da Nang, Vietnam

Carlos Alberto Ochoa
Universidad Autónoma de Ciudad
 Juárez, México
Ciudad Juárez, Mexico

Yaman Parasher
Institute for Communication,
 Information and Perception
 Technologies (TeCIP)
Pisa, Italy

Arsh Javed Rehman
Department of Computer Science
  and Engineering
Gautam Buddha University
Uttar Pradesh, India

Aditya Pratap Singh
Department of Information
  Technology
Ajay Kumar Garg Engineering
  College
Uttar Pradesh, India

Garima Singh
Department of Electronics
  and Communication
  Engineering
Delhi Technological University
Delhi, India

Prabhjot Singh
Research and Development,
  Salesforce.com Inc.
San Francisco, California, US

Priya Singh
Indira Gandhi Delhi Technological
  University for Women
Delhi, India

Arun Solanki
School of Information and
  Communications Technology
Gautam Buddha University
Uttar Pradesh, India

Akanksha Srivastava
Department of Electronics and
  Communication Engineering
Delhi Technological University
Delhi, India

Dimpal Tomar
Department of Computer Science
  and Engineering
Gautam Buddha University
Uttar Pradesh, India

Pradeep Tomar
Department of Computer Science
  and Engineering
Gautam Buddha University
Uttar Pradesh, India

Shweta Yadav
Department of Computer Science
  and Engineering
Gautam Buddha University
Uttar Pradesh, India

Abbreviations

AFC	automatic fare collection
AI	artificial intelligence
AMI	advanced metering infrastructure
AMPS	advanced mobile phone system
ATES	automatic traffic enforcement system
BDMA	beam division multiple access
BECs	building envelope components
BIS	bus information system
BMS	building management system
CA	cultural algorithms
C-AHS	cooperative automated driving highway system
CO_2	carbon dioxide
CRRC	cool roof rating council
DER	distributed energy resources
DGPS	differential global positioning system
DOE	design of experiments
DSM	demand-side management
DSS	decision support systems
EEB	energy-efficient building
ECLA	the economic commission for latin america
EHR	electronic health record
FMIS	farm management information system
FTMS	freeway traffic management system
GCI	green city index
GDP	gross domestic product
GIS	geographic information systems
GSHVAC	green smart heating, ventilation and air conditioning
GSM	global system for mobile
GSSS	green smart sanitation system
GSWM	green smart water management
GWP	global warming product
HANs	home area networks
HIS	hospital information system
ICT	information and communications technology
IDSS	intelligent decision support system
IEQ	indoor environmental quality
IoT	Internet of things
IPM	integrated pest management
ISO	international organization for standardization
ITS	intelligent transportation systems

LCA	life cycle assessment
LED	light-emitting diode
LTE	long-term evolution
MEG	magnetoencephalogram
MIE	medical imaging equipment
MMR	multimedia medical record
MRI	magnetic resonance imaging
NLP	natural language processing
PAN	personal area network
PET	positron emission tomography
PVC	polyvinyl chloride
SBRP	school bus routing problem
SNG	synthetic gas generation
SPECT	single-photon emission computed tomography
SRI	solar reflectance index
UAV	unmanned aerial vehicles
UMTS	universal mobile telecommunication system
VR	virtual reality
VRP	vehicle routing problem
WAN	wide area network
WMS	water management system
ZEBs	zero-energy buildings

1
Green Smart Cities

Akanksha Srivastava, Mani Shekhar Gupta and Gurjit Kaur

CONTENTS

1.1 Introduction 1
1.2 Meaning of Smart Cities 4
    1.2.1 Secure and Safe Society 5
    1.2.2 Mobility in Green Communication 5
    1.2.3 Attain Excellent Education and Health 6
    1.2.4 Real Estate and Building 7
    1.2.5 Technology and Transportation Amplification 7
1.3 A Roadmap of Using Communication for Smart City Development 8
    1.3.1 Green Communication-Enabled Knowledge and Information Sharing 10
    1.3.2 Green Communication-Enabled Forecasts 10
    1.3.3 Green Infrastructure 10
    1.3.4 A Cloud Computing Platform 11
1.4 Potential Communications Technologies to Make Cities Smart 12
    1.4.1 Wireless Technologies 12
    1.4.2 Wired Technologies 13
1.5 Features of Smart Cities 14
    1.5.1 Sensor-Equipped City and Cloud Computing 14
    1.5.2 Future Research Directions 14
1.6 Conclusion 14
Acknowledgment 15
References 17

1.1 Introduction

There has been a remarkable growth in the percentage of the world's population living in urban areas. Presently, almost 55% of the world's population lives in urban areas and it is expected that the urban population will account for 66% of the world's population by 2050 (United Nations 2018). Significant advancements in society and a deficiency of resources in rural areas have led to a continous migration of the rural population to urban areas. Table 1.1 presents the projected growth rate of the world's population from 2020 to 2050, including the urban population percentage. Figure 1.1 shows the percentage of the urban population in the ten most populated countries. This expansion into urban areas requires the optimization of available resources (such as electricity, water, housing and transportation) to meet the needs of citizens. Therefore, emerging technologies such as artificial intelligence, cognitive technology, Internet of Things, machine learning and cloud computing are now being used in a significant manner to convert cities into "smart cities". Energy-efficient green communication and seamless networking are very important pillars of smart city construction, and connect the different essential elements of smart cities. This chapter aims to provide a brief definition of smart cities and to summarize the various emerging technologies which play a vital role in smart cities.

Over the past 50 years, the average growth rate of the world's population has been almost 1.2% per year (Roser et al. 2013). The urban area population has continuously surpassed the rural area population and it is expected that this will continue, with the urban population accounting for 68.5% of the total population by 2050 (Mensah et al. 2019). Therefore, the present scenario is one of greater urbanization. There are many benefits of urbanization but it also brings many challenges. Increased urbanization means increased demand for various resources such as electricity, water, transportation, sanitation and infrastructure, as well as healthcare, public services and education. The urbanization process also plays a key role in environmental degradation on a local, regional and global level.

At the present time, the conversion of digital cities into smart cities is a major issue of concern for many countries. Intelligent and digital cities are those which are technology-oriented but a city becomes "smart" when it is organized, interconnected, self-repairing, self-decision-taking and healthy. In the literature, various models and explanations have been put forward

TABLE 1.1

Projected World Population From 2020 To 2050

Year	World Population	Increase in Population (Yearly)	Change in Yearly Population (%)	Rate of Fertility	Urban Population (%)
2020	7,794,798,739	83,000,320	1.10%	2.47	56.2%
2025	8,184,437,460	77,927,744	0.98%	2.54	58.3%
2030	8,548,487,400	72,809,988	0.87%	2.62	60.4%
2035	8,887,524,213	67,807,363	0.78%	2.70	62.5%
2040	9,198,847,240	62,264,605	0.69%	2.77	64.6%
2045	9,481,803,274	56,591,207	0.61%	2.85	66.6%
2050	9,735,033,990	50,646,143	0.53%	2.95	68.6%

(Mensah et al. 2019)

Green Smart Cities

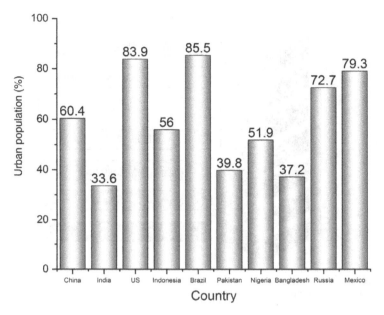

FIGURE 1.1
Urban population percentage of the most populated countries (2017) (Shahidehpour et al. 2018).

for smart city description (Wenge et al. 2014). Figure 1.2 presents a layered architecture model for a smart city. This model consists of seven layers of a smart city, in which each layer plays a very important role. A central role is performed by the third layer (i.e. the communication and networking layer). The first layer is infrastructure, which consists of roads, buildings, the electricity grid and parking.

The communication and networking layer facilitates and supports the different forms of communication such as voice, text, data, video, multimedia and real-time services from a personal area network (PAN) to a wide area network (WAN). A question that emerges is how communication and networking provide support in making a city smart? The answer is that this midpoint layer of architecture connects the upper information and core framework to the infrastructure layer, so this is responsible for the transfer of services and features over one or numerous networks. This layer enables interrelated objects to communicate more efficiently with each other. However, the proper establishment of communication is a very important factor in applying easily integrated engineering to the design of smart cities.

The smart city is one of the emergent topics of research within the past decade. In the first stage of such research, smart cities were scrutinized according to their domain, technological zone and economic state. The focus of research is now turning to how green communication is used to develop urban growth and public administration.

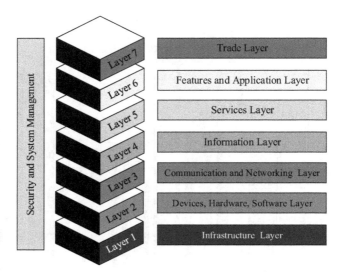

FIGURE 1.2
Layered architecture model for a smart city.

1.2 Meaning of Smart Cities

A smart city can be defined as follows:

> An urban zone that uses various technologies such as sensors or the Internet of Things to gather data and then processes these data to manage resources and assets in an optimized manner.

The technologies in smart cities gather data from people and situations (traffic, climate, crowd etc.), which are then analyzed to achieve proper maintenance of the water supply system, transportation and traffic network; waste management; maintenance of the information network, colleges, libraries, shopping malls, hospitals, power plants and other public areas; and to assist in crime prevention;

The concept of a smart city incorporates communication and various infrastructure-level devices connected to the Internet of Things (IoT) system to improve the efficiency of the services and operation of the city. Efficient communication establishes a real connection between the government and citizens, and also improves the performance, quality and activities of urban services. However, the applications of smart cities are based on the real-time feedback of citizens. Each letter of the word "SMART" contains real meaning for any city, as illustrated in Figure 1.3; the different components will be discussed in the following sections.

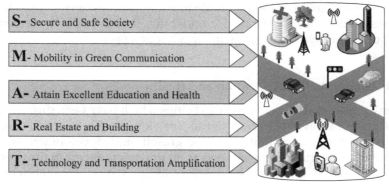

FIGURE 1.3
Composition of the word "SMART".

1.2.1 Secure and Safe Society

In modern society, safety and security is a very challenging issue. These challenges result from terrorism, cyber-crime and natural disasters that impact millions of lives across the world. These problems are the highest-ranked items on the agendas of national and international organizations. In the present interconnected world, security and safety challenges are becoming more complex (Scott et al. 2018). With the increase in globalization and the usage of hyper-connected techniques, new security and safety challenges are emerging and increasing in number day by day (Sookhak et al. 2018). Presently, dealing with security and safety challenges represents a major problem. To deal with such issues, citizens should be required to scale their security adequately (in smart cities, data can be collected from applications such as smart transportation, the smart environment and CCTV cameras). Law implementation should be well organized.

1.2.2 Mobility in Green Communication

Mobility in communication has significant advantages and is very beneficial for modern society. In the process of urbanization, mobility in communication has been a crucial issue and an important engine for progress and growth. It facilitates anyone to use any service anywhere and anytime. Mobility in communication first emerged due to the fact that radio frequency waves can travel easily through the air (Theoleyre et al. 2015; Sethi et al. 2016). Now people can watch a cricket match on their mobile phone at any shopping mall or while traveling in vehicles. The word "mobility" can be divided into two types:

- user mobility;
- device mobility.

User mobility provides users with the freedom to access any communication services in any place, while device portability facilitates the movement of a communication system device with or without a user.

1.2.3 Attain Excellent Education and Health

Among the several elements of smart cities which have been discussed previously, education is the most significant element. This section discusses how education is contributing to the growth and development of smart cities. Education has been observed as an important factor in the economic well-being of any country. The progress of any nation is dependent upon the increase in technical innovation and productivity; these factors are strongly linked to the quality of education provided. Intellectual capital and skills are important for the present economy, and quality education at the school and college level plays an essential role in key skills development (Zhuang et al. 2017). A high-quality education is the base of novel innovation, discoveries, entrepreneurship and knowledge that activate prosperity and growth of the individual as well as the city and nation. The following sections present approaches that can be implemented to bring about quality education.

1.2.3.1 Technology-Based Learning and Teaching Approach

Digital learning is an important concept that needs to be included by educators in today's classrooms. It involves using online websites, programs and services as learning and teaching tools (Wolff et al. 2015). It could also involve use of educational apps, educational webinars, online courses and social networking platforms as tools to generate digital projects. Some recent technologies include:

- machine learning;
- cloud computing;
- use of tablet computing;
- mobile learning;
- wearable technology.

1.2.3.2 Personalized Teaching and Learning

Additional curricular activities such as attending workshops, seminars, conferences and competitions etc. related to education from outside the classroom help students to acquire more information and achieve objectives. Outside classroom learning can improve a student's educational experience by demonstrating real-life applications of theories that they are learning at school. This learning also encourages students to develop problem-solving, critical thinking and decision-making skills. A report published by Ofsted (2008) shows that outside classroom learning contributes significantly to improving pupils' emotional, social and personal development.

1.2.3.3 Healthcare System

The main aim of a smart city is not only to provide the best healthcare solutions for its people but also to strengthen an ecosystem that mitigates the risk of infections and disease. The government should start various healthcare campaigns so that citizens learn how to live a more healthy lifestyle (Cook et al. 2018). On the way to becoming a smart city, it is essential to promote the importance of wellness and healthcare as part of the smart city design. A smart city is said to be a smart city only when its citizens adopt a healthy lifestyle and follow a fit and healthy daily routine; this can only be achieved with purposeful planning.

1.2.4 Real Estate and Building

In this era, urban innovation will be a product of technological advancement. Now our cities are becoming smart and a greater number of technologies (such as autonomous vehicles and drones) are already being used in our modern society. In history, urban sanitation can be sketched back to the Indus Valley Civilization, the remnants of which showed evidence of the development of early waste and drainage systems, making waterways and streets more hygienic. Now progress is based in the digital domain, where emerging technology such as the Internet of Things, artificial intelligence and cloud computing facilitates the emergence of smart city features. Real estate has played an important role in smart city development (Silva et al. 2018). The incorporation of smart buildings with numerous elements will bring many aids to the real-estate trade, and redefine many long-standing industries and fundamentals. The real-estate industry provides plenty of solutions to shape the smart city.

1.2.5 Technology and Transportation Amplification

A city is considered to be a smart city when, through using smart technologies, everything can be connected. The goal of the smart city is not only to provide smart features, services and facilities to its people which can make their lives easier but also to establish a proper channel of communication between citizens and governance (Schaffers et al. 2011). Using this channel, citizens are free to give their feedback to the government about the city. This goal can only be achieved by using the latest technology. Some emerging technologies include the following:

- Internet of Things: every smart solution in smart cities depends on the IoT (Nam et al. 2011), which provides an appropriate platform to connect a device to make a decision and then take action according to the situation. For example, traffic and crowd monitoring of a particular road is very helpful for emergency vehicles (ambulance, fire truck etc.) to follow a different path to reach the destination without delay.
- Artificial Intelligence: a smart city generates digital data in bulk. This raw data can be converted into useful information through proper analysis and processing using artificial intelligence (Naphade et al. 2011). The huge quantity of data generation brings the role of artificial intelligence to the fore, which is capable of making sense of raw data.
- Information and Communications Technology: ICT enables bidirectional communication between the government and citizens in a smart city. It helps the government to evaluate the citizens' demands and thus generate useful resources to address the requirements.
- Sensors: sensors are concealed but universal elements of the urban area. Sensors are crucial components of any intelligent system (Perera et al. 2014). The accessibility of massive sensors and continuously growing technology permit various applications that were not easily approachable in the past due to the high cost and inadequate availability.

A smart city consists of multi-layer transportation, smart parking and smart traffic lights. The city transportation system is a very important pillar of the city for citizens. Private and public road transportation is the basic model of transportation in most cities. Metro and local trains, which are present in some megacities, are the backbones of the transportation system.

1.3 A Roadmap of Using Communication for Smart City Development

This section discusses the importance of appropriate and seamless communication for smart city development. Enabling technologies based on wired and wireless communication are also discussed; these are adopted in smart city design. Communication has a vital role in smart cities since it provides a platform to accumulate information and statistics from the field. This also improves understanding of how the city is operating in terms of use of resources, lifestyles and services, as represented in Figure 1.4.

Green communication performs various roles, which are very significant for maximizing the performance and attaining the goals of a smart city. The conversion process of a city into a smart city involves the steps represented

Green Smart Cities

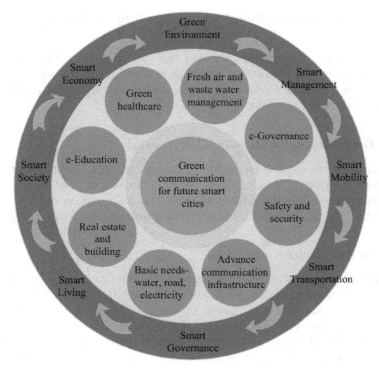

FIGURE 1.4
Essential field elements of future smart cities with the utility of green communication.

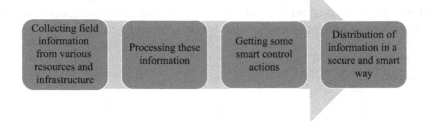

FIGURE 1.5
Effective communication for smart city design.

in Figure 1.5. This process starts with collecting field information from various resources and infrastructure. The data are then processed. After processing of data, some controlled actions will be implemented. Finally, by using these controlled actions, information can be distributed in a secure way.

1.3.1 Green Communication-Enabled Knowledge and Information Sharing

A lack of proper information sharing means that a city cannot become a smart city even if it is fully equipped to respond. With instant and precise information, city administrators can gain awareness about the problem and take action before it worsens.

1.3.2 Green Communication-Enabled Forecasts

Natural disasters require an extensive amount of data devoted to studying patterns, recognizing trends, identifying risk areas and predicting potential problems. Green communication delivers this information more efficiently so that the city can improve its attentiveness and response capability.

1.3.3 Green Infrastructure

"Green infrastructure" is an infrastructure which is environmentally friendly and focuses on sustainable development of the cities. Sustainable development of cities can only be achieved by using solutions that are energy efficient. Power supply is very crucial for the smooth functioning of any city. Usage of renewable energy resources such as the wind, sun and water are managing some of the load of conventional energy resources. Renewable energy is environmentally friendly and cheaper as compared to conventional energy. The main aim behind green infrastructure is to not only obtain economic benefits but also ecological ones. This approach can be brought into reality by applying the concept of "integration" supported by the green infrastructure concept to the progress of a "smart city environment", as shown in Figure 1.6. The green infrastructure idea recommends a practical or corporeal integration of green places into the development of a normal city.

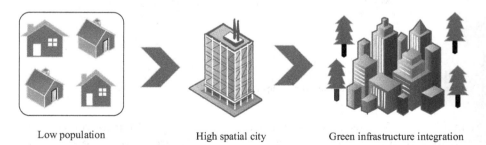

Low population   High spatial city   Green infrastructure integration

FIGURE 1.6
Green infrastructure integration.

Green Smart Cities

For example, the incorporation of green infrastructure can be achieved through using green roofs of houses to reduce surface water overflow in extremely commercial zones, using sky parks in housing zones to supplement the dearth of groundwater-level green spaces, planting trees alongside roads to decrease CO_2 emissions or using small parks such as those associated with schools to provide biodiversity.

1.3.4 A Cloud Computing Platform

A cloud computing-based architecture is considered to blend the methods through which control centers in different places are monitored and their requirements are diagnosed. Cameras are mounted at each and every zone for constant monitoring.

The green communication system is erected on a cloud platform that can be accessed by official bodies through a wide range of smart advanced mobile devices. Concerned official bodies are automatically informed of extreme deviations in the control center, as shown in Figure 1.7. This represents a cloud computing-based platform that enables green communication to make a city smart. This cloud-based architecture utilizes a two-level approach (Lee et al. 2014). The first level covers damage to property, public safety and human activity. This method focuses on zones with high traffic flows such as public parks and crowded streets. The second level covers the standardization of the system.

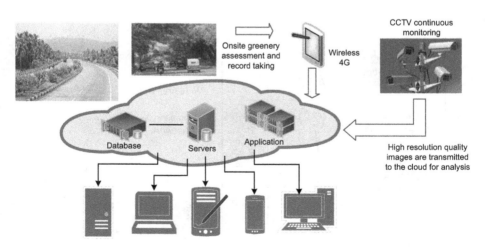

FIGURE 1.7
A cloud computing platform enables green communication to make cities smart.

1.4 Potential Communications Technologies to Make Cities Smart

Smart cities are responsible for providing various services to citizens, and infrastructure is dispersed unevenly in the city. Various communication technologies will be used for service provision. The popular technologies that are presently used and which will have a significant impact on the information communication system framework are presented in Figure 1.8. They are broadly divided into wireless communication (absence of direct connection between transmitter and receiver) and wired communication (direct connection between transmitter and receiver).

Another classification is indoor communication, where transmitter and receiver are present in a small area such as a building, house and room, and outdoor communication where there is a large distance between transmitter and receiver. Wireless communication is now the most used technology, which has been upgraded from 1G to 5G. The wireless communication system facilitates the benefits of mobility and is easy to deploy anywhere. However, a wired communications system is also advantageous in some places where high-speed communication is required in smart cities.

1.4.1 Wireless Technologies

The wireless communication system is currently the fastest growing system, with new technologies, novel standards and announcements

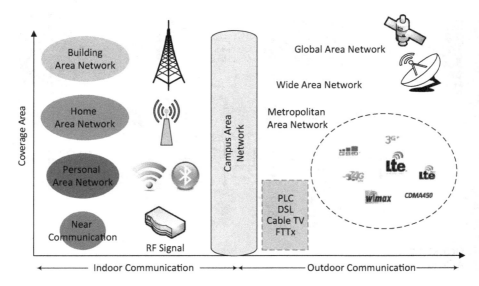

FIGURE 1.8
Emerging technologies of ICT.

Green Smart Cities

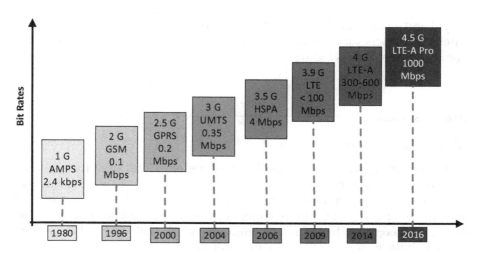

FIGURE 1.9
Advancement in wireless technologies.

TABLE 1.2

Upgrading of Wireless Generation Technologies

Technology	1G	2G	3G	4G	5G
Power Level of Mobile Station	Low	GSM 850/900:35 dBm	20–32 dBm	23 dBm	High
Frequency Band	800 Mhz	900 Mhz	2100 Mhz	2.6 Ghz	30–300 Ghz
Access Technology	AMPS	GSM	UMTS	LTA-A	BDMA
Power Level of Base Station	Low	Macro 46 dBm	32–38 dBm	44–49 dBm	High
Data Rate	2.4 Kbps	100 Kbps	30 Mbps	1.5–3 Gbps	10–50 Gbps

released every year. The first generation of this technology started with the advanced mobile phone system (AMPS) and has evolved to the latest release of long term evolution (LTE) and LTE-Advanced, which is an advancement of LTE (Raza 2016; Yigitcanlar et al. 2015). These systems belong to the fourth generation (4G) of the wireless communication system, as shown in Figure 1.9. The data rates of the mobile communication system are growing drastically; presently, 4G is transmitting at a speed of 300 Mbps in cities. The upgrading of wireless generation technology is represented in Table 1.2.

1.4.2 Wired Technologies

In the framework of a smart city, wired communication is an important component. This offers a high bit rate, protection against interference, stable output and a minimum bit error rate, unlike wireless communications where wireless channels suffer from different channel losses and the signal suffers from problems such as fading, refraction and reflection.

1.5 Features of Smart Cities

In the upcoming scenario, the goal of the Internet is not only to meet the requirements of citizens but also to focus on content. It is used for many applications such as in traffic management systems, parking systems, environment monitoring etc. Internet media contains real-world resources as intelligent people exchange information and work with each other, and it helps with the business development of enterprises.

1.5.1 Sensor-Equipped City and Cloud Computing

Recently, the Internet of Things has emerged as a widely used technology which supports the connection of the Internet with the physical environment. In some aspects, it was initially considered as machine-to-machine communication where ICT systems are connected with sensors through a wireless or wired link. However, the IoT also uses Internet protocols to connect different systems with sensors (Rajab et al. 2018). This guarantees high-quality connectivity and the interoperability of devices from various manufacturers. Figure 1.10 presents an overview of IoT architecture, including different application areas.

1.5.2 Future Research Directions

The major research challenges are outlined in Table 1.3. A widespread effort is required from academia and industry in the areas listed to contribute to greener smart cities.

1.6 Conclusion

The conversion of cities into "smart cities" is one of the key initiatives taken by different countries with the aim to change society in an efficient manner. The key focus is to optimize resources and infrastructure. Real estate is the key component of urbanization and infrastructure development is totally dependent on it. Growth will be in the digital domain, where evolving technology such as the Internet of Things, artificial intelligence and

Green Smart Cities

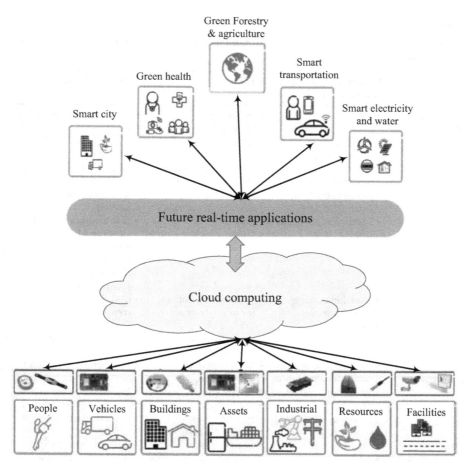

FIGURE 1.10
Overview of IoT architecture, including different application areas.

cloud computing is enabling the development of smart city features and advancements in information and communications technology. It also helps in improving and developing the quality of life of people.

Acknowledgment

The authors would like to thank the Women Scientists Scheme-A under the Department of Science and Technology, Government of India for its financial support of this work under File No: SR/WOS-A/ET-154/2017.

TABLE 1.3

Future Aspects and Key Features of Green Smart Cities with Their Research Challenges and Needs

Application Domain	Specific Applications and Features	Key Needs	Research Challenges
Security and secrecy	• Smart education system • Healthcare • Smart transportation • Smart electricity • Smart schools • Smart surveillance • Intelligent data • Protection of people	• Smart detection of black holes in IoT applications • Strong encryption • Cost-effective tools • De-identification process for maintaining people's privacy • Energy-efficient data transmission	• People's involvement in various security-related services to make the city smart • Lack of standard safety resolutions with no negotiation on data integrity • Safe incorporation of cloud and Edge and their distribution • On-time recognition of outdoor and indoor security threats
Data reconstruction	• Decision making • Event detection • Healthcare	• Optimized solutions for reconstruction • Intelligent reconstruction techniques	• Computational complexity • Quality assurance of reconstructed data
Data prioritization	• Disaster management • Smart transportation • Smart surveillance • Healthcare • Real-time activities monitoring	• Intelligent data prioritization • Energy-efficient system • Minimization of semantic gap • Personalized fusion	• Real-time prioritization of data • Events detection • Multiple data captured in smart cities • Complexity
Data dissemination	• Smart industry • Low-cost communication • Smart surveillance • On-time data transmission	• Detection of communication points • Intelligent spectrum sensing techniques • Cost-effective and reliable data transmission • Upgrading in IoT devices • Ensuring low-cost communication • Extending the battery lifetime	• Extending the IoT devices' battery life • Reliable and secure data communication • Secure data broadcasting • Energy-efficiency measurement
Big data	• Actionable intelligence • Smart transportation • Smart surveillance • Healthcare • Data prioritization	• AI-assisted intelligent tools • Actionable intelligence • Awareness of public about IoT applications in smart cities • Cloud-based big data processing centers	• Ensuring people's privacy • Cost-effective indexing • Lack of intelligent tools • Fast reclamation procedures for anticipated contents extraction

References

Cook, D. J., Duncan, G., Sprint, G. and Fritz, R. L. 2018. Using Smart City Technology to Make Healthcare Smarter. *Proceedings of the IEEE* 106(4): 708–722.

Lee, J. H., Hancock, M. G. and Hu, M.-C. 2014. Towards an Effective Framework for Building Smart Cities: Lessons from Seoul and San Francisco. *Technological Forecasting and Social Change* 89: 80–99.

Mensah, C. M. 2019. Reviewing the Narrative Concerning the Impact of Population Growth in Africa. LeXonomica. Retrieved from: www.worldometers.info/world-population.

Mensah, C. M. 2019. Reviewing the Narrative Concerning the Impact of Population Growth in Africa. LeXonomica. Retrieved from: www.worldometers.info/world-population

Nam, T. and Pardo, T. A. 2011. Conceptualizing Smart City with Dimensions of Technology, People, and Institutions. In *Proceedings of the 12th Annual International Digital Government Research Conference: Digital Government Innovation in Challenging Times*, 282–291. ACM: USA.

Naphade, M., Banavar, G., Harrison, C., Paraszczak, J. and Morris, R. 2011. Smarter Cities and Their Innovation Challenges. *Computer* 44(6): 32–39.

Ofsted. 2008. *Learning Outside the Classroom: How Far Should You Go?* Ofsted: London.

Perera, C., Zaslavsky, A., Christen, P. and Georgakopoulos, D. 2014. Sensing as a Service Model for Smart Cities Supported by Internet of Things. *Transactions on Emerging Telecommunications Technologies* 25(1): 81–93

Rajab, H. and Cinkelr, T. 2018. IoT Based Smart Cities. In *IEEE International Symposium on Networks, Computers and Communications (ISNCC)*, 1–4. IEEE: Italy.

Raza, A. 2016. LTE Network Strategy for Smart City Public Safety. In *IEEE International Conference on Emerging Technologies and Innovative Business Practices for the Transformation of Societies (EmergiTech)*, 34–37. IEEE: Mauritius.

Roser, M., Ritchie, H. and Ortiz-Ospina, E. 2013. World Population Growth. *Our World in Data*. Retrieved from: https://ourworldindata.org/world-population-growth.

Schaffers, H., Komninos, N., Pallot, M., Trousse, B., Nilsson, M. and Oliveira, A. 2011.Smart Cities and the Future Internet: Towards Cooperation Frameworks for Open Innovation. In *The Future Internet Assembly*, edited by J. Domingue et al. Springer: Berlin, 431–446.

Scott, T. and Sokwoo, R. 2018. Smart and Secure Cities and Communities. In *IEEE International Science of Smart City Operations and Platforms Engineering in Partnership with Global City Teams Challenge (SCOPE-GCTC)*, 7—11. Retrieved from: https://ieeexplore.ieee.org/document/8490099

Sethi, R. K. 2016. *The Role of Telecommunications in Smart Cities*. GlobalLogic Inc., San Jose. Retrieved from: www.globallogic.com/wp-content/uploads/2015/12/The-role-of-telecommunications-in-smart-cities.pdf

Shahidehpour, M., Li, Z. and Ganji, M. 2018. Smart Cities for a Sustainable Urbanization: Illuminating the Need for Establishing Smart Urban Infrastructures. *IEEE Electrification Magazine* 6(2): 16–33.

Silva, B. N., Khan, M. and Han, K. 2018. Towards Sustainable Smart Cities: A Review of Trends, Architectures, Components, and Open Challenges in Smart Cities. *Sustainable Cities and Society* 38: 697–713.

Sookhak, M., Tang, H., He, Y. and Yu, F. R. 2018. Security and Privacy of Smart Cities: A Survey, Research Issues and Challenges. *IEEE Communications Surveys & Tutorials* 21(2): 1718–1743.

Theoleyre, F., Watteyne, T., Bianchi, G., Tuna, G., Gungor, V. C. and Pang, A.-C. 2015. Networking and Communications for Smart Cities Special Issue Editorial. *Computer Communications* 58: 1–3.

United Nations. 2018. Revision of World Urbanization Prospects. Retrieved from: www.un.org/development/desa/en/news/population

Wenge, R., Zhang, X., Dave, C., Chao, L. and Hao, S. 2014. Smart City Architecture: A Technology Guide for Implementation and Design Challenges. *China Communications IEEE* 11(3): 56–69.

Wolff, A., Kortuem, G. and Cavero, J. 2015. Towards Smart City Education. In *Sustainable Internet and ICT for Sustainability (SustainIT)*,1–3. Retrieved from: https://ieeexplore.ieee.org/document/7101381

Yigitcanlar, T. 2015. Smart Cities: An Effective Urban Development and Management Model. *Australian Planner* 52(1): 27–34.

Zhuang, R., Fang, H., Zhang, Y., Lu, A. and Huang, R. 2017. Smart Learning Environments for a Smart City: From the Perspective of Lifelong and Lifewide Learning. *Smart Learning Environments* 4: Article 6.

2
Green Smart Town Planning

Yaman Parasher, Prabhjot Singh and Gurjit Kaur

CONTENTS

2.1 Introduction 19
2.2 Vision of Green Smart Town Planning 20
2.3 Strategies and Planning Procedures for Green Smart Town
    Planning 24
2.4 Influential Factors and Policy Instruments 28
2.5 Basic Vital Elements for Green Smart Town Planning 29
2.6 Carbon-Limiting Strategies 34
2.7 Challenges to Enforce the Concept of Green Smart Town Planning 36
Conclusion 37
References 38

2.1 Introduction

According to a recent forecast by the United Nations (UN) (United Nations, Department of Economic and Social Affairs, Population Division 2018), more than half of the world's population lives in cities at the present moment, which is estimated to rise to two-thirds by 2050. The whole American continent includes the most urbanized regions in the world, with 80% of residents living in cities. In contrast, Europe lags behind by a slight margin at 70% and Asia at 40% (Govindarajan 2014). With the rapid infrastructural and technological development, the concept of smart towns has been shown to have enormous environmental consequences both at the local and global level. It was estimated in a report by UN Habitat (UN Habitat. 2016) that nearly 70% of the world's greenhouse gas emissions are produced by residents who live in these smart urban spaces. Widespread urban development has already taken over vital green spaces and arable lands that are an essential part of any ecosystem (Haaland et al. 2015).

With the expansion of these modern smart towns, pressure on water and energy resources, along with sewer, waste management and transport networks, is increasing day by day. Therefore, to avoid lasting damage to the ecosystem and to improve the well-being and health of billions of residents in these areas, concerned authorities need to devise solutions in the planning mandate that can ensure environmental sustainability along with economic development and improved quality of life (Dixon and Wood 2003). This is why the green agenda is a necessary concept that must be included in smart town planning strategies for socioeconomic and environmental sustainability (Deal et al. 2013). It is essential for town planners to devise green technology-based urban planning strategies and solutions that can put forward some eco-sustainable ways to move the whole ecosystem towards more healthy and self-reliant living, where all these issues can be handled in more progressive and effective ways (Beatley 2011).

The concept of the green town typically represents an urban planning idea that takes into account sustainable eco-services, strategies and policies that help in bringing benefits to the whole community (Wheeler 2013). In a typical green infrastructure, one can find an interconnected network of green areas and hydrographic elements, contributing to the preservation, enhancement and maintenance of biodiversity processes within urban ecosystems (Benedict and McMahon 2006). This chapter highlights the vitality of the green technology concept in modern-day planning of smart and sustainable townships. It reflects some of the concerns that need to be addressed with appropriate planning procedures and strategies that need to be enforced in a balanced way. A holistic view of fully fledged green smart town planning can be seen in Figure 2.1.

2.2 Vision of Green Smart Town Planning

The idea of an environmentally friendly smart town has generally grown from the consequences of the undesirable effects of conventional smart systems in urban settings. Green smart town planning therefore serves as a catalyst in understanding pollution-causing areas, decreased biodiversity levels, the urban heat island phenomenon and so on. Green buildings, which serve as an essential part of green smart town planning, are built with a sustainable mandate and are used in a more responsible manner with regard to the environmental consequences during their total life cycle. Some of the main features necessary for a green technology-based smart town development are illustrated in Figure 2.2.

The whole concept of smart town development usually involves design, construction, use, maintenance, rehabilitation and demolition of the unwanted structures that usually inhibit an approach towards an economical and sustainable ecosystem. The foremost precondition for the development of a green residential infrastructure in a smart township is choosing an appropriate location, together with strategies and policies that can then

Green Smart Town Planning

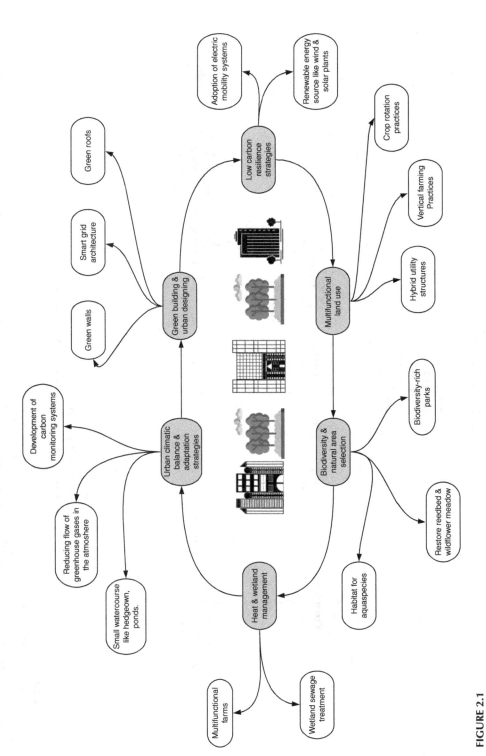

FIGURE 2.1
Holistic representation of green smart town planning.

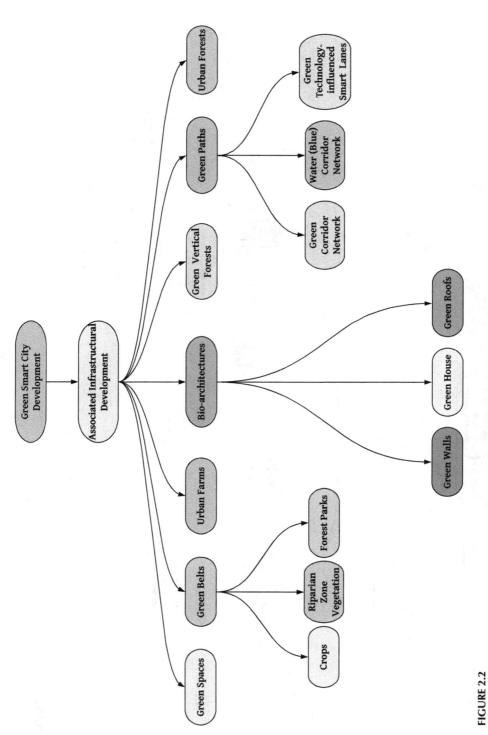

FIGURE 2.2
Features of a green smart town.

be implemented in the chosen area (Birch and Wachter 2008). These policies usually include the energy-efficient usage of renewable resources to promote cost-effective, sustainable solutions to the current hazardous situation that has been caused by the widespread exploitation of resources to their threshold.

The factor of energy efficiency therefore plays a vital role in green town development. This is evident from the fact that most of the green buildings around the world consume 40% less energy than conventional pre-existing ones (Chel and Kaushik 2018). In order to achieve this performance level, architects are devising certain solutions that help in regulating the costs associated with energy supply in a much better way. These solutions usually take into account the building shape and orientation in order to take advantage of natural light, provisions for passive systems of energy, together with automatic and energy-efficient lighting systems, optimum thermal insulation (for instance, cork in the Iberian Peninsula, natural wool in Romania), the use of warm colors and the use of alternative, clean energy.

The construction costs of such a green building, however, are higher initially, but energy savings can compensate against higher building costs. The general advantage of this type of construction is that it allows almost complete recycling of water resources and energy. For example, the building can be provided with facilities meant to capture rainwater for toilet use and for the irrigation of green areas.

For a typical green town development, it is necessary to locate the potential areas that have the scope for further sustainable development (Roo et al. 2011). Therefore, it is necessary to bring low-energy structures forward within the smart towns, in place of conventional ones that usually have a lot of detrimental effects on the overall population residing in that particular environment. Apart from locating and embedding the low-energy solutions, it is imperative to identify the green and blue regions. The green areas generally represent the fundamental components of the urban landscape, including spaces with trees, flowers and shrubs within and around these towns. Blue regions include all the natural and anthropogenic watercourses crossing urban areas (Benedict and McMahon 2012).

The adoption of a geothermal-based power supply can help to reduce the strain on conventional non-renewable energy resources which are on the verge of extinction. Besides the scarcity of these non-renewable conventionally adopted energy sources, their adverse effects have also raised a serious need to switch to clean energy initiatives that are currently in development in most parts of the world. Thus, it is necessary to lay down a mandate for energy use for each type and size of infrastructure that lies in the territory of a particular smart town. Also, putting a limit on automobile usage due to growing pollution levels could prove to be another vital step forwards. Considering means of sustainable mobility systems could be a better alternative to relying on the older conventional systems that we have been using recklessly, oblivious to the adverse consequences on our lives.

The use of a porous-type pedestrian path that can reduce rainwater runoff and reusing sewage water by treating it effectively can help in tackling the scarcity of water for a number of purposes other than drinking. Preparation of compost from kitchen waste to build up methane gas energy reserves could be a better alternative to non-renewable energy resources.

A green smart town generally guarantees its inhabitants the following set of essential features:

- strengthens innovation, economic productivity and the job market;
- increases participation in social interactions and provision of collaborative long-term training exercises;
- facilitates much more involvement in administration-based functioning and services related to smart governance;
- ensures accessibility to modern information and communication technologies (ICT) and sustainable and safe transportation systems;
- controls pollution, together with sustainable management of vital resources; and
- provides a better quality of health, safety, structures and social cohesion.

2.3 Strategies and Planning Procedures for Green Smart Town Planning

To harness the benefits around the green smart town planning strategies and procedures, the development of greener business models and environmental innovations must help the whole township to use vital resources efficiently and effectively. These solutions could possibly help present and future generations to obtain access to clean water and air, abundant recreational natural green spaces, fluid transportation systems and efficient management of waste, among other factors.

To provide a holistic perspective of a typical green smart township in the current smart infrastructure setup, certain quantitative and qualitative types of analysis method are generally employed by various township planners to locate the potential areas for improvement within modern-day towns. Modern investigative tools such as geographic information systems (GIS) and remote sensing techniques have been found to be extremely useful in some cases where sustainable economic development is necessary (Pauleit 2011). These tools generally help in identifying the endangered habitats for biological and aquatic species that are usually present around the smart town infrastructure. A careful analysis of the areas for improvement provides the planners with strategies that can help the whole ecosystem of the smart town in a much more economical and sustainable way. Returning to the analysis, from the set of tools, remote sensing is responsible for identifying and analyzing green areas from different satellite images. On the other hand, the GIS tool is useful for the analysis of spatial areas of relevant green spaces with overlapping layers of information and elements of interest. Both tools are often employed in analyzing suburban and urban areas of the forest.

These maps are usually developed in a GIS environment based on LANDSAT imagery (Stathopoulou and Cartalis 2007) and processed with some of the well-known geospatial analytic tools. The process of managing and creating these high-quality green raster images located within the urban space usually involves implementing specific, appropriate measures which help to reinvent the values associated with the urban ecosystem, sustainable waste management and various energy-saving practices. Based on the analysis, a typical green smart town is usually segregated into two areas during the design phase by the town planner (Carter 1993). These two areas are generally categorized into green and blue regions, which define both the ecological and aquatic habitats separately.

In general, the fundamental elements of a modern-day smart urban infrastructure are mainly represented by green areas. These green areas generally represent the fundamental components of the urban landscape that essentially encompasses spaces with trees, flowers and shrubs within and around these towns. They are usually distinguished further into three categories which include composites, strips and patches. The main reason behind the conservation of these green areas lies in the enormous benefits they offer to living species inhabiting these spaces. They not only control pollution and soil erosion, but also result in conserving water, mitigating the adverse effects of the urban climate and improving the quality of urban living (Manea and Mihail 2007). Depending on their scope and the activities carried out within their perimeter, green areas can be divided into several categories: public green areas (square gardens, parks, plantations along main roads, urban national parks, urban natural parks and urban forests); green areas with limited access (sports parks, private gardens, gardens belonging to institutions or factories); and specialized green areas (botanical gardens, zoological gardens, parks for exhibitions, and garden cemeteries) (Blanc 2008). According to their location, one can speak of urban green areas (including various green area formations), lying within the built-up area of the town, and suburban green areas, encompassing urban forests, recreation areas and other categories of suburban green spaces. These green corridors are not only planned and well-managed regions that have the potential to be useful for multiple purposes but are also areas that hold aesthetic and cultural roles, compatible with the sustainable use of the resources available (Ahern 2002).

Regarding the blue regions, the blue corridors include all the natural and anthropogenic watercourses crossing urban areas, making up genuine urban hydrological systems (Mell 2010). The concept of a blue–green corridor could prove to be useful in integrating green areas with water surfaces as part of the strategic planning paradigm followed in sustainable development. This hybrid concept could prove useful to the cities which are crossed by rivers or to those lying within the proximity of watercourses or canals, such as Amsterdam, Saint Petersburg and Ottawa. Besides the ecosystem benefits (oxygenation, humidification, habitat for various species etc.), these corridors invigorate the urban aesthetics and harmonize people's needs with those of nature.

To give an accurate picture of the benefits of green smart development, examples of four different cities (Vienna, Frankfurt, Katowice and Copenhagen) will be provided. Despite their spatial extension and overwhelming functional complexity, these cities still succeed in serving as ideal models to be followed when it comes to the planning of urban greenness in their respective townships. In each of these four cases, the enormous anthropogenic pressure has been partly compensated by oxygen-producing areas represented by urban forests – the green element that these four metropolises have in common (Miller et al. 2015).

Vienna, with a population of around 11 million (according to the 2011 census), has taken advantage of its proximity to the old Wienerwald (Beatley 2011). Here the term "Wienerwald" indicates the "Vienna Woods", which are forested highlands that form the northeastern foothills of the Northern Limestone Alps in the states of Lower Austria and Vienna. However, in addition, the combination of four different programs that emerged within Vienna has helped to reshape this region in the direction of a sustainable development approach. These programs can be defined as the Vienna Climate Protection Programme (KliP), Eco Procurement Vienna (ÖkoKauf Wien), Eco Business Plan Vienna, and Repair and Service Centres (RUSZ). It is perhaps only due to these programs that Vienna was able to stand out from other places in Austria.

Frankfurt, with a population of around 6 million (according to the 2011 census) in its built-up area, has the largest urban forest in Germany. Due to its various strategies and policies for eco-sustainability, Frankfurt has recently been awarded for being the most sustainable city in the world. The main reasons behind this are as follows:

- Its policies on reducing CO_2 emissions by 10% every year. The main goal of Frankfurt is to reduce CO_2 emissions by 50% by 2030.
- The creation of massive solar panel grids across the roofs of different buildings across the city, which offers a yield of nearly 300,000 kW.
- The creation of a green belt area comprising 200,000 trees.
- The introduction of a policy of cash bonuses on reducing electricity consumption by at least 10%.
- Providing large industries and companies with access to energy and environment management systems, which ultimately results in the preservation of natural resources along with a reduction in detrimental effects on the environment as a whole.

Turning now to Katowice, this European city was considered as a city of coal for a long time. However, the enormous pollution levels that resulted from the coal industry compelled the authorities to take action to address its ecological imbalance and rejuvenate the ecology of the city. As a result, forest parks were developed across the whole city, which ultimately has helped the whole region in revitalizing the concept of urban sustainability. According to Bartniczak and Raszkowski (2018), more than 40% of the city's area is covered by forest, which is expected to increase further to 60% in a decade.

Similarly, Copenhagen has also made pioneering efforts in this regard to expand its green economy. This place has become well known for its promotion of green growth through public–private partnerships that are at the core of its approach for sustainable practices. Since launching its first climatic action plan in 2009, Copenhagen has aimed to become the world's first green smart carbon neutral city by 2025. The region has also adopted a climatic adaptation plan to meet challenges such as heavier rainfall, floods, storms and rising sea levels. In terms of waste management, landfill disposal in Copenhagen has reduced from 44% to merely 2% in the last couple of years, which is a significant improvement in regard to preserving the fertility of the land (Beatley 2012). As Thorpe (n.d.) comments, currently, the whole region of Copenhagen recycles almost 58% of its total waste, and the remaining 40% is generally used to serve as the fuel for the district's heating network. With special restrictions on heavy diesel-powered vehicles, lorries and buses in Copenhagen, the polluted particle emissions have also fallen by 60%. Copenhagen has been awarded the European Green Capital Award, in recognition of the following indicators:

- its contribution to global climate change, transport, green urban areas/sustainable land use, nature and biodiversity;
- quality of local ambient air and acoustic environment;
- waste production and management, water consumption, waste-water treatment; and
- eco-innovation and sustainable employment, environmental management of the local authority, and energy performance.

Analysis of a number of similar initiatives in many townships around the world has shown that most initiatives continue to focus only on a single aspect, such as energy efficiency in buildings or smart grids. In order to bring a fully fledged modern-day smart township into reality, there is an urgent need to take these initiatives and practices into account holistically rather than concentrating only on particular or selective areas of interest. Planning policies, therefore, must adopt a systematic and integrated approach that makes the best use of all of the assets that are vital for the existence of humans and other living beings within the whole town.

2.4 Influential Factors and Policy Instruments

To incorporate green infrastructure in our towns, several factors such as correct functionality, political viability and economic feasibility play an important role. To raise the funds for this green technology, it is crucial that these constituencies come together and apply political pressure for the completion of planning and thereby securing the commitment of elected officials. Due to certain funding approaches such as single-use agenda or planning and process design carried out in a piecemeal format, many design and planning professionals have failed to obtain the full range of benefits that are possible from this green infrastructure. Therefore, to achieve the potential of appropriate planning, there is a need to create a strong team consisting of common people, elected officials, municipality people, business heads and subject experts. Experts from different domains such as finance, ecology, planning and landscape architecture will serve to resolve the technical issues revolving around funding, planning and locating this green advancement (Austin 2014).

To move the vision towards realization, it is important to create master plans along with the first critical steps of setting up a vision and goals supported by subdivision, development ordinances and zoning. The development goals and growth drivers involved in the smart township planning were subdivided into each of the distributed territories of the township planning structure to enable proper enforcement to the best level possible. This entire process of establishing green technology which must be socially, economically and environmentally viable is the outcome of advance planning that may take years or sometimes decades.

Another very important aspect when moving towards a sustainable smart approach for a town is smart governance. It has been observed in various cases that most smart governance tools, such as digital democracy, citizen empowerment and open governance, have impacted hugely on the development of a typical smart infrastructure. Out of the three tools mentioned above, the concept of digital democracy is solely based on implementing modern-day ICT for the service of vital political and governmental processes which indirectly serve as a driving instrument for the growth of a smart infrastructure (Parasher et al. 2018). If implemented in an appropriate manner, it results in much wider participation of the inhabitants, leading to more open and transparent administration. It therefore serves as a critical factor in increasing the common responsibility of the government and residents in the development of their towns.

Open governance, which is the second vital factor, usually lets people keep a check on the policies and strategies which the government is implementing for their welfare and development. It helps the citizens to keep track of the programs introduced by the government for the sustainable growth of their regions. The third concept of citizen empowerment gives the people a sense of belief that they are actively contributing to the development of their town or region in a much safer, economical and sustainable way.

The main idea behind using smart governance for green smart towns is based on the fact that community members are experts in the related urban problems to which they already have knowledge and solutions. Bringing the community members into contact with the government directly allows professionals to identify their actual needs and utilize knowledge and local human resources, rather than imposing the much-needed changes from above.

2.5 Basic Vital Elements for Green Smart Town Planning

With the demand to accommodate energy-efficient sustainable solutions for modern smart towns, many urban setups worldwide have directed their development strategies towards being solely based on sustainable buildings, roofs, material, transportation, streets, lighting, terraces etc. The main objective behind these energy-efficient, environmentally friendly smart solutions is to redefine the quality of life of their residents by offering socio-economic and cultural growth in a safe, dynamic and healthy environment.

Having provided a comprehensive overview of the essential features of the green township concept worldwide in earlier sections, this section of the chapter aims to provide an updated definition of a green smart town development, together with information on the essential elements that form it. Based on the analysis of the most interesting initiatives implemented by different towns in this regard, six essential elements that serve as the heart-beat of a green smart township are described in detail.

2.5.1 Green Buildings

Within the industrial domain, the construction industry can be considered as an important element with the potential to achieve the objective of smart sustainable growth of future smart towns. It has been found that of the world's total energy usage, nearly 40% is consumed by different types of buildings in the European Union (EU), which generally comprise houses, offices, shops, educational institutions and so on (Beatley 2000). To counter this problem, it has been observed in different studies (Kibert 2016) that the building industry holds the potential to lower this energy consumption to 30%, together with applying useful interventions that can help to cut the associated costs significantly to a much more affordable range for inhabitants.

Another significant prospective solution provided by pre-existing building structures is to cover all unused roof spaces and other parts of buildings with solar panels. This is an attempt to meet the EU's 40% energy demand, which can be fulfilled if all roof spaces and other parts of buildings start to be covered with solar panels.

In addition, it has been observed that in Europe nearly 90% of the whole population spend their time inside buildings due to the widespread cold climatic conditions (Brasche and Bischof 2005). This continuous exposure to heating systems inside buildings not only accounts for hazardous effects on people's health but also results in greater expenditure on heating solutions. The following list presents a few methods, measures and practices that can help upgrade the energy sustainability in existing building structures to meet the needs of modern-day smart townships:

- adoption of measures to reduce energy dispersion from the building envelope;
- promoting the use of cleaner and renewable sources for construction purposes; and
- continuous monitoring of the efficiency of the heating systems.

2.5.2 Green House Material

Despite a number of technologically advanced green alternatives that are available for the development of a green infrastructure (Mell 2010), old types of building materials can be sometimes used in towns and be quite useful in the long run. One example is using cob for building green energy concept houses. Cob is a mixture of sand, water and loam and is used in building diverse architectures; it not only provides strength but also conserves the energy within the structure. Such types of houses automatically adjust moisture levels and maintain heat in winter and coolness in summer (Goodhew and Griffiths 2005). If the roof is large enough and the foundations are high (to prevent water stagnation near the walls), these houses have proved to be extremely durable in such situations. These alternatives would fit better in towns, where cheaper means of construction are necessary to move towards a sustainable approach. It could definitely turn out to be a new form of housing, blending energy efficiency with cultural aesthetic values.

2.5.3 Green Roofs

Unlike the conventional smart townships, where urban greenness is only one of the urban structural elements, the green regions in present-day smart towns are regarded as a central point around which the whole planning is laid out. These traditional green areas are embedded in new planning concepts, meant to interconnect residential areas, green areas and water surfaces through unconventional green area infrastructure. The element of novelty is found in the introduction of vegetation to a green building in the form of green roofs and terraces, green walls and green hedges made of trees or shrubs (Dunnett and Kingsbury 2008).

Thus, a green building can help in increasing comfort by bringing the benefits of urban green spaces closer to the user. In comparison to conventional roofs, which get overheated during the day, green roofs mitigate thermal contrasts and add a pleasing appearance to an urban landscape (Cantor and Peck 2008). From the standpoint of quality of life, the green roof acts as an insulator, thus helping to reduce power consumption both in summer and winter. An example of this good practice is Chicago, where hundreds of residential blocks and other buildings, including the town hall, have landscaped terraces and gardens.

2.5.4 Vertical Green Forests

Another extravagant project promoting green values is the concept of the "Vertical Forest", also known as "Bosco Verticale", in Milan, Italy. In simple terms it is a complex of multilevel residential buildings with tree and shrub vegetation growing on the roof and on a number of asymmetric terraces. In these cases, the functionality of the buildings does not prove to be a barrier when it comes to implementing green architecture and technologies in these spaces (Kaur et al. 2018). This is perhaps the main reason why green architecture has spread from individual residential buildings in smart townships to business premises and multifunctional centers (e.g., Commerzbank in Frankfurt, Taipei 101 in Taipei and Green Gate in Bucharest). In this respect, it is therefore advisable to assess the ecological potential of the township territory, which is generally based on the mean length of vegetation season and the ratio between temperature and moisture, when planning to construct such a building type (Manea and Leishman 2011). Another useful example in this regard comes from the Johnston Terrace Garden, a beautiful garden situated in Edinburgh, and the Serre de la Madone in Menton, France, containing broad-leaf tree species with the ability to retain large amounts of sedimentable particles, purifying the urban air to the upmost level. It is preferable to have a high diversity of such species because some of them are far more environmentally efficient than others due to their better filtering capacity (de Roo 2011).

2.5.5 Green Streets and Transportation

In a typical green smart town, it is vital that the transport infrastructure should have minimum adverse effects on the vital components of nature: the living beings, soil and vegetation. An omnipresent environmental problem that any large town is confronted with is the physical and chemical pollution induced by motorized transport systems. This phenomenon becomes more serious when the area is larger and the number of inhabitants is higher than average.

When the ground is covered by impervious materials such as concrete or asphalt laid by big trucks, soil properties suffer. Imperviousness increases the risk of flooding, generates water deficits, contributes to global warming and ultimately affects biodiversity. Green town architecture therefore should envisage using environmentally friendly materials and landscaping techniques (pervious concrete slabs, porous asphalt, natural stone, recyclable rubber tiles, tartan boards etc.) (Kahn 2007).

Three decades ago, with the same thing in mind, the town of Curitiba in Brazil introduced a sustainable public transport system, the Bus Rapid Transit (BRT), in order to decrease vehicle use on their roads. As a result, about 70% of the commuters in Curitiba currently use the BRT to go to work (Netland 2012); consequently, streets have become decongested and the air is less polluted, which is an advantage for the 2.2 million residents of this particular landscape. Another essential element of the urban transport infrastructure of a green smart town are the bicycle lanes, which promote sustainable urban mobility. In Europe there are places where this mode of travel is quite common (e.g. Amsterdam, The Hague, Copenhagen, Stockholm, Helsinki, London, Paris, Vienna, Berlin, Barcelona, Hamburg, Freiburg etc.).

Eco-friendly sidewalks serve as a viable choice over existing concrete pavements. Being more flexible, they have a lesser impact on the trees and their root systems. The rubber sidewalk modular paving system allows water to percolate the soil. At the same time, rubber is a soundproofing material and is less risky for users (it cushions the shock of falls and reduces injury risk, especially for children and elderly people). These sidewalks also take advantage of the used tire recycling principle. Another ecological improvement in this regard is the use of natural stone (granite, limestone, basalt, cobblestones etc.), which is more resistant to freezing and high pressure than concrete and asphalt. On the other hand, it has been observed that the sustainability of grassy tiles is nearly double that of concrete sidewalks, reducing the maintenance and repairing costs to much lower levels (Ahern 2002). Grassy tiles ensure the benefits of permeability and substratum oxygenation are realized, along with enabling preservation of the associated aesthetic value.

An omnipresent environmental problem that any large town is confronted with is the physical and chemical pollution induced by motorized means of transport. This phenomenon becomes more serious when the area is larger and the volume of inhabitants is higher. In Vauban, an ecological district in Freiburg, bicycles are the main transportation vehicle. The residents located in this town do not have personal cars or parking lots. Similarly, the town of Hamburg has an environmental strategy that is included in its "Green Network Plan". This document (Fischer 2016) supports the creation of bicycle and pedestrian lanes meant to connect all the green areas of the town – parks, playing grounds, gardens and cemeteries.

With the expansion of urbanization and the overcrowding of large cities, phenomena such as physical, chemical and biological pollution, as well as the continuous increase in building density and waste volume, are inevitable and difficult to manage. Under the circumstances, it is necessary to reconsider the principles of conventional urban planning. With current knowledge, the smart green town represents a viable strategy that combines all the conceptual characteristics of the town of the future.

2.5.6 Smart Grid and Lighting

In order to reduce energy consumption on the urban scale in modern-day smart townships, the intervention of smart grid architectures is necessary. This not only improves energy efficiency, but also keeps a check on its flow based on energy requirements. Another significant advantage of implementing such structures is the reduction in the overall operational costs. To incorporate this into the existing smart town setup, outdated lamps must be replaced with energy-savvy light-emitting diode (LED) lamps, and a clean, renewable energy source must be introduced to make them operate effectively.

One such initiative was implemented in Buenos Aires in 2016, where Philips and SAP installed intelligent lighting systems which not only reduce energy consumption to 50%, but also bring in numerous other connected advantages, including real-time monitoring of traffic, available parking lots and other necessary features such as surveillance for that particular region (Smart Cities: Understanding the Challenges and Opportunities, Philips 2018).

To keep a town moving effectively, sustainable, smart and safe mobility solutions are considered as vital aspects in bringing about economic prosperity in a given smart township (Tomar et al. 2018). In particular, the concept of green smart mobility brings several useful solutions, which are summarized as follows:

- It employs modes of public transport that rely on low-emission combustion engines with hydrogen or electric motors.
- It promotes the use of hybrid electric vehicles with proper smart charging systems.
- It promotes carpooling practices.
- It is an introduction to the early warning systems that will help to convey traffic- and parking-related information via a smartphone.
- It enables the digitalization of public transit systems through smart panels and applications that provide information about the mobility path, waiting time and so on.
- It introduces intelligent traffic lights which take into account a real-time count of car flows to monitor possible congestion or bottlenecks.
- It is an introduction to intelligent streetlights which are capable of automatically modulating lighting according to the time of their usage, resulting in a reduction in unnecessary energy wastage.

2.6 Carbon-Limiting Strategies

This section aims to present an overview of the global impact of low-carbon and climate-resilient smart urban development on modern-day smart urban ecosystems, with a focus on their planning and services. Climate has always been a vital ecosystem which is greatly affected by urbanization. From its inception, our planet has always been highly susceptible to climate change. Its exposure to various climate changes has often resulted in an increase in natural disasters across multiple regions. According to Kim (2017), it is estimated that by 2050, almost 70% of the world's population will be living in urban spaces located in various townships around the world. This means that there is a vital need to implement effective climate-resilient planning procedures and policies to prevent potential widespread damage to the whole ecosystem.

Specifically chosen case studies show that the design of a low-carbon smart green town usually involves concepts such as energy conservation and environment and climate change impact assessments to ensure effective resilience planning. However, it has been conclusively proven by various studies that conventional urban planning does not usually meet most of the requirements and mandates of a safe and comfortable world.

This ultimately presents the need for continuous innovations in town planning strategies and procedures with the implementation of more effective technologies and policies. Due to the widespread hazardous effects associated with carbon accumulation within towns, effective comprehensive and detailed carbon-focused (3Cs) planning practices should be adopted as a mainstream approach in town-wide climate resilience planning procedures. Therefore, to protect smart towns from the adverse consequences of these environmental degradation factors, there is an urgent need to address global environmental challenges.

The potential adverse effects on the natural ecosystem of the standard urbanization setup and services are evident in the ecological and climate implications which all the major townships around the world are suffering from. With the onset of town development, many factors come into play as the result of the high-density habitation of humans across many urban spaces. These include isolation, loss, fragmentation and pollution of natural habitats, along with the change in climatic processes and accumulation of waste products.

Annually, air pollution accounts for 4.2 million of the world's deaths. (World Health Organization n.d.). In terms of water, high levels of contamination are observed when population density is high and sanitation standards are fairly low. It has been observed that cities and towns together produce 1.7–1.9 billion tons or 46% of global waste (Singh et al. 2014). Many factors are responsible for this problem, but a few of them have led to massive impacts on the socioeconomic and environmental structure in a number of smart towns around the world:

- continuous exposure to climate change;
- high volume of vulnerable populations; and
- insufficient allocation of basic urban services to each and every member of society.

It is therefore necessary for the modern smart townships to include climate-driven responsive planning in their urban mandate. The conventional planning paradigm based on static climatic conditions must be reformed to take into account the complex ecosystem changes. Some of the essential approaches that can help to mitigate these climatic changes to ensure low carbon production are as follows:

- adoption of integrated policies and strategies to localize energy, food and vital component production within the realm of the urban public;
- adoption of organic farming practices;
- bringing improvement in water quality with low environmental impact;
- bringing into existence zero-pollution zone regions in urban setups;
- maintaining a balance between socioeconomic and environmental sustainability in line with economic progress;
- promoting high-density, eco-friendly transit options, together with mixed land use paradigms; and
- implementing water-sensitive urban structure designs in planning layouts.

Since climate change is accountable for a number of adverse effects on the living population of a smart town, it is therefore imperative to develop necessary action plans to counter this situation and to integrate these into the local, regional and national planning of the whole smart urban setup. The short- and long-term plans for low-carbon smart towns have not found their place so far in any policy implementation around the world. Therefore, it is again necessary for planners to develop a comprehensive action plan that can help to halt adversities such as growing air pollution, global warming and other related health problems in a much shorter timeframe.

Green smart urban planning must include a mandate to reduce the overall carbon footprint of the whole town by integrating a number of measures into the planning procedures, including planting trees and the creation of wetlands (Van Veenhuizen 2006).

2.7 Challenges to Enforce the Concept of Green Smart Town Planning

The continuous rapid increase in the global population, pollution, climate change and the decrement in energy sources are some of the main problems facing modern-day smart towns on a regular basis. Therefore, most of the smart towns in the modern world are now focusing on implementing green technology-based structures into their pre-existing infrastructures, which primarily includes the concept of green buildings, roofs, streets, lighting solutions and so on. Because of the adoption of these structures, the towns are able to reduce their greenhouse gas emissions and energy consumption to significant levels. However, the main contrasting feature in this adoption is that no single country has taken into account each and every factor which is actually responsible for the detrimental effects on the lives of many people. As a result, the concentration of carbon is still on the rise, which ultimately results in a global temperature increase.

It has also been observed that buildings in the EU account for 40% of the total power usage, together with 35% of CO_2 emissions and nearly 40% of particulate matter (PM) 10 and PM 2.5 (Lavalle et al. 2009). Despite this devastating scenario, the present mobility system has also resulted in increasing pollution levels by 30% within a couple of years (World Health Organization n.d.), even in urban areas, which usually remain uncrowded even in peak hours. It has been observed that continuous exposure to hazardous chemicals, UV radiation and climate change has increased illnesses to very serious levels.

These problems are expected to surge to more significant levels with the rapid increase of the population of the world, which is estimated to reach 9.5 billion by 2050 (Al Jazeera News n.d.). This situation will worsen further because in the following years the population residing in these urban smart spaces located inside the townships will increase to account for two-thirds of the world's population.

With the onset of such peril, modern smart townships need to become self-sustainable in handling such socioeconomic mutations and other environmental factors that are going to be a hindrance in their sustainable development. A green smart town essentially holds the capability to improve the quality of life of its citizens, with the opportunity for socioeconomic and environmental growth in a much safer, healthier, dynamic and stimulating environment (Carter and Turnock 1996).

In today's world, all of this has become possible only due to the cutting-edge innovation and sustainable technological development that has taken place over the last couple of years. This in turn is playing a major role in changing aspects of the modern-day smart town development, through its functioning, by providing renewable energy sources, advanced ICT infrastructure, transit systems, healthcare and vegetation, along with sustainable waste management practices.

It is estimated that by 2023 the market related to these types of smart cities and townships will be valued at about 27.5 billion USD (Housing and Land Rights Network India 2017), which ultimately means that the research and development in sustainable technologies is going to play a huge role in refining the definition of smart urban spaces.

To transform conventional townships into modern-day green smart townships, the sustainable technologies needed in town planning must be able to add environmentally friendly and energy-efficient components to the pre-existing building structures and smart mobility designs, together with features such as renewable energy sources. It has been observed that all these initiatives discussed have proven to be an effective way to address the environmental objectives and socioeconomic goals that are indeed an essential part of the development of the sustainability concept as a whole.

The ongoing detrimental effects of continuously varying climate crises resulting from 19th-century industrialization have already strained much of the equilibrium between humans and nature to the core. Thus, by adopting these green sustainable measures, we can avert further ecological and climactic variations while reaching the objective of achieving self-reliant, sustainable, safe, eco-friendly smart towns for future generations.

The only important challenge that these green smart towns may face in the future could arise from the integration of sustainable energy-efficient sensors and other ICT technologies with the conventional technical infrastructure; this could very well harness the hidden potential behind the existing infrastructure to much greater levels. Proper interoperability between them will definitely help to deliver the requisite benefits of the green smart town concept to inhabitants, thus contributing to improving their quality of life to a much better level.

Thus, to accomplish these developments, active research and development is needed to explore further ways in which the current infrastructure can offer better services and a platform to a large number of people inhabiting the relevant space in a much smarter and more sustainable way. It is therefore necessary to connect urban planning with solutions that incorporate energy conservation, use of renewable energy and ecological restoration (Tîrlă and Cocoş 2014). It can be concluded, essentially, that there is an urgent need to formulate and implement climate action plans that incorporate these techniques at all levels, spanning from global to local, to achieve our goals for sustained prosperity for ourselves and all future generations.

Conclusion

This chapter has attempted to present the merits of green technology-based smart planning procedures that help in rejuvenating the lost soul of the modern-day smart towns that have deteriorated due to the harmful actions of humans on the environment.

Starting at the initial level, the concept of segregating the whole smart town architecture using smart GIS and remote sensing tools was discussed. The main benefit of this segregation process is that it will prove helpful in categorizing separate habitats for different species such as aquatic life and plants that reside within the town network. Furthermore, several methods that highlight pollution-causing areas and activities that reduce biodiversity levels were discussed and introduced in a brief and concise manner. Also, solutions such as the usage of renewable energy sources and the adoption of advanced infrastructural structures like green roofs, vertical garden forests, green streets and so on were also explored. To conclude, strategies and methods to curb the increasing carbon footprint were addressed, followed by the potential challenges that accompany the integration of the new technologies with the conventional systems.

References

Ahern, J. F. 2002. *Greenways as Strategic Landscape Planning: Theory and Application*. Doctoral thesis. Wageningen University.

Al Jazeera. 2019. UN: World Population Expected to Rise to 9.7 Billion in 2050. Retrieved from: www.aljazeera.com/news/2019/06/world-population-expected-rise-97-billion-2050-190618085808201.html.

Austin, G. 2014. *Green Infrastructure for Landscape Planning: Integrating Human and Natural Systems*. Routledge: New York.

Bartniczak, B. and Raszkowski, A. 2018. Sustainable Forest Management in Poland. *Management of Environmental Quality: An International Journal* 29(4): 666–677.

Beatley, T. 2000. *Green Urbanism: Learning from European Cities*. Island Press: Washington, DC.

Beatley, T. 2011. *Biophilic Cities: Integrating Nature into Urban Design and Planning*. Island Press: Washington, DC.

Beatley, T. 2012. *Green Urbanism: Learning from European Cities*. Island Press: Washington, DC.

Benedict, M. A. and McMahon, E. T. 2006. Green Infrastructure: Smart Conservation for the 21st Century. In *Sprawl Watch Clearinghouse Monograph Series* (report). Island Press: Washington, DC.

Benedict, M. A. and McMahon, E. T. 2012. *Green Infrastructure: Linking Landscapes and Communities*. Island Press: Washington, DC.

Birch, E. and Wachter, S. (eds.). 2008. *Growing Greener Cities: Urban Sustainability in the Twenty-First Century*. University of Pennsylvania Press: Philadelphia.

Blanc, P. 2008. *The Vertical Garden: From Nature to the CityTown*. W. W. Norton and Company: New York.

Brasche, S. and Bischof, W. 2005. Daily Time Spent Indoors in German Homes—Baseline Data for the Assessment of Indoor Exposure of German Occupants. *International Journal of Hygiene and Environmental Health* 208(4): 247–253.

Cantor, S. L. and Peck, S. 2008. *Green Roofs in Sustainable Landscape Design*. W. W. Norton and Company: New York.
Carter Jr, E. J. 1993. Toward a Core Body of Knowledge: A New Curriculum for CityTown and Regional Planners. *Journal of Planning Education and Research* 12(2): 160–163.
Carter, F. W. and Turnock, D. (eds.). 1996. *Environmental Problems in Eastern Europe*. Psychology Press: London.
Chel, A. and Kaushik, G. 2018. Renewable Energy Technologies for Sustainable Development of Energy Efficient Building. *Alexandria Engineering Journal* 57(2): 655–669.
de Roo, G. 2011. The Ecosystem Approach – Complexity, Uncertainty, and Managing for Sustainability. *Planning Theory. Special Issue: Urbanisms and Worlding Practices*. Sage Publications: Thousand Oaks, CA, 92–95.
Deal, B., Pallathucheril, V. and Heavisides, T. 2013. Ecosystem Services, Green Infrastructure and the Role of Planning Support Systems. In *Planning Support Systems for Sustainable Urban Development*, edited by S. Geertman, F. Toppen and J. Stillwell. Springer: Berlin, 187–207.
Dixon, A. B. and Wood, A. P. 2003. Wetland Cultivation and Hydrological Management in Eastern Africa: Matching Community and Hydrological Needs through Sustainable Wetland Use. *Natural Resources Forum* 27(2): 117–129.
Dunnett, N. and Kingsbury, N. 2008. *Planting Green Roofs and Living Walls*. Timber Press: Portland, OR.
Fischer, T. B. 2016. Health and Hamburg's GrünesNetz (Green Network) Plan. In *Green Infrastructure and Public Health*, edited by C. Coutts. Routledge: London, 286–295.
Goodhew, S. and Griffiths, R. 2005. Sustainable Earth Walls to Meet the Building Regulations. *Energy and Buildings* 37: 451–459.
Govindarajan, V. 2014. A Critique of the European Green City Index. *Journal of Environmental Planning and Management* 57(3): 317–328.
Haaland, C. and van den Bosch, C. K. 2015. Challenges and Strategies for Urban Green-Space Planning in Cities Undergoing Densification: A Review. *Urban Forestry and Urban Greening* 14(4): 760–771.
Housing and Land Rights Network India. 2017. *India's Smart Cities Mission: Smart for Whom? Cities for Whom?* Housing and Land Rights Network: New Delhi.
Kahn, M. E. 2007. *Green Cities: Urban Growth and the Environment*. Brookings Institution Press: Washington, DC.
Kaur, G., Tomar, P. and Singh, P. 2018. Design of Cloud-based Green IoT Architecture for Smart Cities. In *Internet of Things and Big Data Analytics Toward Next-Generation Intelligence*, edited by N. Dey, A. E. Hassanien, C. Bhatt, A. S. Ashour and S. C. Satapathy. Springer International Publishing: Cham, Switzerland, 315–333.
Kibert, C. J. 2016. *Sustainable Construction: Green Building Design and Delivery*. John Wiley & Sons: Hoboken, NJ.
Kim, K. G. 2017. *Low-Carbon Smart Cities: Tools for Climate Resilience Planning*. Springer: New York.
Lavalle, C., Birgit, G, Fons, J., Holger, R., Goodstadt, V., Bullet, M., Granberg, A., Berrini, M., Joyce, A., Borsboom, J., Folkert, R., Ludlow, D. and Banos, E. 2009. *Ensuring Quality of Life in Europe's Cities and Towns – Tackling the Environmental Challenges Driven by European and Global Change*. European Environment Agency: Copenhagen.

Manea, A. and Leishman, M. R. 2011. Competitive Interactions between Native and Invasive Exotic Plant Species Are Altered Under Elevated Carbon Dioxide. *Oecologia* 165(3): 735–744.

Manea, G. and Mihail, M. 2007. Vulnerabilitatea terenurilor cu funcții de spațiiverzi în Municipiul București. Factoricauzalișiconsecințe. [Vulnerability of Land Functioning as Greenspace in Bucharest CityTown. Causative Factors and Consequences]. *Comunicări de Geografie* 11: 501–506.

Mell, I. C. 2010. *Green Infrastructure: Concepts, Perceptions and its Use in Spatial Planning.* Doctoral thesis. Wageningen University.

Miller, R. W., Hauer, R. J. and Werner, L. P. 2015. *Urban Forestry: Planning and Managing Urban Greenspaces.* Waveland Press: Long Grove, IL.

Netland, T. 2012. Public Transportation That Works: The Curitiba Case. Retrieved from: https://better-operations.com/2012/04/01/public-transportation-that-works-the-curitiba-case/.

Parasher, Y., Kedia, D. and Singh, P. 2018. Examining Current Standards for Cloud Computing and IoT. In *Examining Cloud Computing Technologies Through the Internet of Things*, edited by P. Tomar and G. Kaur. IGI Global: Pennsylvania, 116–124.

Pauleit, S., Liu, L., Ahern, J. and Kazmierczak, A. 2011. Multifunctional Green Infrastructure Planning to Promote Ecological Services in the City. *Handbook of Urban Ecology*, edited by I. Douglas, D. Goode, M. Houck and D. Maddox. Routledge: London, 272–285.

Philips. 2018 [White Paper]. Smart Cities: Understanding the Challenges and Opportunities. Retrieved from www.smartcitiesworld.net/resources/resources/smart-cities---understanding-the-challenges.

Roo, M., Kuypers, V. H. M. and Lenzholzer, S. 2011. *The Green City Guidelines: Techniques for a Healthy Liveable City.* Elsevier: Wageningen, Netherlands.

Singh, J., Laurenti, R., Sinha, R. and Frostell, B. 2014. Progress and Challenges to the Global Waste Management System. *Waste Management & Research* 32(9): 800–812.

Stathopoulou, M. and Cartalis, C. 2007. Daytime Urban Heat Islands from Landsat ETM+ and Corine Land Cover Data: An Application to Major Cities in Greece. *Solar Energy* 81(3): 358–368.

Thorpe, D. n.d. What Has Being Europe's Green Capital Done for Copenhagen? And What Might Bristol Expect? Retrieved from www.smartcitiesdive.com/ex/sustainablecitiescollective/what-has-being-europes-green-capital-done-copenhagen-and-what-will-bristol-expe/1032456/.

Tîrlă, M. L., Manea, G., Vijulie, I., Matei, E. and Cocoș, O. 2014. Green Cities – Urban Planning Models of the Future. In *Cities in the Globalizing World and Turkey: A Theoretical and Empirical Perspective*, edited by R. Efe, N. Sam, R. Sam, E. Spiriajevas and E. Galay. University Press: Sofia, Bulgaria, 462–479.

Tomar, P., Kaur, G. and Singh, P. 2018. A Prototype of IoT-Based Real Time Smart Street Parking System for Smart Cities. In *Internet of Things and Big Data Analytics Toward Next-Generation Intelligence*, edited by N. Dey, A. E. Hassanien, C. Bhatt, A. S. Ashour and S. C. Satapathy. Springer International Publishing: Cham, Switzerland, 243–263.

UN-Habitat. 2016. *World Cities Report 2016: Urbanization and Development–Emerging Futures.* UN-Habitat: Nairobi.

United Nations, Department of Economic and Social Affairs, Population Division. 2018. *World Urbanization Prospects: The 2018 Revision, Methodology.* Working Paper No. ESA/P/WP.252. United Nations: New York.

Van Veenhuizen, R. 2006. *Cities Farming for the Future – Urban Agriculture for Green and Productive Cities.* RUAF Foundation, IDRC and IIRR Publishing: the Philippines.

Wheeler, S. M. 2013. *Planning for Sustainability: Creating Livable, Equitable and Ecological Communities.* Routledge: London.

World Health Organization. n.d. Air Pollution. Retrieved from: www.who.int/airpollution/en/.

3
Green Smart Buildings for Smart Cities

Dushyant Singh Chauhan and Gurjit Kaur

CONTENTS

3.1 Introduction 44
3.2 Green and Smart Building Trends 45
    3.2.1 Zero-Energy Buildings (ZEBs) 45
    3.2.2 Building Enabled by the Internet of Things 47
3.3 Features of the Green Smart Building 49
    3.3.1 Life Cycle Assessment (LCA) of a Green Smart Building 49
    3.3.2 Structural Design Efficiency of a Green Smart Building 49
    3.3.3 Energy Efficiency in a Green Smart Building 50
    3.3.4 Water Efficiency in a Green Smart Building 50
    3.3.5 Waste Reduction in a Green Smart Building 50
3.4 Passive Design Technologies 51
    3.4.1 Building Envelope 51
    3.4.2 Building Envelope Technologies 54
    3.4.3 Design Factors for Passive Cooling Technologies 58
    3.4.4 Passive Lighting or Daylighting 61
    3.4.5 Passive Heating Technologies 64
    3.4.6 Benefits of Passive Solar Heating 65
3.5 Active Design Technologies 65
    3.5.1 Green Smart Heating, Ventilation and Air Conditioning
          (GSHVAC) 66
    3.5.2 Electrical Lighting 66
    3.5.3 Energy-Efficient Elevators 66
    3.5.4 Plug and Process Load Management 67
3.6 Green Smart Building Standards and Regulations 67
    3.6.1 Building Standards 67
    3.6.2 Green Building Codes 68
    3.6.3 Green Product Certification 70
3.7 Challenges in the Adoption of Green Smart Buildings 70
    3.7.1 Limited Awareness 70
    3.7.2 Insufficient Government Policies 70
    3.7.3 Extra Approvals and Clearances 71
    3.7.4 Poor Incentives to Motivate Adoption 71

3.7.5 Overpriced Equipment 71
3.7.6 Unskilled Manpower.. 71
Conclusion 71
References... 72

3.1 Introduction

Smart buildings in general denote structures that utilize automation to control the building's operational processes such as ventilation, air conditioning, heating, security and lighting according to user requirements. To achieve this, newly emerged sensors, actuators and chips are now used in many smart buildings to collect and manage data according to a particular type of service. Figure 3.1 shows the features of a typical green smart city that are essential for its sustainable development. These types of infrastructure usually help operators and owners to improve the reliability and performance of the assets located within the building structure with minimal energy consumption, space usage and minimal detrimental impacts on the building as a whole (Zuo and Zhao 2014).

Green building can be defined as a process of building structures that utilize energy-efficient resources and methods, accountable to the environment during the complete lifecycle of a building. The economy is one of the main parameters in the development of conventional as well as modern green smart buildings, and is essential to take into consideration. The difference between smart and green buildings comes from the concept of energy efficiency. Figure 3.2 shows the similarity between smart buildings and green buildings.

Sustainable building is considered as the most crucial step in the development of green buildings. It is generally considered to involve the implementation of sustainable methods that take into account not only the environmental conditions of the building site, but also the best use of resources during the subsequent building stages. Figure 3.3 shows various stages of the green building lifecycle. These green smart buildings are mainly useful for reducing pollution and environmental degradation, which not only helps in improving the residents' well-being and productivity but also protects their health in a much more holistic way (Hernandez et al. 2014).

During the last few decades, with the rapid agglomeration of smart cities around the world, much attention has been given to the unfavorable relationship between normal economic growth and over-exploitation of natural resources. Stress on the environment as a whole has increased due to the production of harmful waste materials from land shedding, pollutants and the production of building materials such as asbestos, lead and radioactive sources.

Green Smart Buildings for Smart Cities

FIGURE 3.1
Features of green smart cities.

3.2 Green and Smart Building Trends

The growth of smart cities around the world has been accompanied by the widespread acceptance of various green technology-based building approaches. Some of these green smart technologies and trends in building structures will be described in this section.

3.2.1 Zero-Energy Buildings (ZEBs)

ZEBs are energy-efficient commercial or residential buildings that achieve greater energy efficiency through the renewable energy generated at the site.

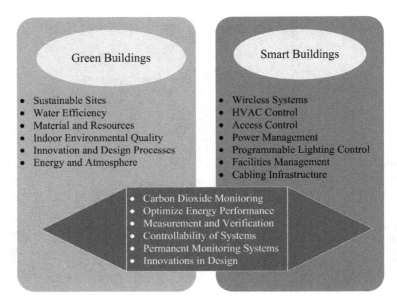

FIGURE 3.2
Similarity of smart buildings and green buildings.

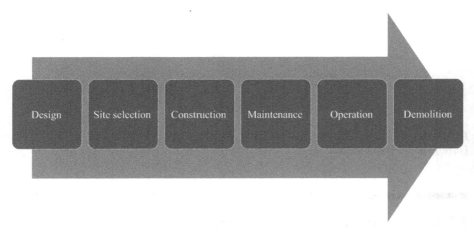

FIGURE 3.3
Various building life cycle stages of a green building.

On an annual basis, these buildings have generally used less energy than, or equal to, the level of renewable exported energy. A ZEB produces a sufficient amount of renewable energy to meet its requirements; therefore, the usage of non-renewable energy can be reduced in the building. These buildings use most of the cost-effective techniques to reduce energy usage through

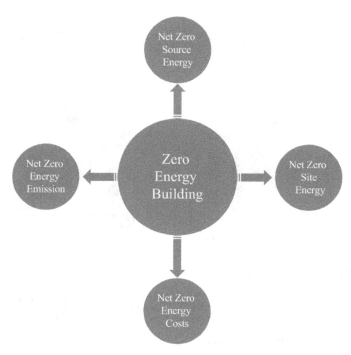

FIGURE 3.4
Terms associated with ZEB for green smart buildings.

energy efficiency. There are numerous long-term advantages of adopting these ZEBs, including lower operating and maintenance costs, lower environmental impacts, better capacity to recover quickly from natural disasters and improved energy security (Bucur and Moga 2018). A ZEB can be defined by the terms shown in Figure 3.4. These terms are useful for a set of buildings that consume less energy.

3.2.2 Building Enabled by the Internet of Things

In recent times, the number of sensors and devices linked to the Internet has increased at a rapid rate, resulting in the development of a much more connected and smarter world. With the onset growth of this technology on a broad spectrum, the concept of the Internet of Things (IoT) is on course to play an important role in the overall development of smart cities in the near future. IoT-based systems have been increasingly adopted in the building sector in order to make conventional buildings much more efficient and smart. For example, a huge number of buildings have to date been consuming a lot of energy for their various operations. One of the main objectives behind using smart IoT-based solutions in conventional building structures is to monitor,

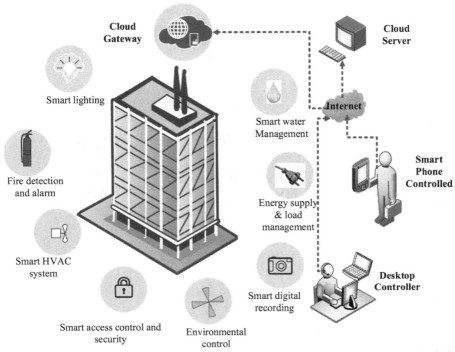

FIGURE 3.5
IoT-enabled smart building.

minimize and manage the consumed energy of the whole building architecture in a more efficient way. It has been observed from various examples that heating, ventilation and air conditioning (HVAC) systems usually result in the consumption of a large amount of energy within buildings. In addition, plug loads and lights are also responsible for high energy consumption. Hence, various types of IoT sensors and devices can be installed inside mechanical systems such as the HVAC systems of smart buildings to make these systems adaptive and intelligent, enabling much more energy-efficient usage of the different associated operations (Pöhls et al. 2014). Another significant advantage that comes with this is the accumulation of a large amount of vital data collected from the smart embedded sensors through the connected controllers mounted in different spaces inside the smart building structures. The data acquired from these sensors is usually collected, filtered and analyzed for the use of smart building analytics that ultimately help the concerned authorities to regulate and exercise different essential functions through projections. Therefore, it can be concluded that there will be increased interest in using big data management analytics in smart building development for future green smart cities. Figure 3.5 shows an IoT-enabled smart building.

3.3 Features of the Green Smart Building

The concept of green smart building development mainly initiated from the aspiration for environmentally friendly and more energy-efficient construction practices. Alongside this, a variety of motivations are responsible for the development of green smart building structures in today's world. Most of the motivations are linked to the social and environmental benefits which can be derived from sustainable design approaches. In general, these approaches incorporate green practice and the green building life cycle into the design purpose, with different methods being used alongside each other in the development (Jadhav 2016).

Green smart building therefore involves an enormous collection of techniques, skills and practices which are implemented to minimize and eventually eradicate the negative effects of buildings on human health and the environment (Edwards 2006). Moreover, it emphasizes taking advantage of renewable resources, for example, wind through proper ventilation and sunlight through solar equipment. Some green smart buildings use very low-impact building resources or porous concrete material over conventional concrete to boost the renewal of groundwater. The tradition and methods associated with green building are thus growing steadily and may vary from one region to another. The following sections present the essential principles of a green smart building.

3.3.1 Life Cycle Assessment (LCA) of a Green Smart Building

This is an organized set of processes for inspecting and compiling the outputs and inputs of materials. The last few years have seen a shift away from the conventional approaches in terms of green smart building development, which assumes that specific approaches are better for the environment.

3.3.2 Structural Design Efficiency of a Green Smart Building

The concept and design phases comprise the primary roots of any construction plan. The concept phase includes the most essential and significant steps in the project life cycle which help to provide an overview of the objectives for the development. Moreover, this phase has a primary impact on the overall performance and associated costs. In designing smart green buildings, the primary goal is to reduce the overall environmental impact associated with various life cycle stages of the building project. In addition, the building includes much more complex products composed of an assembly of materials constituting several design variables which need to be defined at the design stage (Magno et al. 2015). During the building life cycle stages (raw material extraction, manufacturing, construction, operation and maintenance, demolition and disposal, reuse or recycling), variation in each design variable holds the potential to affect the environment in many ways.

3.3.3 Energy Efficiency in a Green Smart Building

Green smart buildings are generally comprised of features that reduce energy consumption to a much greater extent as compared to conventional buildings. Embodied energy is required to process, transport and install building materials, while operating energy is necessary to maintain heating and lighting equipment. Much greater importance has been assigned to embodied energy, which accounts for as much as 30% of all energy consumed throughout a building's lifetime (Ruuska and Häkkinen 2014). The US Life Cycle Inventory Database Project (Trusty and Meil 2002) shows that buildings mainly built with wood will have lower embodied energy than those built mostly with brick, concrete or steel. Designers can reduce the operating energy usage by using features that minimize air leakage through the building envelope. They can also specify extra insulation in walls and high-performance windows, floors and ceilings that are vital for a range of applications.

In low-energy homes, another strategy known as passive solar building is frequently employed. Trees should be grown in such a way that windows, porches and walls can be shadowed by them during the summer. Moreover, more natural light can be achieved by effective window placement, reducing the necessity for electric lighting in the daytime. Additionally, heating water through solar technology minimizes the cost of heating for a number of day-to-day activities. The environmental impact of a building can therefore be significantly reduced by the generation of renewable energy sources through the power of wind, solar, water and biomass. Generally, power generation in the building is supposed to be the most expensive feature; therefore, energy efficiency is the main feature that needs to be taken into consideration when designing a green smart building.

3.3.4 Water Efficiency in a Green Smart Building

The main objectives of a sustainable building can be attained by reducing the on-site water consumption and safeguarding water quality. One of the most severe issues in many areas of water consumption is that demand exceeds the ability to refill the groundwater. By designing dual plumbing techniques, the conservation and protection of water in a building can be accomplished. By utilizing water-conserving fixtures, the waste of water may be reduced.

3.3.5 Waste Reduction in a Green Smart Building

During the process of construction, the wastage of water, energy and materials used can be minimized by seeking green architecture. A survey by the California Integrated Waste Management Board (1999) showed that around 60% of unwanted materials come from commercial buildings. One of the vital goals during the process of construction should be to minimize the matter going to landfill sites. Well-designed buildings also help in reducing the waste material generated by the residents by providing suitable solutions to the residents on-site, such as bins to compost waste materials going to landfill. The buildings are typically demolished and hauled to landfill sites when they become old. With the assistance of a method known as deconstruction, the harvesting of waste can be reclaimed into useful building materials. Structures which use construction materials such as wood make the restoration of structures much easier.

3.4 Passive Design Technologies

The carbon footprint of any building can be improved by utilizing passive design technologies that help by functioning with ambient conditions. These passive design technologies not only consume less energy, but they also minimize overall energy usage in the building. These technologies are therefore considered an essential aspect of designing green smart buildings for future green smart cities. However, these technologies are not actively controllable, so are not viewed as being active enough to tackle the requirements and variable demands of buildings. Once a building design is finalized, passive design features cannot be changed to meet specific operating needs (Mazzucchelli et al. 2018).

3.4.1 Building Envelope

The physical barrier between the interior and exterior environments surrounding a structure is the building envelope. This envelope generally encompasses a series of systems and components that protect the inside space from the effects of various environmental factors such as wind, precipitation, humidity, ultraviolet radiation, temperature and so on. The atmosphere inside a building envelope is generally composed of the occupants, furnishings, lighting, building materials, machinery, equipment and the HVAC system (Shahin 2019). Figure 3.6 shows the building envelope of a green smart building.

A building envelope has many functions, which can be separated into three categories:

1. Support: the building should guarantee rigidity and strength and must provide structural support against external and internal loads and forces.

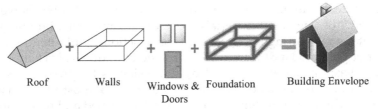

FIGURE 3.6
Building envelope for green smart buildings.

2. Control: the building must be able to control the exchange of water, air, condensation and heat between the exterior and interior of the building.
3. Finish: the finish of the building should be appropriate while still performing support and control functions. This feature concerns the associated aesthetic value of the building.

Building Envelope Components (BEC) may be categorized into two components: transparent and opaque components. The transparent components of a building envelope comprise ventilators, windows, doors, skylights that are more than one-half glazed and glass block walls. The roofs, walls, basement walls, slabs and opaque doors are examples of opaque components. Physical protection from weather conditions, indoor air quality, climate, durability and energy efficiency account for the conventional methods of effectiveness associated with the building envelope. The consumer should be both psychologically and environmentally satisfied (Sadineni et al. 2011). Psychologically, external or outside views are very important in a building development. Environmentally, buildings should be able to respond to radiation from the sun's heat and be capable of minimizing heating loss and noise to a much larger extent than conventional buildings. Figure 3.7 shows the components of the building envelope and physical processes.

3.4.1.1 The Role of Determinants in Building Envelopes

Concerns relating to the internal and external loads from heat, as well as the advantages of daylight, must be taken into consideration in the BEC design. To achieve workability, the highest efficiency and sustainability, BECs are always designed with an objective to achieve environmental, technological, functional and aesthetic design determinants. Numerous models and equations have already been developed based on these dependent variables. An expression of satisfaction with the thermal environment is generally known as the thermal comfort. However, this concept usually depends upon environmental parameters, such as behavioral factors, physiology and psychology. The vital factor that plays a major role is thermal comfort. Figure 3.8 shows the variables capable of measuring comfort levels.

Green Smart Buildings for Smart Cities

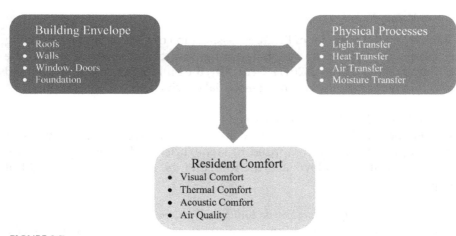

FIGURE 3.7
Components of the building envelope and physical processes.

FIGURE 3.8
Comfort level of residents measured by various variables.

Types of Building Envelope

To interpret the behavior of BECs, a classification of various kinds of building envelope is required to maintain appropriate climate control within the building (Dascalaki et al. 2011). Generally, the building envelope is classified into different parts, the two most essential of which are as follows:

a. Single-Skin Facade Building Envelope
In this, the building is enclosed by a single-skin envelope. These types of building are generally comprised of stone/brick walls with openings for the placement of windows and the roof, and the provision of a skylight, if required.

b. Double- or Multiple-Skin Facade Building Envelope
The performance of a building can also be improved by the concept of a double-skin facade. In these systems, extra skin is applied to a structure with a cavity between the outside facade and external wall. A double-skin facade envelope has an advantage over the traditional building facade system in terms of the working and operational system as a whole. The requirements of mechanical service systems and service installations in these buildings can be minimized by optimizing the cavity between the two facades.

3.4.2 Building Envelope Technologies

These are the techniques through which the heat loss in the building can be reduced. In general, the building envelope functions as a partition between the inside and outside environment of a building. Some of the technologies that make these buildings greener are summarized below.

3.4.2.1 Cool Roofs and Coatings

As per the report presented by the Cool Roof Rating Council (CRRC), black surfaces usually become hotter than white reflective surfaces with the heat of the sun. The system life span shortens due to continuous exposure to heat. Roofs which become hotter due to an increase in internal building temperatures require the cooling systems to work harder. A heat island effect observed in such cases mainly arises due to trapped heat inside the roofs, in which the temperature of the surrounding roofs rises fast and remains even hotter after the sun has gone down. As a result, there is an increase in energy demand, greenhouse gas emissions and air pollution.

On the other hand, a cool roof is one that absorbs less heat and has been designed to reflect more sunlight than a traditional dark color roof. A particular type of white pigment is used for cool roof coatings that reflect light from the sun. Roof coatings are very thick paints that can shield the surface from chemical damage and some offer protection from water. A cool roof can benefit nearly any type of building, but the climate and other factors have to

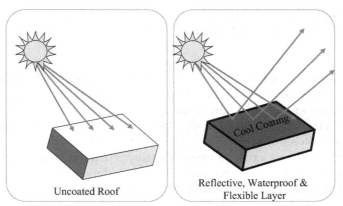

FIGURE 3.9
Roof of building coated with reflective material.

be considered before the development of cool roof coating structures. A cool roof under the same environment could stay cooler and save money and energy by using the air conditioning system less frequently. Figure 3.9 shows the roof of a building coated with refractive material (Altan et al. 2019).

Effectiveness of Cool Roofs
The effectiveness of cool roofs can be measured by using the following:

1. Solar reflectivity: this is the ability of a roof to reflect heat and sunlight away from a building. Solar reflectivity is measured on a scale from zero to one. An effective cool roof is the one that absorbs a small percentage of solar energy inside the structure and reflects more than 65% of it. However, traditional roofing materials have been observed to reflect only 5–15% of the energy received from the sun.
2. Solar Reflectance Index (SRI): this is the ability of the material to reflect heat from the sun. It is essential in determining how well a material and its color will work. In general, SRI is measured on a scale from 0–100%. For dark surfaces, zero refers to a high temperature, while for white surfaces a value of 100% represents a cool temperature.
3. Thermal emittance: this refers to the ability of a roof to release absorbed heat. Thermal emittance is also measured on a scale from zero to one. In warmer areas, a high value of thermal emittance is desirable so that the heat is not held within the roof. By properly adopting all these measures the building usually becomes cool.

Types of Cool Roofs
When various sectors (e.g. residential, commercial and industrial) are considered, different roof types can be used for different structures. There

are multiple types of cool roof techniques to suit different types of building designs. The various methods that are used for different types of roofs are outlined below:

1. Low-sloped roofs: these types of roof have an extremely flat roofline and are generally used for draining needs with a slight incline. These types of roof are used in industrial, commercial and institutional buildings. Because of their vast roof surface areas, they are supposed to be the best alternatives to standard cool roofs. Heat enters a building through the roof, and a cool roof can significantly help in lowering energy costs and heat gain. Cool roof techniques for low-sloped roofs generally include the following points during the development:
   - Coated roofs: to help enhance a roof's durability and adhesion while simultaneously reducing the growth of bacteria, roofs are glazed with a paint-like finish. To implement various cooling technologies, white paint is not used directly; instead, a variety of colors can be pigmented onto the roof. Most coatings in such roof types are rated according to different energy star ratings.
   - Foam roofs: these roofs are covered by a material such as foam for isolation. Over the last few years, foam roofing has become known for being a long-lasting, dependable and inexpensive technique for cool roofing. The foam is usually made with the help of liquid chemicals that combine to form a flexible, solid and lightweight physical substance that attaches seamlessly. It has proven itself to be sustainable by creating minimal waste and minimal maintenance requirements.
   - Single-ply membranes: where more extensive repair is required, single-ply membranes are used. These membranes are prefabricated sheets that are independently applied to the top of the roof. The principal types of single-ply membranes are single-ply thermoplastics and single-ply ethylene propylene diene monomer.
2. Steep-sloped roofs: these types of roof are generally seen in residential areas with an inclined roofline. Asphalt shingles, tiles, shakes materials and metal roofing are the materials used for steep-sloped roofs. Various techniques for cool roofing are used for these kinds of roof because of their various materials and structures.

3.4.2.2 High-Performance Glazing

Bay windows, doors, conservatories and so on are zones where heat is unable to escape. The weakest thermal component in the building envelope is uncoated single-glazed windows that transmit large amounts of heat in and out of a building. With the help of high-performance glazing, the performance of buildings can be improved, which can help in reducing the

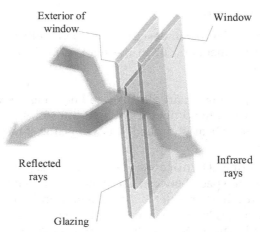

FIGURE 3.10
Effect of glazing materials on the windows of a green building.

energy consumption and lowering carbon dioxide emissions to much more appropriate levels (Raji et al. 2018). Figure 3.10 shows how the sun's rays are reflected back due to glazing materials placed on the windows.

Characteristics of High-Glazing Energy Performance

The following characteristics can fully specify a window system:

1. U-value: this value defines the heat flow rate due to radiation, convection and conduction through a space causing a temperature gradient between the outside and inside. In winter, as this factor is high, more heat is lost through the window.
2. Smart green solar heat gain coefficient (SGSHGC): this specifies the quantity of heat transmitted through the window when the solar energy strikes the window fully. An increase in the SGSHGC value causes the solar gain potential to also increase.
3. Visible transmittance: sunlight is a form of electromagnetic energy exchange between the earth and sun. This value indicates the percentage of the visible portion of the solar spectrum that is transmitted through a given glass product. It comprises a range of electromagnetic (EM) wavelengths, collectively known as the solar spectrum.

Nowadays, the latest technologies for enhancing the performance of double-pane windows and standard single-pane or glazing units are being explored and made available. These latest technologies are multiple-pane glazing, low emittance glass coatings, inert gas fills, adaptive glazing and selective transmission films.

Types of Glazing Materials

Glazing plays a crucial role in designing an energy-efficient building (EEB). As various advancements occur in the field of EEBs, a variety of glazing materials are used.

1. Switchable glazing: smart glass is also known as a switchable glass or glazing in which the properties of light transmission are altered when light, voltage or heat is applied. It is a versatile solution for businesses and homes.
2. Vacuum glazing: if two or more glass panes are separated about 0.22 mm apart, and the space between the panes is discharged or vacuum, to minimize the heat transfer by conduction and convection, then this type of glazing is known as vacuum glazing. As a result, the U-value is reduced to 1 watt/m^{2}K. Also, the innermost glass pane is coated with a low emissive thin film.
3. Suspended particle devices (SPDs): in SPDs, very small particles are suspended in a liquid and inserted between two glass panes. These nanoparticles are aligned on applying a voltage; similarly, the intensity of light can be controlled by regulating the applied voltage.
4. Electrochromic devices: some special materials alter their opacity when a certain electric potential is applied to it. These materials are in general often seen to change their state from colored to transparent. This change in state occurs from the edge and usually takes a few minutes depending on the window size.

3.4.2.3 Natural Ventilation and Cooling

These types of ventilation systems use air from the outside environment to cool the building environment without using cooling from mechanical instruments present inside the structure. Air supply is provided through open windows or air intake louvers, either automatically linked or manually controlled by a building management system (BMS) (Raji et al. 2019). The features needed for natural ventilation to be more efficient and effective are as follows:

1. Good air quality to achieve temperature stability, which is appropriate for the proposed purpose of the space; and
2. Bringing sufficient air inside the spaces to replace the accumulation of heat and harmful gases.

3.4.3 Design Factors for Passive Cooling Technologies

In recent times, architects and building designers have been building their trust in using natural ventilation systems for those spaces where there is no continuous human occupancy. Examples of such spaces include stairwells or

Green Smart Buildings for Smart Cities

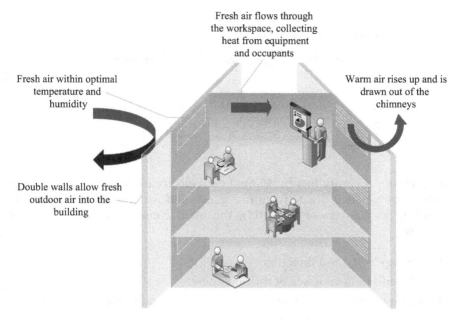

FIGURE 3.11
Cooling a building using natural ventilation.

corridors. However, the specific approach for designing natural ventilation systems will differ based on the type of building and local climate conditions that apply to a given structure. Figure 3.11 shows how a passive cooling technique like natural ventilation helps in cooling the building (Al-Sallal and Abdelhamid 2018). To increase the natural ventilation effect in buildings, the following technologies could be used.

3.4.3.1 Orientation and Massing for Natural Ventilation

There are many vital design aspects that need to be considered for passive cooling, specifically for natural ventilation. Wind speed is greater at higher elevations, meaning that taller buildings hold the potential to improve natural ventilation. In addition, taller buildings can reduce sun exposure if they are oriented correctly in a precise manner. With the help of good orientation and massing, the sustainable benefits associated with the building can be maximized. Table 3.1 shows the Beaufort scale used to classify wind speed for natural ventilation in buildings.

3.4.3.2 Openings for Cross-Ventilation

To encourage cross-ventilation, windows must be located on opposite walls. Also, the kinetic energy of airflow must be lost each time an obstacle comes

TABLE 3.1

Beaufort Scale to Measure Wind Speed

Beaufort Scale	Type of Wind	Wind Speed (m/s)
1	Calm, light air	0–1.5
3	Gentle breeze	3.4–5.4
5	Fresh breeze	8.0–10.7
7	Near gale	13.9–17.1
9	Strong gale	20.8–24.4

within close proximity to a building. Low-velocity airflow inside buildings can effectively halted due to furniture, goods and walls within the building. According to Gautam et al. (2019), the main constraints on the airflow levels are:

1. wind velocity and direction;
2. inlet and outlet surface of the openings;
3. a temperature difference between the outdoor and indoor environment;
4. relative wind shadowing; and
5. relative position of the openings of the building.

3.4.3.3 Wing Walls

With the help of wing walls, most of the natural air is allowed to enter the home. By utilizing a cross-ventilation system with a pairing of two wing walls (one larger and one smaller), airflow in the building can be improved. These architectural structures have often proved to be helpful in navigating winds for improving natural ventilation. The associated pressure gradient is generally responsible for pulling the air from the outdoor to the indoor environment on a regular basis. Therefore, these wing walls are often considered the best option where the outdoor wind velocity is less and where the direction of the wind can vary significantly. The main advantages of these wing walls are that there are no maintenance costs, it is an effective natural cooling system and the walls are easy to install (Nejat et al. 2018).

3.4.3.4 Stack Ventilation

As the air gets hot, it becomes less dense and lighter. There is a tendency for warm air to rise, and this effect of hot air is used to ventilate buildings naturally. Cool air from the outside is drawn inside the building at a lower height. This air is warmed by the heat generated from sources within the building (such as equipment, people, heating and solar gain), and then rises up to vent out through the building. In summer, to avoiding mechanical cooling, stack ventilation can be considered an effective mechanism for passive cooling in a building. In winter, however, it has a negative effect (Song et al. 2018).

The effectiveness of stack ventilation is subject to:

1. the overall effective open area;
2. the stack height;
3. the temperature gradient of the top and bottom of the stack; and
4. the pressure gradient outside the building.

3.4.3.5 Solar Chimney

A solar chimney is an example of both a passive cooling and heating system. The temperature and ventilation of a building can therefore be controlled by a solar chimney. The solar chimney directly gains warm air and causes it to rise while collecting cool air from the bottom. CO_2 emissions, energy use and pollution can be reduced by using a solar chimney, which helps in benefiting the passive cooling and natural ventilation policies of buildings (Chaichan et al. 2018) and hence contributes in moving towards a much greener building structure.

3.4.4 Passive Lighting or Daylighting

When daylight is used to brighten indoor spaces instead of lighting from artificial sources, it is known as passive lighting. The energy consumption can be significantly reduced in buildings which are primarily in operation during the daytime. Natural light or daylight have many associated psychological and physiological advantages. However, it is important to differentiate between daylight and sunlight. Sunlight is powerful and results in surplus heating of spaces. Daylight is not powerful and can be used to light up the inside spaces of a building (Stritih et al. 2018).

3.4.4.1 Methods to Efficiently Use Daylighting

The following are techniques that use daylight effectively (Jakica 2018) to brighten the indoor spaces of a building:

Light Shelves

A light shelf (Figure 3.12) is a horizontal platform situated near windows to reflect light deep into a room. They are usually placed above eye level on high windows and can be present either indoors or outdoors. Light shelves often reduce glare close to a window and increase the availability of natural light away from windows. In many cases, they may be paired with adaptive electronic lighting that dims when natural light is available. In theory, light shelves could dynamically control their reflective properties to respond to indoor temperatures or to allow people to control lighting levels. In practice,

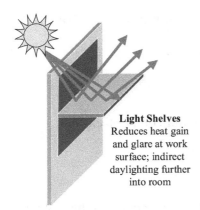

FIGURE 3.12
Benefits of incorporating light shelves into a green building.

they are more often paired with other control mechanisms such as smart windows.

Translucent Concrete

Translucent concrete is a type of concrete that allows light to pass through it. It is usually based on a fine concrete that encases optical fibers designed to let light move from one side to the other. The fibers are typically visible on both sides of the concrete and may be placed to create a pattern of light. Objects on the light source side of the concrete may appear as shadows on the other side. Due to this unique feature, it is used in facades, interior walls and ornamentation.

Lightwells

A lightwell is an open space within the volume of a building designed to provide light and ventilation to interior spaces. Traditionally, lightwells are unroofed spaces open to the sky, and windows line the walls facing the lightwell. These windows provide a degree of airflow and light to the deep interiors of large buildings. The open space provided by a lightwell may also be usable as a small outdoor area or garden.

Sunlight Transport

Sunlight transport is a passive system that captures sunlight at a source, such as a roof, focuses it, transports it by a medium such as fiber optic cables and diffuses it into a room. It is typically used to get natural light into interior rooms and to reduce energy consumption. Sunlight transport is typically paired with adaptive lighting that dims when natural light is available.

Light Tubes

Light tubes commonly have a lens that captures light, a pipe of reflective material and an element for diffusing light into a room. They also can be paired with a dynamic electronic lighting system that dims when natural light is available. In some cases, a light tube also provides controls so that it can be turned off or dimmed.

3.4.4.2 Mirror Ducts and Light Pipes

Light pipes or mirror ducts are structures that work as optical waveguides for transferring and dispensing light for brightness. They are often known as sun pipes or sun tubes due to their use as daylighting. Light pipes normally involve sunlight being brought into the interior by the phenomenon of reflection inside the pipe, and therefore the light ray supply becomes the lighting function inside the building (Fakourian and Asefi 2019). Figure 3.13 shows the working principle of light pipes and mirror ducts.

The light guided through the pipe is, respectively, from two physical phenomena:

1. the total internal refraction of a glass pipe; and
2. the reflection of a mirror.

Components of a Mirror Duct

There are two essential components of a light pipe and mirror duct:

1. a dome that collects light is typically installed like a skylight, but a hemisphere is used to bring light from different angles; and
2. a hollow tube that is typically made of metal that reflects light and helps in passing light over the length of the tube attached at one end.

Benefits of the Mirror Duct System

There are several benefits of incorporating a mirror duct system into a green building. Some of those benefits are listed as follows.

3.4.4.3 Transmission of the Natural Light to Any Area

The light duct can introduce natural light into the building in two ways: one is the horizontal type, where light is captured at the exterior wall; the other is the vertical type, where light is admitted via the roof and rooftop. The vertical type seems ideal when the building structure has depth and where light access is difficult.

3.4.4.4 The Effect of Energy Conservation

After installation, the mirror duct system requires no energy to run or maintain it. It will continue to bring in natural light throughout its lifetime of service, enabling considerable energy consumption saving. In the actual case of an office building, the saving in power consumption for lighting was calculated

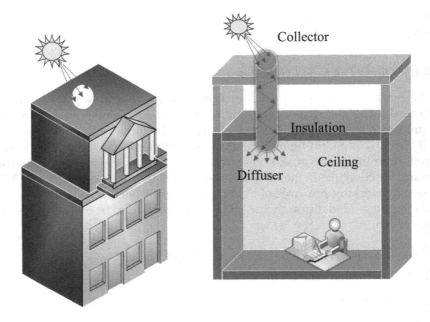

FIGURE 3.13
The working principle of light pipes and mirror ducts used in a green building.

to be around 65% of conventional consumption before the installation of the mirror duct system (Nikken Sekkei n.d.). In addition, the CO_2 discharge from lower lighting power consumption has been shown to reduce drastically.

3.4.4.5 Comfortable and Healthful Light

Natural light is part of the natural world that supplies a balanced mix of nutrients on which various life-forms depend. It also provides healthy portions of UV rays that serve a sanitizing function by eliminating harmful microorganisms.

3.4.5 Passive Heating Technologies

These are the technologies that utilize the rays from the sun to warm a living space by directing sunlight to the area along with minimizing the energy usage by mechanical systems such as HVAC. Sunlight can warm a space even through absorption of heat in the walls or roof and can heat the interior surfaces through the long wavelength of sunlight, called infrared radiation, which is also commonly known to exist in the form of heat in general. Passive solar buildings benefit from how the sun rotates and moves throughout the day to warm living spaces, without the requirement of any high-efficiency system or fuel to do so (Albayyaa et al. 2019).

The five important factors of passive solar designs typically include absorbers and thermal mass to absorb and retain the heat, an aperture for the sunlight to pass through, distribution for heat circulation and a fixed control that will provide shade during the summertime.

3.4.6 Benefits of Passive Solar Heating

Passive solar heating has several benefits: the most important factor, however, is reducing energy bills. Apart from this, the whole passive solar heating system is considered a cleaner alternative to non-renewable variants. Further, there are no associated operational costs or maintenance requirements.

Since there are no greenhouse gas emissions or pollutants released into the air, passive solar heating is considered an effective environmentally friendly solution. On the other hand, most other methods for heating a building use fossil fuels for power, for example, furnaces, electric resistance or space heaters, which release harmful pollutants and greenhouse gases into the air.

Designs utilizing passive solar heating have numerous benefits: one of the most important is the absence of noise from furnaces and radiators when the passive solar heating process is used. Besides improvements and modernization in solar panels, various new technologies allow solar power to be stored in solar panels. A large area of commercial and residential buildings offer creative solar solutions by utilizing these solar-embedded products. Some of these products are:

1. Solar windows: these windows look like ordinary windows, and as sunlight passes through, energy is generated by the phenomenon of quantum dot coating in the glass. These represent an exciting technology that emerged on a bigger scale in 2018.
2. Solar shingles: this integrates innovative solar technology into modern roof shingles. One company, Tesla, offers owners a way to incorporate solar energy creation to their roofs without any noticeable change in the previous roof with their textured, smooth, French slate glass tiles. Solar shingle tiles are more robust and provide better insulation than ordinary roofing materials.

3.5 Active Design Technologies

With the help of active design technologies, all the energy requirements and various operating conditions can be dynamically controlled. Since more energy is consumed by active design technologies, energy efficiency is an essential aspect that is needed in active design technologies. The energy efficiency of any green smart building is the ratio of the output energy delivered to the supplied energy.

Active design technologies may be seen in the following areas.

3.5.1 Green Smart Heating, Ventilation and Air Conditioning (GSHVAC)

In buildings, through equipment known as GSHVAC, residents' comfort may be achieved in a more controlled way. In warm climates, more cooling than heating is required in any building. In addition, in humid and tropical weather, dehumidifying the air to a comfortable level is required. Inside the green smart building however, human comfort depends on several factors.

Numerous factors often need to be considered to achieve the thermal comfort of residents in green buildings (Jazaeri et al. 2019). Some of these are as follows:

1. the air temperature surrounding the residents;
2. moisture content in the air, also known as relative humidity;
3. the rate of air movement or air velocity;
4. the temperature radiating from the surrounding areas;
5. generated energy from the human body; and
6. isolation due to the clothes the person is wearing.

The purpose of the active GSHVAC technologies is to attain optimum thermal comfort at the cost of minimum energy expenditure.

3.5.2 Electrical Lighting

Another key feature of green and smart buildings, apart from thermal comfort, comes from visual comfort. In every building, active or artificial electrical lighting is required to deliver the best visual comfort over the course of a full day apart from that provided by daylighting techniques. The aim of a green smart building must be to provide optimal visual comfort, while seeking to minimize energy expenses.

3.5.3 Energy-Efficient Elevators

As the numbers of multistory buildings are rising, the requirement for vertical movement in a building structure becomes more important. Elevators alone consume about 10% of the net energy in residential and office buildings. Therefore, taking into consideration energy-saving possibilities in elevators is essential. A simple step to minimize elevator energy consumption is to encourage and motivate the residents to use the stairs. It also helps in improving the health of the residents. However, in tall buildings, this is not always practical, and hence energy-efficient elevator technology is used, which helps to convert the very tall building structures into green smart buildings.

3.5.4 Plug and Process Load Management

Significant portions of the overall energy usage of a building generally comprises plug and process loads. These loads are often not directly correlated with general electrical lighting or cooling systems and usually do not offer comfort to the residents in any form. These plug loads arise from equipment that is plugged in and used by the residents, such as microwaves, refrigerators, printers, computers, wall-projectors, vending machines, mobile chargers, audio equipment and so on. Therefore, there are enormous opportunities to understand and manage these type of loads. The key challenge in dealing with plug loads can arise from the changing needs of the residents, and the usage patterns are often unknown.

3.6 Green Smart Building Standards and Regulations

In the developed countries of the world, governments have started to care about both sustainability and energy efficiency in the building sector. As a result, several standards, regulations and codes related to green smart buildings have been formed. There has been a rapid increase in the number of rating, certification and standards programs on the market to guide the delivery of high-performance green and smart buildings. There are approximately 700 green product certifications that are required for different applications. Different rating programs are used all over the world, and these programs differ in their methodology with some optional credits and outlining requirements (Vitunskaite et al. 2019). Figure 3.14 shows the various standards, codes and certification of a green smart building.

3.6.1 Building Standards

A building can be tested against a set of rules and criteria known as building standards. Some of the smart building practices standards are established through agreement procedures with various organizations such as the American Society for Testing and Materials (ASTM) and the American National Standards Institute (ANSI). The International Standards Organization (ISO) helps in instituting the governance of these standards and certifications to develop worldwide standards forming the foundation of industry norms that often then materialize as law.

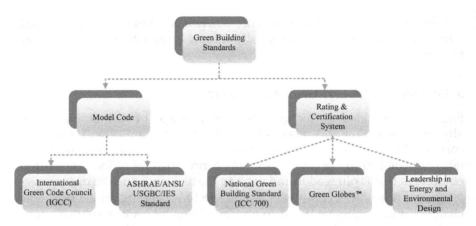

FIGURE 3.14
Various standards, codes and certification of a green smart building.

Some of the requirements found in building standards are usually based on stating expectations of end results, also known as performance-based standards. The standards formed by a formal and voluntary agreement process are known as agreement-based standards. These standards are demonstrated by an open process with international support, government support and an immediate buy-in. As per the law formed by the National Technology Transfer & Advancement Act (NTTAA), private sector, voluntary, consensus-based standards should be followed by federal agencies instead of forming proprietary, non-consensus-based standards. Today there are several proprietary or non-consensus-based standards for green buildings that have been established outside the International Organization for Standardization (ISO) and ANSI process.

The presence of ANSI/ASHRAE/USGBC/IES Standard 189.1 is a milestone addition to the International Green Construction Code (IGCC), which is a standard made for the design of high-performance buildings. It is a group of technically stringent necessities which, like the IGCC, includes criteria for indoor environmental quality, water use efficiency, materials and resource use, energy efficiency, and the building's impact on its community and the site. This standard was designed by building experts from across the building industry, who contributed particularly to the field of building construction.

3.6.2 Green Building Codes

A legal and official statement that highlights building safety features is defined in the building codes. Building energy codes are an essential policy to improve energy efficiency in existing and newly developed buildings. They are a set of rules and minimum standards to ensure the safety, welfare and health of the people living in the building for a particular period of time.

Codes are the statements that address all aspects of a building such as life safety, construction, structural, mechanical, electrical and plumbing aspects. Throughout history, the regulation of building construction can be traced back as far as 4,000 years.

The built environment has both indirect and direct impacts on the environment. Therefore, various green rating systems were established which help in reducing these impacts through the recognition, measurement and encouragement of the performance of sustainability.

The initial green rating system of the UK, formed in the 1990s, was the Building Research Establishment (BRE) rating system. The Building Research Establishment Environmental Assessment Method (BREEAM) was then followed by the Leadership in Energy and Environmental Design (LEED) rating system of the US in 2000. BRE announced the further expansion of BREEAM in partnership with, building-wise, the award-winning US-based LEED certification consultancy on 9 June 2016, to create BREEAMUSA. Today there are around 600 green certification systems all over the world. Table 3.2 shows some of the green rating systems that are commonly used for a green smart building. There are several benefits of a green rating system. Some of them include:

1. the objectives for the sustainable growth of the environment can be set;
2. the performance on the environment can be verified and explained to other parties;
3. enhancement can be demonstrated and measured;
4. education on green technologies can be promoted; and
5. marketing can be generated positively.

The primary difference between building rating systems and building codes is that codes are compulsory while the rating system is not. Their impact can alter the surroundings rapidly and extensively if green codes become widely accepted. In a building project, whether it is a restoration or a new construction undertaking, there is a need to check for a local green code, which will provide the most appropriate directive for proper usage.

TABLE 3.2

The Green Rating Systems That Are Commonly Used

System No.	Rating Agency	Country
1.	BCA Green Mark Scheme	Singapore
2.	BREEAM	UK
3.	CASBEE	Japan
4.	Green Globes	USA
5.	Energy Star	USA
6.	Pearl Rating System for Estidama	UAE

IGCC is proposed as a legal and municipal building code for major restorations and new building construction. It is a document of codes that can be readily used by designers, contractors and manufacturers that was created with the intention to be managed by code officials and accepted by the concerned government bodies.

3.6.3 Green Product Certification

The practice of verifying that a certain product has qualified or passed various performance tests is known as product certification. These certifications are proposed to plan and approve a product meeting a specific standard and offering benefits for the environment. Numerous certification programs and product tags verify products on the basis of lifecycle parameters; they are programs with multiple features. These parameters comprise recycled content, air, water emissions and energy usage from disposal and also manufacturing. Another highlight usually depends upon a single attribute, for example, chemical emissions, water and energy that directly influence indoor environmental quality (IEQ).

When a third party other than a developer is responsible for conducting a test and presenting a certificate, then green product certification is highly desirable. The meaning of third party is that they have no relationship with the contractor, manufacturer and designer.

3.7 Challenges in the Adoption of Green Smart Buildings

Green smart building practices are gaining growing acceptance among designers due to the feasible solutions they offer for meeting the growing demand for environmentally friendly and self-sustaining smart buildings (Ejaz and Anpalagan 2019). Some of the challenges that prevent the widespread adoption of these green smart building technologies are discussed in the following sections.

3.7.1 Limited Awareness

Even today, a very large group of residents are unaware of green smart buildings and their persistent benefits. Also, users who know little about green smart buildings view them as an expensive and economically unachievable option.

3.7.2 Insufficient Government Policies

Although governments have set ambitious targets for a green smart building agenda, sadly this is not generally complemented with government rules and regulations to encourage growth. There is a lack of appropriate government rules and guidelines to impose large-scale implementation of green smart building norms.

3.7.3 Extra Approvals and Clearances

Builders, designers and developers have to go through a very monotonous process when it comes to clearance and approvals. Adding to this burden is the list of approvals required for green building agreements, which serves as one probable factor that discourages the rapid adoption of green smart buildings.

3.7.4 Poor Incentives to Motivate Adoption

As of now, there are few incentive plans available to motivate green smart building adoption. In addition, the ones that exist are also not uniform across different cities. Mostly, green smart building incentives are in the form of added floor-area-ratio (FAR), which is followed by a rebate on the tax imposed on property and other similar schemes. However, throughout the world, this encouragement and incentives have often failed to motivate any large-scale adoption of green smart building practices.

3.7.5 Overpriced Equipment

The equipment and products used in the development of green smart buildings are too expensive compared with conventional ones. Many builders are concerned that adopting green features into their buildings will involve high costs; hence they are reluctant to invest in them.

3.7.6 Unskilled Manpower

One of the biggest factors holding back green smart building adoption is the lack of skilled workers. From policymakers to architects, engineers to contractors and workers, none of the groups possess the adequate knowledge and skills needed for green smart building construction.

Conclusion

In this chapter, smart green building technologies have been discussed. Since these technologies are new, the rate of widespread adoption is currently very low. These smart technologies must therefore be incorporated and implemented in the early stages of the building life cycle in order to move towards implementing greener smart building solutions for future smart cities. The performance of buildings has been shown to be improved through the use of passive design technologies such as proper ventilation, daylight for lighting and passive cooling methods for cooling a building, which result in a significant reduction in the energy consumption of a building. After considering passive designed technologies, active design technologies were also discussed as effective methods for improving the performance and efficiency of the whole building structure. In addition, another useful concept in this regard, the building automation system (see IoT-enabled smart buildings in section 3.2.2), was explained. Moreover, the whole functionality of this system, in using smart sensors and devices to improve the performance of the building, was briefly explained. The data collected from the sensors ultimately highlights the scope of optimizing the energy usage in any building using smart analytics. The IoT can also be integrated with green technologies, which can significantly reduce the electricity, water and energy usage in the building, thus making it greener and smarter in comparison with conventional buildings. Building standards and rating agencies were also discussed, which give certification to high-performance buildings.

References

Al-Sallal, K. A. and Abdelhamid A. R. 2018. Natural Ventilation in Hot Seaside Urban Environments. In *Seaside Building Design: Principles and Practice*, edited by K. A. Al-Sallal and A. R. Abdelhamid. Springer: Cham, Switzerland, 27–41.

Albayyaa, H., Hagare, D. and Saha S. 2019. Energy Conservation in Residential Buildings by Incorporating Passive Solar and Energy Efficiency Design Strategies and Higher Thermal Mass. *Energy and Buildings* 182: 205–213.

Altan, H., Alshikh, Z., Belpoliti, V., Kim, Y. K., Said, Z. and Al-chaderchi, M. 2019. An Experimental Study of the Impact of Cool Roof on Solar PV Electricity Generations on Building Rooftops in Sharjah, UAE. *International Journal of Low-Carbon Technologies* 1: 1–10.

Bucur, A. and Moga L. 2018. Common Thermal Insulations vs. Nano Insulations: A Comparative Analysis Regarding the NZEB Targets Fulfillment. *International Multidisciplinary Scientific GeoConference: SGEM: Surveying Geology &Mining Ecology Management* 18: 65–72.

California Integrated Waste Management Board. 1999. *Statewide Waste Characterization Study: Results and Final Report*. December 1999. p. ES-2: Commercial and Self-Haul Commercial Values Combined.

Chaichan, M. T., Abass, K. I. and Kazem H. A. 2018. Dust and Pollution Deposition Impact on a Solar Chimney Performance. *International Research Journal of Advanced Engineering and Science* 3(1): 127–132.

CRRC (Cool Roof Rating Council). n.d. Retrieved from: http://coolroofs.org/.

Dascalaki, E. G., Droutsa, K. G., Balaras, C. A. and Kontoyiannidis, S. 2011. Building Typologies as a Tool for Assessing the Energy Performance of Residential

Buildings – A Case Study for the Hellenic Building Stock. *Energy and Buildings* 43(12): 3400–3409.
Edwards, B. 2006. Benefits of Green Offices in the UK: Analysis from Examples Built in the 1990s. *Sustainable Development* 14(3): 190–204.
Ejaz, W. and Anpalagan, A. 2019. Internet of Things for Smart Cities: Overview and Key Challenges. *Internet of Things for Smart Cities*, edited by W-S. Gan, C.-C. Jay Kuo and M. Barni. Springer:Cham, Switzerland, 1–15.
Fakourian, F. and Asefi, M. 2019. Environmentally Responsive Kinetic Facade for Educational Buildings. *Journal of Green Building* 14(1): 165–186.
Gautam, K. R., Rong, L., Zhang, G. and Abkar, M. 2019. Comparison of Analysis Methods for Wind-Driven Cross Ventilation through Large Openings. *Building and Environment* 154: 375–388.
Hernandez, L., Baladron, C., Javier, M. and Carro, B. 2014. A Survey on Electric Power Demand Forecasting: Future Trends in Smart Grids, Microgrids, and Smart Buildings. *IEEE Communications Surveys & Tutorials* 16(3): 1460–1495.
Jadhav, N. Y. 2016. *Green and Smart Buildings: Advanced Technology Options*. Springer: Cham, Switzerland.
Jakica, N. 2018. State-of-the-Art Review of Solar Design Tools and Methods for Assessing Daylighting and Solar Potential for Building-Integrated Photovoltaics. *Renewable and Sustainable Energy Reviews* 81: 1296–1328.
Jazaeri, J., Gordon, R. L. and Tansu, A. 2019. Influence of Building Envelopes, Climates, and Occupancy Patterns on Residential HVAC Demand. *Journal of Building Engineering* 22: 33–47.
Magno, M., Polonelli, T., Benini, L. and Popovici, E. 2015. A Low Cost, Highly Scalable Wireless Sensor Network Solution to Achieve Smart LED Light Control for Green Buildings. *IEEE Sensors Journal* 15(5): 2963–2973.
Mazzucchelli, E. S., Aelenei, L., Gomes, M. D. G., Karlessi, T., Alston, M. and Aelenei, D. 2018. Passive Adaptive Façades: Examples from COST TU1403 Working Group 1. In *Facade 2018 – Adaptive, Proceedings of the COST Action TU1403: Adaptive Facades Network Final Conference*, Lucern, Switzerland, November 26–27, 2018, 63–72.
Nejat, P., Jomehzadeh, F., Hussen, H., Calautit, J. and Abd Majid, M. 2018. Application of Wind as a Renewable Energy Source for Passive Cooling through Windcatchers Integrated with Wing Walls. *Energies* 11(10): 2536.
Nikken Sekkei. n.d. Incorporating Natural Light into Windowless Rooms by Introducing a Mirror Duct System. Retrieved from: www.nikken.co.jp/en/expertise/mep_engineering/incorporating_natural_light_into_windowless_rooms_by_introducing_a_mirror_duct_system.html.
Pöhls, H. C., Angelakis, V., Suppan, S., Fischer, K., Oikonomou, G., Tragos, E. Z. and Mouroutis, T. 2014. RERUM: Building a Reliable IoT upon Privacy-and Security-Enabled Smart Objects. *IEEE Wireless Communications and Networking Conference Workshops* (WCNCW) (report). Springer: Piscata Way, NJ.
Raji, B., Tenpierik, M. J., Bokel, R. and van den Dobbelsteen, A. 2019. Natural Summer Ventilation Strategies for Energy-Saving in High-Rise Buildings: A Case Study in the Netherlands. *International Journal of Ventilation* 18(4): 1–24.
Raji, B., Tenpierik, M. J. and van den Dobbelsteen A. 2018. Temperate Climate. *A+BE: Architecture and the Built Environment* 19: 129–186.
Ruuska, A. and Häkkinen, T. 2014. Material Efficiency of Building Construction. *Buildings* 4(3): 266–294.

Sadineni, S. B., Madala, S. and Boehm R. F. 2011. Passive Building Energy Savings: A Review of Building Envelope Components. *Renewable and Sustainable Energy Reviews* 15(8): 3617–3631.

Shahin, H. S. M. 2019. Adaptive Building Envelopes of Multistory Buildings as an Example of High-Performance Building Skins. *Alexandria Engineering Journal* 58(1): 354–352.

Song, J., Fan, S., Lin, W., Mottet, L., Woodward, H., Davies Wykes, M. and Aristodemou, E. 2018. Natural Ventilation in Cities: The Implications of Fluid Mechanics. *Building Research & Information* 46(8): 809–828.

Stritih, U., Tyagi, V. V., Stropnik, R., Paksoy, H., Haghighat, F. and Joybari, M. M. 2018. Integration of Passive PCM Technologies for Net-Zero Energy Buildings. *Sustainable Cities and Society* 41: 286–295.

Trusty, W. B. and Meil, J. K. 2002. The US Life Cycle Inventory Database Project. *Journal of Advanced Science* 13(3): 195–198.

Vitunskaite, M., He, Y., Brandstetter, T. and Janicke, H. 2019. Smart Cities and Cyber Security: Are We There Yet? A Comparative Study on the Role of Standards, Third Party Risk Management and Security Ownership. *Computers & Security* 83: 313–331.

Zuo, J. and Zhao Z.-Y. 2014. Green Building Research – Current Status and Future Agenda: A Review. *Renewable and Sustainable Energy Reviews* 30: 271–281.

4
Green Smart Environment for Smart Cities

Yaman Parasher, Gurjit Kaur and Pradeep Tomar

CONTENTS

4.1 Introduction 75
4.2 Green Environment 77
4.3 Green Products and Services 78
4.4 Role of Green Technologies in the Environment 79
4.5 Crucial Objectives of Green Science and Technology 80
4.6 The Six Environmental Spheres 80
    4.6.1 The Hydrosphere 82
    4.6.2 The Atmosphere 84
    4.6.3 The Geosphere 85
    4.6.4 The Biosphere 86
    4.6.5 The Anthroposphere 87
    4.6.6 The Sociosphere 88
Conclusion 88
References 89

4.1 Introduction

From the beginning of their existence, human beings have always faced challenges, in one form or another, for survival. However, the exploration of different science and technology domains over time has meant that humans became capable of addressing these challenges. As a result, diseases which would previously have wiped out a mass population have now have been either eradicated or are under control. Similarly, evolution in modern agricultural practices and water extraction from scarce sources have come about as a result of the development of humankind. However, despite the remarkable overall growth of human civilization, the relationship with the environment has worsened compared to a century ago.

The net global carbon uptake has increased significantly by about 0.05 billion tonnes of carbon per year, and that global carbon uptake doubled, from 2.4 +/- 0.8 to 5.0 +/- 0.9 billion tonnes per year, between 1960 and 2010

(Ballantyne et al. 2012). The risk of very harmful consequences resulting from emission of greenhouse gases has therefore increased, which already have had devastating effects on the overall climate of the earth. The continuous growth in environmental pollution has also raised grave concerns about the overall health of the ecosystem. Though a number of steps and measures have already been taken in the form of different action policies and plans to alleviate the environmental destruction, the problem has to date not been resolved to any significant extent. All of the policies and measures implemented in this regard have always been taken with a view to eradicate a problem rather than to prevent it. It is therefore necessary to implement sustainable measures to prevent further exploitation of resources to their maximum limit (Manahan 1997).

Regarding the measures to deal with such situations, one useful approach could be the adoption of green science or technology practices designed to carry out various functions while using few resources and with limited environmental impact and the highest degree of sustainability. The Population Reference Bureau (2012) estimated that there were approximately 7.06 billion people in the world at the end of 2012, which is expected to rise to 10 billion by 2100 (World Population Estimates n.d.). As a result, there will be massive demand for clean, basic amenities such as water, food and a safe environment for all. However, taking into consideration the current scenario, it seems that it will be quite difficult to sustain all resources for future generations.

According to a global nutrition report in 2011 (FAO 2011), around 30% of the food that the world buys goes to waste every year. In contrast to this, 40% of children in Africa below the age of five are malnourished. Such statistics simply underline the need to devise policies that can meet the demands of the whole world more comprehensively and securely instead of exploiting resources to the fullest. With a similar view, the then UN Secretary-General, Ban Ki-moon, launched the Zero Hunger Challenge at the United Nations Conference on Sustainable Development in 2012, where all countries were requested to work to better the future along with ensuring adequate nutrition for each of their citizens. Also, organizations like the World Health Organization (WHO), the United Nations Environment Programme (UNEP) and others are focusing their attention on food waste reduction and launched a global campaign – "Think. Eat. Save" – to reduce the unnecessary waste footprint as the theme of World Environment Day in 2013. Alongside all these initiatives, many other programs have been brought into the mainstream to increase the focus on promoting sustainable food systems for food nutrition and the security of the emerging green smarter world.

In regard to the vision of achieving a smarter green environment for future smart cities, a few solutions are listed below that can help to channel the development of humankind in a much better way:

- the adoption of appropriate ecological landscape design procedures and mandates in city planning;
- the development of energy- and fuel-efficient transit systems;

Green Smart Environment for Smart Cities

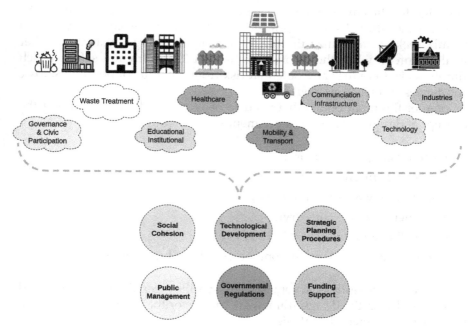

FIGURE 4.1
The key growth drivers of a smart city.

- the adoption of sustainable manufacturing practices;
- the development of energy-efficient ICT equipment;
- the adoption of sustainable waste collection, management and disposal practices; and
- the development of particular enzymes for detection and treatment of emerging pollutants.

The proper adoption of all these strategies will help the natural environment by maintaining sustainability among different interdependent environmental spheres. Drivers of the growth of a typical smart city are illustrated in Figure 4.1.

4.2 Green Environment

The concept of design for the green environment usually takes into account all the environmental impacts of a typical product development life cycle at subsequent stages of its growth (Billatos 1997). In general, the whole concept of design for the environment is categorized into two main broad areas: the first is concentrated on the principle of design for sustainability, which mainly seeks to minimize the usage of material, energy, water and other vital resources, while the second uses design for health and safety, which simply seeks to reduce any potential risk from any type of pollutants, toxic substances and waste as a whole.

Some of the main points that need to be considered while adopting the concept of design for green environment are as follows:

- eco-efficient resource use;
- eco-efficient processes;
- minimum waste and pollution;
- cleaner products and services;
- recycling and reuse;
- less pollution and waste from the use of products and services.

To realize the above areas, several related, specific green design policies need to be included in the design framework meant for the establishment of sustainable environmental strategies.

It will be more beneficial if readily available materials from recyclable and renewable sources are taken into account. In addition, the usage of energy throughout the cycle of system development must be minimized. Depending on the nature of end-use such as recyclability or disposal, the product must be made with biodegradable and combustible materials so that minimum harm is caused to the environment when seeking to achieve the objectives of the whole product.

4.3 Green Products and Services

In simple terms, a green product can be defined as a commodity or thing that utilizes lower amounts of hazardous material, with a vision to lower the potential exposure of the environment and people to hazardous substances; such a product produces less waste in its use and disposal (Hall et al. 2000). Green services, on the other hand, can be defined as the services that fulfill environmental criteria in a much more appropriate and sustainable way.

For example, hybrid electric and fuel automobiles can be considered a green product while a basic, well-utilized transportation system based on such automobiles can be considered a green service. Both green products and services play a vital role in improving profitability due to the lower requirement for hazardous material and the lower associated cost of disposal and environmental cleanup.

These green products are generally considered as highly durable, re-manufactured, reparable and reusable items which usually are more economical to transport. Electric batteries can also be considered a green product if recycling facilities are properly maintained for their collection.

Therefore, it is necessary that the planning strategies for the development of green technology-based smart towns and cities take into consideration such products and services that can help them to eradicate the pre-existing environmental problems.

4.4 Role of Green Technologies in the Environment

In general terms, technology simply refers to the way in which human beings apply knowledge and techniques and construct things using different materials and energy, while always focused on achieving a practical endpoint. However, technically, the whole concept of technology is defined as the product of engineering that is built on various essential scientific principles (Dangelico and Pujari 2010). It is only because of human intelligence that human beings have been able to survive for so long on earth compared to other living species that usually have a short time frame for survival. However, with the advancement of technology, humans have also disturbed the equilibrium with nature in a very harmful way. This can be perceived through the depletion of the ozone layer and accumulation of non-biodegradable waste and other chemical components, exposure to which is fatal for humans. These problems could very well be blamed on the peril brought about through factors such as the emergence of modern-day chemical industries, the growing number of automobiles and deforestation (Ahvenniemi et al. 2017). Thus, to address this, green technology primarily based on sustainable practices and measures needs to be deployed in all environmental conservation strategies and policies in a far more integrated way.

In simple terms, green smart technology defines the way in which conventional environmentally hazardous technologies can be replaced with eco-friendly measures that can mitigate the downfall of the natural ecosystem as a whole (Zygiaris 2013). With the evolution of human–environment relationships, conventional technological practices can become more sustainable (Barrionuevo et al. 2012). This can be achieved through the adoption of more renewable energy alternatives, efficient resource utilization, along with the development of smart grid structures, green parks and so on. In order to bring this agenda forward, the concept of this green smart technology can be brought into the mainstream via proper planning procedures enforced by concerned governments (Viitanen and Kingston 2014).

The main focus of green technology is always on the adoption of renewable energy sources, biofuels, sustainable agricultural practices and bioremediation of contaminants to bring about sustainability of vital resources in the environment on a much wider scale (Sridevi and Thangavel 2015). Widespread use of all these eco-friendly alternatives will ultimately result in the reduction in the accumulation of waste and pollution and will promote sustainable growth for the future.

4.5 Crucial Objectives of Green Science and Technology

The concept of human development generally requires producing defined strategies and objectives that can harness the hidden potential of conventional systems for the benefit of humankind. The foremost objective is to sustain the human species on earth for a long time to come. Major global climate change, global warming, depletion of water, land and resources and similar effects can, however, reduce the earth's capacity to support human life by rendering a large part of it unsuitable for habitation. This could be mitigated by adopting green smart technologies and services.

The second objective that is important for the vitality of human existence on earth is the concept of sustainable development. Therefore, it is considered essential to devise ways to help restore adequate water, mineral and fuel resources, along with measures to repress the associated hazardous effects (e.g. deforestation, over-exploitation of bio reserve etc.) on a regular basis.

The third essential objective is the maintenance of biodiversity. With the onset of hazardous effects from rapid development, different species inhabiting a region can face fatal consequences due to various factors that mainly include loss of water quality and availability, deforestation and change of land use. Therefore, it is necessary for the concerned authorities to work to protect the natural habitats of the affected species.

The final objective in our list is the maintenance of aesthetic richness. Factors such as water, air and uncontrolled land usage usually have the potential to damage aesthetics. In a typical congested smart urban city, one of the biggest detractors in terms of aesthetics is photochemical smog, which can reduce visibility. Oil spills in offshore areas can also ruin the aesthetics by destroying the lives of millions of species that surround the nearby beach areas.

4.6 The Six Environmental Spheres

There are six main overlapping and interacting spheres of the environment which affect each other by interchanging matter and energy (Maczulak 2010).

Traditionally, environmental science has always taken into consideration water – the hydrosphere, air – the atmosphere, earth – the geosphere, and life – the biosphere as its prime components. But when undesirable human activities such as pollution are considered alongside these spheres, there is a need to define another sphere – the anthroposphere – which, in simple terms, defines the things humans make and do on a regular basis which affect the whole environment in a detrimental way. The anthroposphere can be regarded as an integral part of the environment when humans modify their related environmental activities as much as they can.

The atmosphere, which is another crucial part of the environment, is a very thin layer which is mainly responsible for carrying various atmospheric gases a few kilometers above sea level. Among these gases is oxygen, which is responsible for providing life to living organisms; carbon dioxide, which provides support to carry out the photosynthesis process in plants; and nitrogen, which is considered an inert gas responsible for helping organisms to make protein. On top of this, the layer of the atmosphere also serves the vital function of protecting the earth from highly energetic UV radiation from the sun; had it been absent, it might have impacted a large proportion of all living organisms on earth.

A particularly important part of the atmosphere is the stratospheric ozone layer which, because of its ability to absorb infrared radiation, helps the whole earth to stabilize its overall surface temperature. Also, it is only due to the atmospheric layers that the solar energy, which usually falls with great intensity on the equatorial regions, is distributed away.

After the atmosphere, another vital part of the environment that plays an important role is the hydrosphere. The hydrosphere defines the area of the earth's water surface, which is estimated to contain 97% of seawater from the oceans and the remaining 3% from the water entrapped in the form of ice in glaciers and polar ice caps, in addition to the freshwater in lakes, streams and rivers. A small fraction of water also exists in the form of vapor in the atmosphere. Water is also often present on the surface of the earth, in reservoirs, streams, lakes or groundwater, which is mainly used for a variety of applications ranging from agricultural to industrial use.

Another very significant part of the whole environment is the geosphere, which is the solid part of the earth that mainly includes minerals and rocks. A very important part of the geosphere is soil, which is mainly responsible for providing plant growth and is the basis of food for all the living species on the planet. Another very important layer of the geosphere that interacts very well with other environmental spheres and human beings is the lithosphere. This is a relatively thin, solid layer of the earth's surface, up to a depth of 100–250 km, that contains lighter silicate-based materials.

The second to last essential sphere of the environment is the biosphere. The biosphere consists of all living organisms that reside mainly in the geosphere and atmosphere.

The last vital sphere, the anthroposphere, mainly comes into existence due to its strong interaction with the environmental sphere discussed earlier. By taking part in numerous agricultural activities, human beings have modified both the geosphere and biosphere to a great extent. Humans have often been seen to exploit natural water flow by contaminating and returning it to the hydrosphere without looking into any future hazardous consequences. The emission of large, hazardous particles and gases into the atmosphere also results in disturbing visibility and other characteristics of the atmosphere, along with introducing instability regarding the temperature of the earth as a whole. Apart from this, the anthroposphere also disturbs various essential biogeochemical cycles that exist as a part of various environmental processes.

Similarly, the emission of large quantities of carbon dioxide to the atmosphere through combustion of fossil fuels may be modifying the heat-absorbing characteristics of the atmosphere to the extent that global warming is almost certainly taking place. For an in-depth view of the interrelated relationships of the chemistry of different spheres of the environment, see Figure 4.2.

4.6.1 The Hydrosphere

Water is one of the basic and foremost essential resources required by all living species in the world. Concerning its availability, 97.5% of the world's

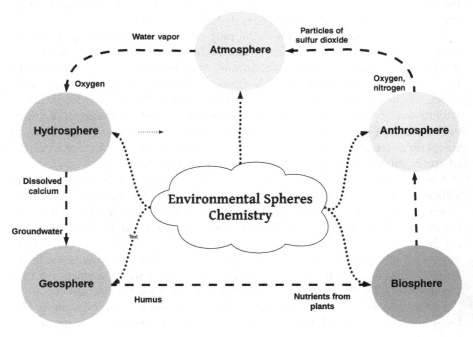

FIGURE 4.2
Interrelated relationships of the chemistry of different spheres of the environment.

water is found in the oceans, leaving the remaining 2.5% as freshwater. Furthermore, approximately 1.7% of this 2.5% freshwater concentration is contained within ice caps and glaciers in polar regions. This leaves merely 0.7% as the only reserve of freshwater that is accessible for use by humans. This freshwater is available to living species in the form of surface water and groundwater (USGS n.d.). The surface water is found in rivers, lakes, reservoirs, impoundments and so on, and is always more susceptible to interference by humans. Groundwater is subject to similar fate and has also been disturbed by human activities. In the environment, water should be saved and used carefully because it is the main resource responsible for the maintenance of sustainability on earth (Bataan 2011). Over a long period, factors such as climatic-induced droughts have been responsible for bringing about variations in the normal water supply, leading to severe problems for humans and other living organisms.

New technologies should be designed to save the environment because global warming has already increased. The destruction associated with droughts and occasional floods is far more detrimental now compared to a few years ago. Since both water quantity and quality are the two most important factors for sustaining healthy and prosperous human populations, the prevention of water pollution has always been a major objective of the environmental movement. Avoiding the discharge of harmful pollutants into water is therefore considered one of the crucial objectives of green technology. Technologies should be designed to eliminate water pollution. Many measures have already been adopted, i.e. humans have endeavored to manage water by building dams, building reservoirs to store water for future use and using dikes and dams to control flooding. In some cases, the construction of reservoirs for providing water to arid regions has been quite useful. However, in the case of prolonged droughts, these methods have been unable to cope with the high demand for water in water-deficient areas.

With the growing demand for water for industry and agricultural practices, there has also been a rapid drop in the water table level. With all of the constraints prevailing in the environment (concerning the dropping of groundwater levels along with its contamination through industrial waste), there is a dire need to devise sustainable solutions that can help to avoid the further depletion of water and replenish it to improved levels.

Some of the practices that need to be adopted around the world to enable green environment are as follows:

- construction of wetlands near points of diversion;
- recharging of groundwater by constructing paving surfaces that are made from porous materials; and
- construction of a basin consisting of a reservoir of water excavated into a porous geospheric material from which water flows into the aquifer.

4.6.2 The Atmosphere

The atmosphere is a layer of different gases or air that surrounds the whole surface area of the earth. Approximately 99% of the atmospheric mass is found within the first 40 km from the earth's surface where different concentrations of various gases exist. Although there is no specific upper limit to the atmosphere, an altitude exceeding 1,000 km is generally considered a practical upper limit to the atmosphere. One of the significant aspects of this layer comes from its ability to act as a blanket which helps in maintaining the stability of the earth's temperature.

Apart from providing oxygen, which is considered an essential component for human survival, it also provides carbon dioxide, which is essential for green plants to carry out the process of photosynthesis. Also, nitrogen is taken up from the atmosphere for the synthesis of protein, which is necessary for human survival. Water vapor, which constitutes another important part of the atmosphere, is typically composed of evaporated water from the oceans that then forms various types of precipitation overland through the hydrological cycle. The whole atmosphere is typically made up of 2–3% water vapor by volume (theweatherprediction.com n.d.); the total is higher in tropical regions and lower in other areas. At an increasing altitude, the air becomes cooler, which enables water vapor to condense into droplets. This whole phenomenon results in the formation of dry air, which is approximately 78.1% nitrogen (N_2), 21.0% oxygen (O_2), 0.9% of the noble gas argon (Ar) and 0.038% carbon dioxide (CO_2). The remaining 0.002% is formed from gases such as helium, hydrogen, krypton, nitrogen dioxide, ozone, xenon, and sulfur (Manahan 2006). Out of the five layers of the atmosphere, the lowest layer is categorized as the troposphere. The troposphere stretches from the earth's surface to a distance of 11 km, where the average temperature is around −56°C compared to 15°C at the earth's surface. The ultra-cold region at the top of this stratosphere is usually called the tropopause and is generally considered one of the most significant areas for sustaining life on earth. It is mainly due to the freezing of water vapor at the tropopause that a protective shield is formed; in another scenario, this could have broken apart, resulting in the formation of a dry planet which would not be able to support any form of life. Above this troposphere lies the stratosphere where the average temperature increases from −56°C to −2°C at the lower and upper limit, respectively.

Heating in the atmosphere arises mainly due to the energy that is absorbed by the air molecules because of the solar radiation that impinges on them. UV radiation from the solar radiation helps in breaking down the molecules of elemental oxygen, which ultimately results in the formation of ozone. This ozone layer helps to absorb portions of extreme UV radiation which otherwise would lead to the destruction of living species from the earth's surface. However, despite its benefits, chemical compounds such as chlorofluorocarbons, which are mainly found in refrigerators, often react in ways that deplete the protective ozone layer.

In addition to the troposphere and stratosphere, the atmospheric mesosphere and thermosphere also exist. However, these are of less importance regarding our brief study of the spheres of the environment in general.

One of the peculiar characteristics of the earth's atmosphere lies in its ability to absorb, distribute and radiate the enormous amount of energy that comes from the sun. This energy is usually absorbed and converted back to heat in the whole atmosphere and at the surface of the earth. When this energy comes back from the atmosphere, it is often delayed due to reabsorption by carbon dioxide, water molecules and other minor species in the atmosphere. All of this results in keeping the earth's atmosphere a little warmer and gives rise to a phenomenon known as the greenhouse effect, which is crucial to sustaining life on earth. However, excessive greenhouse effects have harmful effects on the global climate. Therefore, it is essential to put in place measures in city planning to control the emission of harmful gases into the atmosphere.

Among the different vital factors mentioned regarding the atmosphere, climate is also one of the most important factors for sustainability. This can be clearly seen in agricultural production, where growth and decline vary according to climatic conditions.

4.6.3 The Geosphere

The geosphere is defined as the solid part of the earth, which is responsible for food, vital metal resources, fossil fuels and so on. The outermost solid layer of the earth is the lithosphere, which generally consists of solid rocks with varying thickness, ranging from 100–250 km. Below the lithosphere is the asthenosphere, which contains a viscous liquid rock mantle which is relatively weak and plastic. In general, the geosphere mainly consists of rocks, minerals, soil, sediments and hot layers of molten rock along with an iron-rich inner core. On earth, there are known to exist around 2,000 minerals, each of which is characterized by a definite chemical crystal structure and composition. However, most rocks in the geosphere are only composed of around 25 minerals. The crust of the earth, which is an important part of the geosphere, contains around 25.7% silicon and 49.5% oxygen with little traces of other minerals like carbon, iron, sulfur and aluminum. Only about 1.6% of the earth's crust contains the important resources that are essential for the sustainability of living species on earth (Manahan 2006).

Another very crucial part of the earth's crust is the thin layers of weathered rocks that are formed from decayed organic matter, water and composting soil. Soil is crucial for sustainability as it supports the growth of plant life that provides food to humans and animals for their continued existence. Among the different layers, the topsoil layer is considered the most essential part due to its productive nature. However, adverse climatic conditions, along with poor cultivation practices, have led to the loss of topsoil through water erosion and strong winds. Conservation and enhancement of soil productivity are the key aspects required for the sustainable growth of living species on earth.

With the widespread development of smart cities around the world, large quantities of consumer and industrial waste have accumulated throughout the earth's solid surface. It is, therefore, considered essential to preserve the geosphere due to the large area it covers.

All other environmental spheres are strongly linked tithe geosphere. Any modification or alteration in any one of them can bring hazardous consequences to the associated environmental sphere in one way or the other; this is why it is very much necessary for humans to bring into practice sustainable methodologies to mitigate all of these detrimental effects.

4.6.4 The Biosphere

This part of the environmental sphere mainly includes living species, along with the materials that are produced by them. These living species are generally characterized by a particular class of biomolecules that mainly includes carbohydrates, proteins, lipids, nucleic acids and so on. In the biosphere, usually a classic hierarchical order prevails, which begins at the molecular level and goes straight to the biosphere through stages such as living cells, organs and organisms.

Like the other environmental spheres, the biosphere is also observed to be well connected to all other environmental spheres. It has strong effects on the other spheres and, in turn, is strongly influenced by them. Some of the major effects can be visualized through a number of interdependent applications. For example, the biosphere productivity of a particular region can be determined through the condition of the soil in the geosphere. Similarly, the abundant availability of water in the hydrosphere always has a strong impact on the growth of living species. The biosphere is also influenced by the consumption of large quantities of fertilizers and pesticides produced by the anthropogenic unit, which is mainly used to boost the productivity of farm crops in the biosphere.

Some of the major efforts to control pollution and waste disposal in the anthroposphere come from humans who reside in the biosphere. Thus, it can be concluded that nature and atmospheric activities strongly impact the biosphere and its productivity one wide scale. Like all other spheres, the biosphere plays a key role in achieving sustainability, mainly due to its photosynthesis activity. Despite various essential utilities which the biosphere provides, biogenic material always has the potential to self-produce in a much more sustainable and greener manner than is the case with materials produced by humans.

In contrast, humans have been shown to behave in an unsustainable manner concerning their own existence, which often results in the exploitation and pollution of the environment in various ways. The complex and sustainable ecosystem where living species reside in a sustaining relationship with each other and their surrounding area has given rise to another type of sphere in the environmental sphere classification. The anthroposphere in a general sense can be classified as the infrastructure which is made up of systems, utilities and facilities that serve large sections of people and are considered essential for the proper functioning of the whole community. Thus, by adopting proper methodologies and practices in the biosphere and its associated ecosystem, humans should develop a much more eco-friendly green environment for the healthy and prosperous living of their peers.

4.6.5 The Anthroposphere

The anthroposphere, in general, encompasses the part of the environment that is made and modified by humans through their activities. The visible part of the anthroposphere mainly includes the construction of roads, airports, buildings, factories, power lines and numerous other features that are constructed or built by humans. With the rapid hazardous interventions of humans, the earth is undergoing major transitions in its global climate, due to which the survivability of all living species on earth is at stake. The essential features of the anthroposphere mainly include those processes that are useful for the distribution of energy, fuel and water for the collection, recycling and disposal of waste and sewage water, together with particular interest in sustainability as a whole.

Regarding the anthropogenic activities in transport systems, these mainly include the modification of railroads and waterways for the efficient movement of goods and humans. Cultivated fields for growing crops and the provision of water systems for irrigation are the major contributors to food production that lie within the anthropogenic segment. Similarly, a number of machines including construction machinery and automobiles serve as an essential part of the anthroposphere. The communication sector of the anthroposphere mainly includes radio towers, fiber optic networks and so on. In addition, gas, and oil wells are employed mainly for extracting fuel from the geosphere. A vast portion of these infrastructural facilities mainly consists of physical systems that include communication systems, transport systems and energy generation facilities, along with distribution grids and waste collection, recycling and disposal units. A further essential part of this whole infrastructure also includes the proper working and adoption of regulations, procedures and laws that are essential for the correct functioning of all the associated systems.

These infrastructural facilities of the anthroposphere generally enable the acquisition, processing, conversion and distribution of materials which are to be manufactured. Outdated, obsolete and poorly managed anthropogenic infrastructural facilities are usually responsible for making the whole service production inefficient, negatively impacting the factor of sustainability in the long run. All of these associated problems usually begin due to the improper adoption of design elements and maintenance of various infrastructural facilities from their inception, leading to building of unreliable structures.

4.6.6 The Sociosphere

The sociosphere, or the societal organization of people, is an essential part of the anthroposphere and its infrastructure. This sphere typically includes cultures, families, social traditions, government, laws and so on. The fundamental unit which defines this whole concept simply comprises group of people that interact regularly with each other within a defined boundary of an ecosystem for the growth and prosperity of species that share the same ecosystem with them. Therefore, a well-functioning sociosphere is usually considered useful for human welfare, sustainability and prosperity. An important field that gives a voice to the sociosphere is economics (Daly 1997). Economics in a general sense is the science of describing the production, distribution and use of commodities. The environment is related to the economic system where the different commodities are considered to be developed or are considered futile, depending upon their nature and objective of usage.

It is often the case that only vital resources are regarded as environmental endowments, which tend to give a fruitful return depending upon their usage and management. For example, resources such as grasslands when properly grazed and managed have the potential to maintain their productivity indefinitely. However, when these vital resources are over-exploited through their usage, it generally damages their endowment value. An example could be when grasslands are grazed by too many animals, resulting in a loss of vitality due to removal of the valuable topsoil. It is therefore essential for different spheres of the environment to collaborate effectively to promote the growth of the world's economy as a whole.

Conclusion

This chapter has provided a holistic view of environmental sustainability by taking into consideration design methodologies along with vital green technological practices that need to be implemented in each of the six fundamental environmental spheres. The chapter has also provided concise and brief descriptions of these environmental spheres that form part of the emerging ecosystem. The unique and specific points of each of the environmental spheres have been highlighted to show their contribution towards the sustainability of the whole environment to enable better future for the world.

References

Ahvenniemi, H., Huovila, A., Pinto-Seppä, I. and Airaksinen, M. 2017. What Are the Differences between Sustainable and Smart Cities? *Cities* 60: 234–245.

Ballantyne, A. P., Alden, C. B., Miller, J. B., Tans, P. P. and White, J. W. C. 2012. Increase in Observed Net Carbon Dioxide Uptake by Land and Oceans During the Past 50 Years. *Nature* 488(7409): 70.

Barrionuevo, J. M., Berrone, P. and Ricart, J. E. 2012. Smart Cities, Sustainable Progress. *IESE Insight* 1414: 50–57.

Bătăgan, L. 2011. Smart Cities and Sustainability Models. *Informatica Economică* 153: 80–87.

Billatos, S. 1997. *Green Technology and Design for the Environment*. CRC Press: Boca Raton, FL.

Daly, H. E. 1997. *Beyond Growth: The Economics of Sustainable Development*. Beacon Press: Boston, MA.

Dangelico, R. M. and Pujari, D. 2010. Mainstreaming Green Product Innovation: Why and How Companies Integrate Environmental Sustainability. *Journal of Business Ethics* 953: 471–486.

Food and Agriculture Organization of the United Nations (FAO). 2011. *Global Food Losses and Food Waste – Extent, Causes and Prevention*. FAO: Rome.

Hall, R. E., Bowerman, B., Braverman, J., Taylor, J., Todosow, H. and Von Wimmersperg, U. 2000. *The Vision of a Smart City*. Paper presented at the 2nd International Extension Technology Workshop, Paris, France.

Maczulak, A. E. 2010. *Environmental Engineering: Designing a Sustainable Future*. Vol. 4. Infobase Publishing: New York.

Manahan, S. E. 1997. *Environmental Science and Technology*. CRC Press: Boca Raton, FL.

Manahan, S. E. 2006. *Environmental Science and Technology: A Sustainable Approach to Green Science and Technology*. CRC Press: Boca Raton, FL.

Population Reference Bureau. 2012. World Population Trends 2012. Retrieved from: www.prb.org/world-population/.

Sridevi, G. and Thangavel, P. 2015. *Environmental Sustainability: Role of Green Technologies*. Springer: Boca Raton, FL.

theweatherprediction.com. n.d. Atmospheric Water Vapor. Retrieved from: www.theweatherprediction.com/habyhints/40/

USGS. n.d. The World's Water. Retrieved from: https://web.archive.org/web/20131214091601/http://ga.water.usgs.gov/edu/earthwherewater.html.

Viitanen, J. and Kingston, R. 2014. Smart Cities and Green Growth: Outsourcing Democratic and Environmental Resilience to the Global Technology Sector. *Environment and Planning* A 464L: 803–819.

World Population Estimates. n.d. *Wikipedia*. Retrieved from: https://en.wikipedia.org/wiki/World_population_estimates.

Zygiaris, S. 2013. Smart City Reference Model: Assisting Planners to Conceptualize the Building of Smart City Innovation Ecosystems. *Journal of the Knowledge Economy* 42: 217–231.

5
Green Healthcare for Smart Cities

Prabhjot Singh, Varun Dixit and Jaspreet Kaur

CONTENTS

5.1 Green Healthcare 91
5.2 Green Hospitals for Smart Cities 93
    5.2.1 Green Building Design of Hospitals in Smart Cities 95
    5.2.2 Green Hospital Energy Management 96
    5.2.3 Water Management for Green Hospitals................................ 98
    5.2.4 Waste Prevention and Management in Healthcare............... 99
5.3 Green Medical Devices for Smart Cities... 100
    5.3.1 Medical Imaging 100
    5.3.2 Efficient Biomedical Signal Processing 104
5.4 Green Smart Digital Healthcare System 107
    5.4.1 Electronic Healthcare Information Management System 107
    5.4.2 Telemedicine, eHealth and Telecommunications........................ 112
    5.4.3 mHealth and pHealth Systems 114
    5.4.4 Smart Wearable Technologies for Health Monitoring 116
    5.4.5 eLearning for Green Healthcare 119
    5.4.6 Modeling and Simulation Methodologies........................... 121
    5.4.7 Intelligent Healthcare Services.. 123
Conclusion 125
References.. 126

5.1 Green Healthcare

Rapid industrialization and advancements in the use of technology along with many other factors have led to reckless exploitation of the environment, and the consequences have drastically impacted the way governments, industries and public think. These institutions have realized the importance of sustainable development and environmental protection. As a result, several initiatives and organizations are coming to the forefront with ideas that focus mainly on development policies but also reduce environmental impact and reduce waste.

Healthcare is one of the biggest sectors in the US, accounting for 17.9% of the total US GDP (Cassandra 2013). This sector is further expected to increase to 19.7% of GDP by 2026 (NHE-Fact-Sheet 2019). The healthcare industry significantly impacts the environment. A recent study calculated the total global warming potential (GWP) directly caused by the US healthcare sector to be 254 billion kilograms of CO_2 eq. Approximately 80% of the GWP in the healthcare sector is attributed to carbon dioxide (CO_2), which is one-tenth of the total CO_2 emissions in the US (Cassandra 2013). Greenhouse gases are one of the major causes of global warming and climate change along with other natural factors.

A course correction is needed to save the planet and future generations. New policies and guidelines need to be implemented and acted upon to avoid the devastating effects of climate change. The healthcare sector, which in itself is a significant contributor to changing the climate, also needs to improvise. As published in *American Journal of Preventive Medicine*, the health sector is one of the most trusted and respected sections of society, and it is also one of the largest employers and consumers of energy. This presents both a duty and a window of opportunity to achieve climate-neutrality, efficiency and cost reduction, all at the same time (Dhillon 2015).

To make the healthcare industry cleaner, a holistic overview of its systems with fresh eyes is required, where one should challenge the existing practices that produce a high amount of greenhouse gases, along with having tighter integration between systems. A few basic measures can have an enormous impact on making the healthcare sector more sustainable and environment-friendly.

The term "green healthcare" refers to implementing effective measures and environment-friendly practices in the healthcare sector, to reduce the carbon footprint while providing healthcare services. Green healthcare helps in safeguarding the environment along with providing financial benefits. This allows healthcare professionals to protect and promote health and can be a platform for educating students and members of the public. As shown in Figure 5.1, a healthy environment helps recovery in patients which in turn reduces the pressure on healthcare facilities. This reduced pressure promotes a healthy environment. For example, improved air quality reduces the risk of asthma incidents, bringing fewer patients to hospitals and bringing down the carbon footprint caused by commuting. Further, lower patient intake reduces generation toxic waste from the hospitals thereby improving the health of the environment and quality of life.

The healthcare ecosystem can be subdivided into three categories, namely physical healthcare, medical devices and digital applications that support healthcare. For achieving the desired result of the eco-friendly healthcare sector, impactful changes are required in all three categories without compromising on the quality of services. This mandates the formulation of an ideal ecosystem for green healthcare which aims to be self-sustainable from the environment perspective. As shown in Figure 5.2, the ecosystem of green healthcare can be divided in three broad categories:

Green Healthcare for Smart Cities

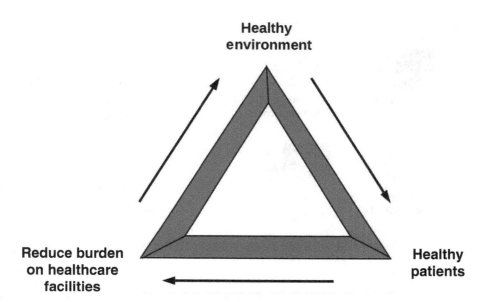

FIGURE 5.1
Iron triangle of green healthcare.

1. Green hospitals for smart cities
2. Green medical devices for smart cities
3. Smart eHealth.

5.2 Green Hospitals for Smart Cities

Green or sustainable building is the practice of designing, constructing, operating, maintaining and removing buildings in ways that conserve natural resources and reduce pollution (Coussens and Frumkin 2007). These guidelines are fully applicable to building hospitals or improving the preexisting infrastructure at the stages of design, construction, operation and management.

A few of the domains where environmental impact caused by healthcare can be reduced are listed as follows:

1. Green building design of hospitals in smart cities
   a. sustainable location
   b. sustainable hospital building design
   c. green/living hospital roof

94   Green and Smart Technologies for Smart Cities

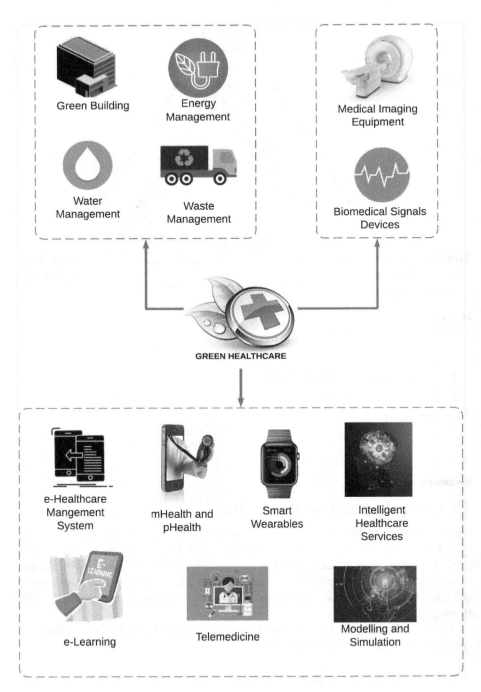

FIGURE 5.2
Ecosystem of green healthcare.

d. green area surrounding the hospital building
   e. green roofs and green spaces.
2. Green hospital energy management
   a. energy conservation
   b. renewable energy resources.
3. Water management
4. Waste prevention and management
   a. legislative and regulatory policies
   b. healthcare waste minimization, reuse and recycling
   c. segregation, storage and transport of healthcare waste.

5.2.1 Green Building Design of Hospitals in Smart Cities

Hospital buildings consume a significant portion of healthcare energy. Designing hospitals with green building design in mind will not only reduce the energy consumption, but will also provide a better environment for patients' recovery without any compromise on the quality.

5.2.1.1 Sustainable Healthcare Locations

Selection of site for building a hospital can play an essential role in deciding its environmental impact. A hospital site should be selected in a way that offers robust transportation methods, preferably walking, biking or easily accessible public transport, and thus minimize the overall amount of greenhouse/toxic gas emissions by the hospital building as well as by the commute to the building. The hospital site should be selected in a way that causes the least natural hazards, like endangering local wildlife. The building should be able to maximize the use of natural resources available. For example, sunlight can provide natural lighting to the hospital as well for powering the solar panels installed on the rooftops and parking lots. A good source of wind can power wind turbines to produce green energy to hospitals. This selection will not only be able to provide energy efficiency for healthcare facilities, but also will be a boom for patients' mental and physical health, as they will be getting plenty of sunlight and fresh air. Although selecting a site that is close to such natural resources is good, it is vital to ensure that the ecosystem of that place is left undisturbed.

5.2.1.2 Sustainable Hospital Building Design

Finding innovative ways to reuse/use the existing materials in a way to solve the purpose needed not only helps reduce waste, but at the same time reduces the extra effort to buy new materials and keep the cost low, to build the structure. As an example, while designing the St. Luke's Magic Valley facility at Twin Falls, Idaho, the designers faced a situation where it was important to manage the underlying layer of lava rock before building the structure. Instead of removing the rocky layer and transporting to other location, authorities came up with a clever idea to use the rock for landscaping and decorative purposes. The material was crushed and used at the site itself. This approach saved both money and effort for the builders along with reducing the production of waste and harmful emissions from transporting the entire rock material to a new place (Going Green: 5 Ways 2011). In another example, Pittsburgh Children's Hospital was built using post-consumer steel and other recycled materials (Top 10 Energy-Efficient 2015).

5.2.1.3 Green/Living Hospital Roof

Green roof is the concept of covering the entire, or part of the building's, roof with vegetation above a waterproof membrane. The vegetation planted over the roof can help absorb rainwater, act as a home for wildlife, provide insulation to the building and also improve the aesthetics of the area. Studies have proven that making the hospital roof green can help reduce stress among the patients by keeping them near to nature. Besides, green plants help improve the quality of air by absorbing harmful/toxic gases, thus adding to the reduction of environmental impact caused by the building (Top 10 Energy-Efficient 2015).

5.2.1.4 Green Area Surrounding the Hospital Building

Tree plantation drives in collaboration with the local colleges and student communities near the hospital building add to the quality of air and consequently improve the healing time of the patients (Healthcare Facilities Using Green 2017). They also help to absorb the greenhouse gas emissions from the hospitals thus improving the quality of air nearby. Torre de Especialidades, Mexico, is a state-of-the-art hospital that has a screen attached outside. This screen is coated with titanium dioxide which not only blocks sunlight for the hospital for reducing energy requirements for air conditioning, but also converts smog in benign chemicals on the contact (Top 10 Energy-Efficient 2015).

5.2.2 Green Hospital Energy Management

Energy management is a key aspect of green hospitals. Compared with commercial buildings, hospitals have 2.5 times the energy intensity and carbon dioxide emissions (Department of Energy Announces 2009). With the rising costs of energy, it is becoming more difficult to provide affordable care (Lutheran 2010). According to the Department of Energy, hospitals produce more than 30 pounds of CO_2 emissions per square foot (Department of Energy Announces 2009).

According to an HCWH report, a typical 200-bed hospital dependent upon electricity generated from coal using 7 million kWh is responsible for more than $1 million per year in negative societal public health impacts and $107,000 per year in direct healthcare costs (Quinlan 2014).

The only logical move is in the direction of both energy conservation as well as sourcing energy from renewable resources.

Healthcare facilities are the most energy-intensive of all commercial buildings in the US because of the unique characteristics of the hospitals. Hospitals need energy 24x7 without interruption. Conserving the energy at such facilities saves not only energy but also keeps the environmental impact of the structure to the lowest possible level. The following are few recommended ways for energy management.

5.2.2.1 Energy-efficient Hospital Products

Usage of energy-efficient electronics in the buildings such as LED instead of the traditional incandescent bulbs is a big step towards saving electricity. Although the upfront cost of such products is high compared to traditional products, in the long run, these products save energy and money (How Energy-Efficient Light Bulbs Compare n.d.). For example, Fresenius Medical Care's production facility in St Wendel retrofitted all 70% of the conventional lamps with energy-efficient lighting, which resulted in 42% reduction in electric power consumption (Top Ten Green Medical Devices 2011).

5.2.2.2 Motion Sensor Electronics

With the advancements in technology, several motion sensor products have come on the market that turn on only when there is movement in the room, for example, motion sensor lights. These lights stay off by default and turn on when somebody moves in the area near to the lights.

5.2.2.3 Environment-aware Hospital Devices

Environmental-aware devices can turn on/off according to the present condition of the room. These devices can be further connected with cloud-based technologies for pattern- and usage-based analytics which can in turn be used to improve the overall efficiency. For example, a heater that turns on when the room temperature goes down to a specific temperature or a cooling system that turns on only when the temperature of the room goes above a set temperature value. These devices can prove to be advantageous when conserving energy in the building (Top 10 Energy-Efficient 2015).

5.2.2.4 Renewable Energy Resources

Hospitals are already inclining towards buying a percentage of energy from green energy plants to minimize their environmental impact. To further strengthen the environmental commitment, healthcare facilities should leverage the space and location to generate on-premise energy using renewable sources. On-site power generation from renewable energy sources not only removes the external dependencies but also minimizes the carbon emissions that are usually incurred due to the transfer of energy.

Solar Energy

Hospitals usually have a lot of open space to capture solar energy, which can then be channeled to meet hospitals' energy needs. Hospitals can install solar panels in the parking lots as well as on the rooftops. Solar panels do not have any significant environmental impact beyond manufacturing solar panels. For example, Mount Elizabeth Novena Hospital, Singapore, has rooftop trees and photovoltaic solar panels which boil water and power rooftop garden lights (Top 10 Energy-Efficient 2015).

Wind Energy

Wind turbines are being designed in a way that can be installed in parking lots and can produce sufficient energy in hospitals which have low sunlight. Companies like EcoVert are building wind turbines for hospital needs which can even generate energy for hospitals in the aftermath of hurricanes when other energy sources fail (Wind Power for Hospitals 2015). Khayelitsha Hospital, South Africa, has a wind turbine on the premises to generate energy (Top 10 Energy-Efficient 2015).

The other forms of renewable energy sources that healthcare facilities can leverage are geothermal, biomass, low-impact hydroelectric sources.

5.2.3 Water Management for Green Hospitals

One of the main threats in sustainable development today is scarce water resources and reckless exploitation of water resources (The Importance of Water Sustainability 2017). One of the significant steps that could help in reducing the impact of healthcare facilities on the environment is an effective water management program. New laws and standards should be introduced and enforced by the policymakers to reduce the wastage of water.

The healthcare facilities should install water meters at strategic locations in order to measure and manage water consumption and detect leaks. Hospitals should invest more in alternative water sources such as harvested rainwater, gray water (waste water directly from kitchen, sinks and bathtubs) and water from condensation units. This water can be treated and reused to grow and maintain green roofs and trees around hospitals.

The usage of water can be better controlled if the water management system (WMS) can make data-driven decisions. For example, data points such as the upcoming rain forecast, current weather and water availability in the storage tank can help WMS to decide when to release water for the lawns around the hospitals. Further, the frequency of flush in hospitals when controlled by WMS can help in further reducing the water wastage. In a case study performed at Asheville Civic Center, North Carolina, a two-minute delay was added to the sensors to assure that urinals do not flush more frequently than every two minutes. This small change in itself resulted in saving of 90% urinal water consumption (Water Management Options n.d.). Installing motion sensing faucets/shower heads also help reduce water wastage. Healthcare facilities can also install in-house water treatment facilities. The water from these facilities can be used in bathrooms as well as in washing clothes.

5.2.4 Waste Prevention and Management in Healthcare

Hospitals are equally responsible for taking care of patients' health as well as that of the environment. In a study done in Pakistan and published by WHO, an estimated 1.28–3.47 kg of waste is generated per hospital bed per day (Safe Management of Wastes from Health-Care Activities. 2014). Maternity homes generate 4.1 kg/patient-day of waste. Metropolitan General Hospital in the United States produces 10.7 kg/occupied bed/day of waste (Safe Management of Wastes from Health-Care Activities. 2014). With a high amount of waste going into the ecosystem, this starts the downward spiral trend of the healthcare system. This negative impact on the ecosystem harms citizens' health, which in turn puts more load onto the healthcare system. Unmanaged waste can have a negative impact and can start a chain of infection (Manual on Healthcare 2011). Therefore, it is critical to introduce measures for more green and efficient waste management.

For effective waste management and prevention, the following actions should be incorporated:

5.2.4.1 Legislative and Regulatory Healthcare Waste Policies

Policymakers should introduce new laws/strictly implement the existing laws, to guide healthcare facilities on categorizing waste and how to manage it, hygiene standards, audit procedures, staff waste awareness mandates and storage guidelines for products.

5.2.4.2 Healthcare Waste Minimization, Reuse and Recycling

3R principally is a globally known and accepted principle for waste management. WHO has published a waste management hierarchy which is based upon 3R's, namely to reduce, reuse and recycle. The best practices will be to avoid waste, followed by recovering and reusing as much as possible. The environment-friendly purchase should be encouraged. For example, mercury-free thermometers should be purchased as when a mercury thermometer breaks, there is a cost associated with cleaning up of mercury which is a hazardous material (Azmal et al. 2014). Latex or nitrile gloves should be used instead of polyvinyl chloride (PVC) which has a toxic effect. Devices and equipment that do not pose a risk should be reused, like blood pressure meters. Procurement of reusable devices/instruments should be encouraged while the non-medical disposable items should be avoided. At an estimate, 41% of Canadian hospitals are able to reuse many types of non-disposable medical devices (Campbell et al. 1987). Hospitals should invest in sterilization techniques like thermal sterilization and chemical sterilization for enhancing the reusability of devices.

5.2.4.3 Segregation, Storage and Transport of Healthcare Waste

Waste generated by the healthcare system should be appropriately labeled into categories like trash, chemical waste, biohazard waste, radiation waste and so on. The best practice is to have containers/trash-cans labeled in which staff can throw the waste material. Transportation of this waste should be done during non-rush hours to prevent exposure as well as fuel efficiency. Hazardous waste should be transported in separate boxes/containers to treatment facilities. Vehicles carrying waste should be labeled to indicate its load type.

5.3 Green Medical Devices for Smart Cities

One of the critical pieces of making healthcare eco-friendly is investing in green medical devices that not only help in lowering energy consumption, but also play a crucial role in the reduction of waste. The US Department of Energy estimates that medical equipment makes up 18% of the facilities energy use (Medical Imaging Equipment Study 2016).

5.3.1 Medical Imaging

Medical imaging can be defined as the use of different technologies that are utilized to view the human body in order for monitoring, diagnosis and exploring treatment options (Medical Imaging 2019). With the recent advancement in the computer vision and multimedia processing technologies, Medical Imaging has been developed into one of the most important fields within computer-aided medical diagnosis.

Medical Imaging and image analytical technologies have numerous applications. They allow researchers to look inside the body of the patient and plan for more personalized treatment options without the need for any invasive medical procedure. It is vital for the detection of cancer at early stages. With high-resolution images and machine learning technologies, the system can help predict cancers at very early stages thereby enabling the possibility of an extensive array of treatments. The system works by creating machine-learning models based on analysis of X-ray and CT scans from databases, and then X-ray and CT scans are matched for spatial patterns for cancer.

Medical imaging equipment (MIE) needs to be worked upon to make them energy-efficient. A separate analysis of each equipment is required along with the ancillary energy-consuming equipment to make a substantial change in energy consumption.

5.3.1.1 Medical Imaging Technologies

The common techniques that modern medicine uses for medical imagery are as follows:

1. general radiography (X-ray)
2. computed tomography (CT)
3. magnetic resonance imaging (MRI)
4. ultrasound imaging/sonography
5. mammography equipment
6. nuclear imaging.

CT Scans

CT is used to generate a 3D computer image using a two-dimensional X-ray when rotated around an axis. X-rays are used in CT to produce cross-sectional images of anatomical structures and spaces within objects. Most energy consumed in the CT scan process is during the X-ray components operations and by computers to generate digital imagery (Medical Imaging Equipment Study 2017). Also, the radioactive waste produced during these techniques needs to be managed carefully.

Magnetic Resonance Imaging (MRI):

A magnetic field is created by passing electric current in the wires, which in turn is used for obtaining high defined images of the subject. MRI is generally made up of superconductive magnets to maintain low resistive losses in the wire. At an estimate, 40% of the energy consumption of MRI machines is used for continuous cryogenic refrigeration of superconducting magnets (Medical Imaging Equipment Study 2017).

General Radiography (X-ray)

X-rays are a type of electromagnetic radiations. X-rays penetrate through bones and skins variably according to the tissue and bone composition. X-rays are one of the most common techniques used in modern healthcare for diagnosis. X-rays are exposed only for few hundredths of a second, but can instantaneously draw 60–80 kWh (Medical Imaging Equipment Study 2016).

Nuclear Imaging

In nuclear imaging short-lived isotopes, which emit radiation, are consumed by the patients. The radiation is then measured by the gamma camera to capture images of the inside of the body. Scintigraphy produces 2D images, while SPECT and PET technologies produce 3D images (Medical Imaging Equipment Study 2017). Molybdenum-99/Technetium-99m are commonly used isotopes in almost two-thirds of all diagnostic medical isotope procedures that occur in the United States (Medical Isotope Production 2009). Due to the nature of Mo-99 decay in transit, the supply is generally 50% above the demand (Radioisotopes in Medicine 2019). Proper radioactive waste management procedures need to be implemented and followed to avoid any unintentional exposure to radioactive material.

Ultrasound

In this technique, sound waves are used to create an image of the inside body. This technique is commonly used for examining causes of pain, swelling and infection, as well as tracking the health of the fetus in pregnant women (ACR, R. 2019).

Mammography

In this technique, a low dose X-rays are leveraged to examine human breasts for early stages of cancer, tumor or cyst.

Along with the medical imaginary devices, a few other ancillary energy-consuming equipment and areas include

1. equipment room
2. multiple monitors
3. equipment cooling apparatus
4. dedicated HVACs
5. technician room
6. electrical room
7. patient change room and waiting room
8. patient consultation space.

5.3.1.2 Innovative Technologies in Medical Imaging

The science and engineering behind the sensors, instrumentation and software that are used in biomedical imaging devices has evolved a lot in recent years. Modern machines used for medical imaging use much faster hardware which helps in drastically reducing the exposure time. With the enhanced image quality, doctors can provide more reliable and accurate diagnoses. These machines have a negligible carbon footprint because of their low consumption of energy, hence supports the green healthcare initiative.

Innovative techniques and technologies also are making the process of data capture faster, at the same time reducing the exposure time to radioactive material to the patients. These combined are helping in large-scale reduction of medical and radioactive waste. A lot of companies are coming up with new designs and research in the area. For example, Siemens Healthcare recently introduced the CT scanner, SOMATOM Definition Flash. This scanner requires only a very small radiation dose in comparison to traditional CT scanners. It can perform a cardiac scan with less than one millisievert (mSv) whereas the average dose would be in the range of 8–20 mSv for traditional machines. This means less radioactive waste generation during CT scans. Per Siemens Healthcare, these machines are very energy-efficient and consume up to 85% less energy than traditional CT scanners in the market, and 45% less than its predecessor. The machine also replaces lead-based counterweights with steel material (Top Ten Green Medical Devices 2011). This would help the hospitals to be more energy-efficient and generate less hazardous waste per patient.

Mammo Diagnost DR is an energy-efficient digital radiography system. The device uses energy-efficient techniques to capture images and then uses image enhancement software to provide excellent quality imaging. The machine integrates hospital management systems, which allow paperless patient management more easily. This allows the hospital to reduce paper waste. Since the films are also replaced by digital technology, there is no extra energy required for film processing. It also helps in reducing silver and chemical pollution in water caused by film processing (Top Ten Green Medical Devices 2011). All the data captured can then be stored in patients eHealth medical records. Further, the energy required per screening is reduced when compared with traditional systems.

The Green Neuroscience Lab uses a new technique called Stanford Synchrotron Radiation Light to analyze the distribution and functions of metals in the human brain, rather than using traditional methods. This reduces the carcinogenic waste produced by the traditional machines by a large quantity, hence making it more environment favorable (SJWG Rapporteur Report 2017).

GE designed Optima MR360, an energy-efficient 1.5TMR system, is another breakthrough innovation in the field of medical imagery. This device with innovative water-cooling technology is claimed to save 60,000 kWh of energy annually (Top Ten Green Medical Devices 2011).

These technologies not only help doctors and medical professionals gather more information about the patient without the need for invasive medical procedures, but also help in reducing energy consumption as well as in waste reduction.

5.3.1.3 Recommendations for Greener Medical Imaging Equipment

Medical imaging equipment still has a long way to go green. What follows are the recommendations to promote and avail opportunities for making medical imaging equipment energy-efficient (Medical Imaging Equipment Study 2016).

Energy Star Specifications for MIE

Energy star specifications should be provided for MIR devices. In research, it has been found that there are significant energy consumption differences in the scanning mode and low power mode (Medical Imaging Equipment Study 2016). Also, along with MI equipment, the ancillary equipment should also be energy rated.

Incorporate Energy Consumption while Making Purchase Decisions

Purchasers should look into the energy consumption required by machines while making the decisions. The cost of medical devices will go up significantly if the buyers include the energy consumption bill for these devices. For example, the operational cost of X-ray alone is $100 when not in use, and around $400 when they are used. Then the additional cost of ancillary energy should be considered too (Medical Imaging Equipment Study 2016).

Both governments as well as buyers should promote energy-friendly machines through legal and purchase policies. The Green Public Procurement initiative in Europe is estimated to result in 80% energy savings for X-Rays and 50% for MRI and CT scans (Medical Imaging Equipment Study 2016).

Develcp Energy Behavior Guidance Directed at MIE Users

Imaging equipment should be either turned off or on low power mode when not in use. The operators of these types of equipment should be trained, and policies should be implemented for the same (Medical Imaging Equipment Study 2016).

Optimizing Energy Consumption of Cooling Requirements for MIE Equipment

Medical imaging equipment needs to be kept at specific temperatures as part of the operational requirements. Optimization of a cooling system can also result in significant energy-related behavior change for these devices.

5.3.2 Efficient Biomedical Signal Processing

Biomedical signal processing involves the analysis of physiological signals such as heart rate, blood pressure, oxygen saturation levels, blood glucose, nerve conduction and brain activity through bio-measurement tools. With more sophisticated tools, it allows physicians to get a greater insight into patient health and hence allow them to investigate a wide array of treatment options.

In traditional settings, these physiological signals are recorded at particular points in the patient's medical history. Physicians need to make treatment decisions based upon these limited subsets of data calling for guesswork. This process requires a medical professional to be available in the geographical proximity of the patient to record these readings at a specific time. Also, in complex cases, these readings might be skewed at the time of recording.

Biomedical signal processing helps eliminates guesswork. It allows the physician to gain better real-time insight into the patient's health and hence provide more personalized healthcare. With the convergence of real-time monitoring and biosignal processing technologies, an alarm is activated for the attention of doctors and medical staff when the patient's signals are not in normal range. These integrations can act as a lifesaver for many.

Telemedicine, biosensors and biosignal processing technologies allow physicians to monitor patients without being in geographical proximity regularly. For patients, it means that they do not have to be in a hospital just for monitoring. For a medical professional, it allows them to monitor the patients even from the comfort of their homes. This allows hospitals to reduce paper waste and consumption of electricity per patient.

5.3.2.1 Types of Biomedical Signals

Of the various biosignals that are measured by medical professionals, some of the major ones are as follows:

1. electromyogram (EMG)
2. electrocardiogram (ECG)
3. electroencephalogram (EEG)
4. magnetoencephalogram (MEG).

Electromyogram (EMG)

EMG is a tool for recording and monitoring the electrical activity of muscles. It measures the electrical potential of the muscle cells when the cells are activated through small electrical currents transferred through electrodes placed inside the muscle. This procedure is invasive in nature as it requires needles to be placed inside the skin into a muscle. It is used to understand the biomechanics of human movement.

EMG signals are also used for smart prosthetics. Research learns about the pattern of a signal generated by a patient when the patient wants the muscle to perform specific actions and programs it to a prosthetic limb. Now whenever the patient wants to move a muscle, the prosthetic limb converts the particular pattern back to an action command.

Many companies and researchers are coming up with new EMG sensors which considerably reduce the energy requirements. These new EMG sensors are capable of processing signals in real-time. These enhancements, in turn, will tend to be significant energy savings. For example, these will reduce the amount of training required for prosthetic limbs, thereby reducing the power consumption per patient per session.

Electrocardiogram (ECG)

This technique is similar to EEG and is used for measuring the electrical activity of the heart. The electrodes are placed near the heart over the skin. Similarly to an EEG, this technique is non-invasive and extremely mobile. There has been a lot of advanced research in this area. For example, the latest version of the Apple watch is capable of generating an ECG for the user in real-time for a very low power compared to traditional CRT monitor-based ECG machines. Bringing ECGs to the fingertips of patients reduces the need to go to hospitals, thereby lowering the carbon footprint per patient.

Electroencephalogram (EEG)

EEG is the monitoring method to record the electrical activity of the brain. It is generally performed by placing two electrodes on the scalp and measuring the voltage fluctuations that occur due to the functioning of neurons. EEGs have a lot of benefits over other techniques to measure brain activity. An EEG is non-invasive unlike other techniques like electrocorticography which require actually pushing electrodes to the surface of the brain, and it does not require bulky instruments which allows it to be extremely mobile and to be used in remote areas. The sensors are simple electrodes; hence the overall cost of hardware is pretty low in comparison to other techniques. This mobility reduces the need for patients to travel long distances for diagnosis. The minimum hardware requirement to perform this technique also reduces waste drastically when compared with other techniques like electrocorticography.

Magnetoencephalogram (MEG)

MEG is a technique to record brain activity by recording the magnetic field that is produced when neurons are fired into the brain, causing some electrical impulses. These magnetic fields are recorded through a very sensitive magnetometer. Currently, MEGs are used to understand the functions of various parts of the brain and the coordination between those parts. An MEG requires a very sensitive magnet which requires a high amount of energy. Nowadays with advancement in quantum sensing, the magnetic field sensors are becoming capable of achieving the same level of precision and sensitivity as the ones that are cryogenically cooled (Boto et al. 2017).

The biosignal field is still at a very nascent stage but is emerging at a very fast rate. With enhancements in low power sensors and upcoming artificial intelligence techniques, biosignal devices are becoming smarter. These devices will then continuously monitor patient/subject from the convenience of their homes without the intervention of medical professionals and will raise alarms in the case of any abnormalities. Doctors can then remotely monitor the patient and patient data and can guide accordingly. This synergistic relationship between diagnosis, alerting and other eHealth tools will provide affordable care to patients in an energy-efficient way.

5.4 Green Smart Digital Healthcare System

The advancement of technology plays a very crucial role in the oncoming of a new generation healthcare system. Cloud-based smart digital health systems empower both medical providers and patients to access services in a much more streamlined and on-demand basis. Digital healthcare can be termed as the ecosystem where digital technologies converge with healthcare to provide affordable, personalized, efficient and easy access to healthcare services.

Digital technologies have significantly strengthened the relationship between patients and medical professionals. Patients can now ask more direct and data-driven questions to medical professionals from the convenience of their homes. Digital healthcare enables healthcare professionals to serve more patients efficiently. These technologies reduce the overall impact of the healthcare industry on the environment by reducing pressure on healthcare facilities, by providing remote healthcare support and saving on patients' commute.

In a country like the United States where the current healthcare system is unscalable due to the high volume of patients and shortage of medical professionals, a smart digital health system provides an alternative where patients can be treated across the globe without having to be in close proximity. These technologies are challenging traditional medical procedures and allowing healthcare to be more environmentally friendly and green.

Green smart digital healthcare can be categorized into the following subcategories:

1. electronic healthcare information management system
2. e-learning
3. smart wearables
4. telemedicine
5. intelligent healthcare services.

5.4.1 Electronic Healthcare Information Management System

The emergence of information and technology has helped much innovation in the healthcare sector. These innovations have paved the way for new processes and technologies to manage patients and hospitals. On the one hand, these systems provide better and more effective management for hospital and patients. On the other hand, they provide an energy-efficient alternative to physical healthcare facilities.

Electronic healthcare management can be broadly divided into three categories:

1. hospital information system
2. clinical information system
3. electronic health record.

5.4.1.1 Hospital Information Systems

A hospital information system (HIS) can be defined as a comprehensive, integrated information system designed to manage all the aspects of a hospital's operation, such as medical, administrative, financial and legal issues and the corresponding processing of services (HISTree (Hospital Information System), 2019). HIS allows healthcare professionals to provide better healthcare by allowing them to access patient information and history at the click of a button. It enables them to securely share and coordinate the patient information both internally and externally with the help of healthcare networks. This also contains patients' medical history as well as clinical information such as visit summaries and laboratory test results for patients such as blood count, X-ray, EEG and ECG.

HIS and healthcare networks are emerging at a fast rate. A lot of healthcare industry leaders and software companies are heavily investing in this area. HISTree is a HIS system developed by Trio-Technologies. HISTree is highly configurable and user-friendly and can be deployed on-premise or can be accessed through the cloud. It allows for tightly linked workflows, more granular access controls, which in turn provide better control over the healthcare data (Solutions. Hospital Management System 2019). eHospital by Adroit Infosystems is a comprehensive and integrated information system that is designed to manage every possible hospital operation. It covers end-to-end management starting from patient information to hospital stock management (Ehospital – Best 2019). Similarly, Mediware HIS focuses on managing the unstructured information and uncontrolled processes in multiple areas in a hospital. It allows for managing the needs and information of homecare patients as well as rehabilitation information management (Digitized Knowledge Processing 2019).

Benefits of HIS

HIS enables the healthcare providers to manage the operations of the facilities in an easy and efficient way. It also provides other benefits like:

a. Provides insights on quality of service HIS allows for integration with the business intelligence module to provide valuable insights into hospital operations and quality of patient care. Using this information, healthcare facilities can take data-driven actions to improve the overall quality of healthcare.

b. *Enables more personalized healthcare* It allows medical professionals to manage the patient information at any point in time, which allows them to provide personalized healthcare.

c. *Reduction in operational costs* These systems store information in digital format which helps in maintaining the information integrity by eliminating the errors introduced due to handwritten notes. It helps to reduce the operational costs for healthcare.

d. *Automation of tasks* HIS allows for automation of management tasks which allows medical professionals to focus more on patient health and safety.

e. *Green healthcare* HIS facilitates saving data in a digital format which was traditionally saved in files using environmental resources such as papers and plastics. Healthcare providers can use the data-driven insights provided by HIS to design more efficient resource management processes and reduce the wastage in facilities.

5.4.1.2 Clinical Information System

Increased prevalence of diseases and decreased availability of physicians along with an effort to reduce healthcare costs have led to the introduction of several clinical information systems (CIS). Traditional clinical visits do not guarantee patient satisfaction and they have limited time available per visit which acts as a hindrance to improved patient-doctor relationships. CIS such as patient portals and e-visit systems aim to address these problems by enabling patients to send an email to the doctor at any time, build a better relationship with doctor and access healthcare information whenever needed.

CIS is a secure healthcare information technology that enables patients convenient, 24-hour access to their healthcare information and allows them to view information like after visit summary, test results, prescriptions, medications, immunizations and so on (Frequently Asked Questions 2019). Some patient portals also allow sending an email to a doctor, scheduling appointments, making prescription refill requests and so on.

Patient portals allow improved communication between a patient and the doctor which is of significant value in the case of chronic diseases. It facilitates secure and easy management of personal health information for a patient. The data in CIS can be shared over the portal which saves on paper.

Benefits of Clinical Information Systems

CIS can act as a very powerful tool in providing more efficient and personalized healthcare. It provides many benefits both to the patients and healthcare providers. Some of them are described below:

a. Ease of access   Patient portals can be accessed 24 hrs a day and allow the user to get respective healthcare information at any time of the day. This ensures that the patient does not have to wait for specific office hours or make an appointment for concerns that can be addressed via email.

b. Improved patient-doctor relationship   Using patient portals acts as a useful tool to strengthen the patient-doctor relationship. This is possible because of the ease with which a patient can send an email message to the doctor and at the same time it is convenient for the doctor to reply to the email whenever the message is received. Also, if a patient is undergoing treatment and is in contact with more than one doctor, then the doctors can post the medical data, instructions or medications relevant to that patient at the central location, which is the patient portal. This ensures that all the doctors are aware of other medical conditions and treatments that the person is undergoing and hence ensures that the patient gets the right medical advice and treatment. This improved and continuous communication also helps reduce the number of admissions in inpatient facilities, hence reducing the carbon footprint and medical waste per patient.

c. Lowered cost of healthcare   The use of patient portals lowers the number of office visits as patients can get medical advice for their concerns by contacting the doctor via email. This has made healthcare accessible to more and more people and at a lower cost. Patients can have their concerns addressed by being in their comfort zone and also, at the same time, making it convenient for the doctor to help more people.

Patient portals act as a big time-saver. In the traditional system where a team was needed to answer the phone calls and make appropriate appointments for the patients, now everything can be done online through the portal at both the patient and doctor's convenience. Making the entire system online has enabled healthcare providers to focus more on the quality of the services that are provided. Patients are also allowed to pay for the services online, hence saving a lot of time.

d. Reminders help keep track of health   Patient portals store all the healthcare information of a patient which can be easily accessed by the patient at his comfort and convenience. Apart from keeping track of the patient's ongoing treatments, it contains information as to which vaccinations the person has had and which are due in future, immunizations, what are his allergies and so on. This way, patient portals act as a reminder and help the patient keep track of his health in a much better way.

e. Helps reduce the negative impact of healthcare on the environment CIS helps to reduce the negative impact of healthcare on the environment in many ways. It reduces the need for admissions to inpatient facilities by enabling doctors to manage the treatment plans remotely. It is also better and more personalized healthcare by enabling doctors to track patients' health more precisely, hence allowing them to be proactive, which in turn helps patients to be healthier and reduce the need for hospital visits. It also allows patients to access all the information from their home, for which traditionally they had to travel to the hospital. Hence, it allows reducing the impact on the environment by reducing the carbon footprint through vehicle emission.

5.4.1.3 Electronic Health Record

A medical record can be defined as the collection of patient health information and medical information. These records consist of information about the symptoms, the process of diagnosis, hospital and doctor visits, prescribed drugs, test results and much more. These records when collected and saved in digital format are known as an electronic health record (EHR) (Gunter and Terry 2005).

Medical records add up the medical history of the patient. They help the doctors and caregivers to communicate more effectively with their patients. Effective communication is essential for identifying both patients' needs and expectations. Medical records consist of patient data and other biomedical signals than mere text, for example images, X-rays, videos, MRI scans and so on, and are called a multimedia medical record (MMR) (Sicurello n.d.). A collection of these EHR and MMR can be termed a clinical database.

The researchers often study the information in these clinical databases in combination with other factors such as demographics to determine the causes and risks associated with a disease. Epidemiology is one such use of the clinical database where the researchers study and analyze the factors and determinants of health and disease conditions related to defined demographics. It includes the cause of the disease, transmission, investigation and the outbreak of the disease (WHO Epidemiology 2019). Clinical and multimedia databases provide the information required for creating epidemiological databases. This information is then transformed and mapped to demographic factors and diseases to create an epidemiological database. Epidemiological databases help to find relationships between demographics and diseases. This information can be used by researchers to understand the causes and factors of the disease outbreak better. It helps them create a risk analysis for epidemic outbreaks based on specific demographic factors.

Advantages of Clinical Databases and Epidemiological Databases

Some of the major benefits of clinical and epidemiological databases are described below:

a. Environment-friendly Traditionally, the medical records were recorded on paper which was hard to manage or transfer and caused a high wastage of environmental resources like trees. With EHR and epidemiological databases, all the information is stored in digital format. Even when moving service providers, there is no need for re-printing patient records. These technologies also help in reducing other medical waste. Digital X-rays are more accessible to share, support and manage, and create much less toxic waste when compared to traditional film-based X-rays.

b. Ease of use EHR is easy to share and can be accessed on the go. This helps the doctors and other healthcare personnel to fetch the information quickly and hence provide better treatment to the patient. It allows easy coordination and movement between service providers. Doctors/patients can easily share data with multiple doctors to get the best treatment options available.

c. Better community healthcare EHR when augmented for research purposes like epidemiology helps in better preparedness and superior care. Epidemiological databases make it easy for researchers to investigate and connect demographic factors to a disease outbreak which in turn allows the healthcare providers and policy makers to monitor the health status of the whole community.

d. Better healthcare policies Epidemiological databases help government and health organizations to create plans to inform and educate the population about the diseases to which they are more prone. This information can also be used by governments to allocate resources and focus their efforts in specific areas. It helps the government to create a competent public healthcare system.

5.4.2 Telemedicine, eHealth and Telecommunications

Telemedicine is a mechanism to perform remote diagnosis, provide support and treat patients using the means of electronic communication and technologies (MHealth 2019). Telemedicine allows and advocates for the use of low power sensors rather than the traditional machines with high power consumption. It also helps doctors to manage and monitor patients in outpatient facilities and even at patient homes which will decrease the hospital admissions. Each hospital admission converts to a large carbon footprint due to the power consumed by hospitals and transportation used by patients to reach the hospital. This carbon footprint is even larger for patients from remote areas. For example, telemedicine is used in Queensland, Australia, to manage 17% of pediatric outpatients with burns. On an estimate over six years, 1,000 video conference consultations eliminated about 1.4 million kilometers of patient travel, which reduced CO_2 emissions by 39 tons each year (Smith 2007).

With the help of telemedicine, doctors can replace hospital admissions with remote monitoring and management for chronic conditions like Parkinson's disease (Types of Telemedicine 2019).

5.4.2.1 Type of Telemedicine Solution and their Benefits

Depending upon the location, patient's need and upon the services available, the following are the three types of telemedicine solutions that are becoming prevalent in the market (Types of Telemedicine 2019).

Store and Forward Telemedicine Solutions

Here the patient's data is stored and transferred to other medical facilities or professionals in an asynchronous fashion (Types of Telemedicine 2019). For example, a remote lab installed in a rural area can take an X-ray or can perform radiology on a patient and can then forward the results/images to a specialist at a later point in time for consultation. In this model, the patient, the consulting specialist and the medical professional need not communicate at the same time. Thus, this model eliminates the need for everyone to be at the same place at the same time, thereby improving healthcare system efficiency. This model also lowers the patient's wait times, provides better outcomes for patients and helps create more optimized schedules for physicians. Physicians and patients can now consider rush hours as to whether to visit their local medical care centers, thus not only improving the overall healthcare system, but also the environmental impact.

Remote Patient Monitoring in Telemedicine

In this model, the healthcare providers can track patients' vital signs along with other healthcare data remotely (Types of Telemedicine 2019). This model, when integrated with the wearable technology, can forward the monitoring data to physicians at regular intervals for better health monitoring. For example, a patient wearing a blood pressure monitoring device can be remotely monitored by the physician. The device can continuously push data to the healthcare system and raise the alarm when the data does not coincide with the usual pattern. The critical part of remote patient monitoring is the efficiency of health monitoring devices. With emerging technologies, the devices are becoming better day by day. This system effectively reduces the carbon footprint as patients need not visit healthcare facilities for low-impact, regular checkups.

Real-time Telehealth

In this approach, both the patient and doctor discuss and communicate in real-time. In this approach, patients interact with the physicians remotely. The video chat is the most common form for this kind of approach. There are many emerging companies such as TeleDoc and DoctorOnDemand that are investing heavily in this space. In this approach patients and their doctor can meet each other at a scheduled time suitable to both of them, and healthcare facility timings and rush hours do not affect the schedule. The commute for patients and doctors is not required. Also, the communication channel is more streamlined, thus putting a stop to the back and forth conversations required in other approaches. Last, proximity to the location of the patient is not a limitation, thus facilitating access to medical professionals across the globe (Types of Telemedicine 2019).

5.4.3 mHealth and pHealth Systems

The Global Observatory for eHealth (GOe) defines mHealth (or mobile health) as medical and public health practice supported by mobile devices, such as mobile phones, patient monitoring devices, personal digital assistants (PDAs) and other wireless devices (Mhealth New Horizons 2019). mHealth focuses on using the basic facilities of voice and text on a mobile device to provide healthcare service, but it also extends to more complex and newer technologies such GPRS, 3G, 4G, and Bluetooth technologies (Mhealth New Horizons 2019). As mHealth makes use of pre-existing and readily available devices, it is easier for patients to adapt to the change. It also eliminates the need to buy/manufacture new devices specific to the service and hence rules out the possibility of generating e-waste. Using mHealth reduces the need of patients to visit healthcare facilities for non-critical issues which can be solved using mHealth tools, hence reducing the negative impact of healthcare on the environment.

The World Health Organization breaks down mHealth initiatives and technologies into five categories (Mhealth New Horizons 2019):

1. communication between individuals and health services;
2. consultation between healthcare professionals;
3. intersectoral communication in emergencies;
4. health monitoring and surveillance; and
5. access to information for healthcare professionals at the point of care.

pHealth (or personalized health) is a new paradigm that encourages the participation of the whole nation in the prevention of illnesses or early prediction of diseases such that pre-emptive treatment can be delivered thus achieving pervasive and personalized healthcare (He et al. 2013, pp. 589–598). pHealth entails monitoring patients with the help of invasive and non-invasive smart wearable sensors. The information is then collected and analyzed for patterns and thresholds. pHealth can be leveraged to raise an alert or perform corrective actions if needed. For example, when a smart wearable device raises an alert to a healthcare facility regarding an abnormality in the patient, the caregiver team can perform some corrective actions, such as releasing the dose of medicine in the patient's body or forwarding a call to emergency services.

5.4.3.1 Benefits of mHealth and pHealth

Some of the critical applications and benefits of mHealth and pHealth system include:

Spreading Education and Awareness about Medical Issues

mHealth has been proven to be one of the most effective ways to spread education and awareness about various conditions and diseases. mHealth application sends alerts and reminders through SMS and notifications to offer information about symptoms, treatment options available for diseases and general health advice for the population.

In developing countries, these technologies have provided healthcare providers with a way to reach the population in hard-to-reach and rural areas where lack of reach of healthcare prevents the population from making informed decisions about their personalized health. The wearable tech industry is growing at a fast pace and is attempting to reach the maximum population.

Project Masiluleke and Text to Change are two projects currently operating in South Africa and Uganda. These projects use SMS technology for information about HIV/AIDS and test and medicine reminders for HIV patients (Mhealth New Horizons 2014). Traditionally they have used pamphlets and booklets to spread this information. Traditional mediums made the information hard to reach the major population and used a lot of resources like paper. This harmed the environment as a whole. Using mHeatlh has not only enabled them to distribute information quickly and efficiently, but has also reduced the negative impacts of the traditional system on the environment.

Patients Remote Data Collection for Personalized Healthcare

In this era of personalized healthcare, data collection becomes a more crucial and essential part of the healthcare systems. It enables healthcare workers and policymakers to gage the effectiveness of treatment options and policies for health management. Traditionally, this information is collected through paper-based surveys during doctor visits by the patients. However, this method was not effective in remote areas where patients do not have access to doctors even in case of severe illness. Also, the traditional method depends on extensive use of environmental resources such as fuel, paper, and so on.

Recently, a lot of mHealth projects have moved the focus to solve the problems in this area through the use of mobile technology and focus more on making the whole system environmentally friendly. These initiatives make use of mobile phones, PDA and SMS technology to conduct surveys and questionnaires for the patients (Mhealth New Horizons 2014). This allows the doctors and healthcare providers to collect and process the information in real-time and customize the corrective actions accordingly.

Remote Monitoring

One major focus and area of growth with this initiative is remote monitoring. Remote monitoring allows doctors to monitor patients in their homes and opens the possibility of new and better treatment options without keeping patients in hospitals. It allows healthcare providers to ensure adherence to medicines and regimen which increase the survival rates for patients in case of a disease such as AIDS or diabetes (Mhealth New Horizons 2014).

In the case of remote areas where hospital beds or rooms are not accessible, this opens up a lot more treatment options which were not possible traditionally. Doctors can now monitor patients with motor disabilities or chronic conditions without needing them to be in the hospital. With the help of the integrated and wearable sensor, doctors can monitor advanced parameters Traditionally, this kind of monitoring required large machines.

With one of the initiatives in Thailand, TB patients were given mobile phones, and healthcare workers would monitor their medicine schedule and regimen. The rates of patients who tested positive in the first test and tested negative in the second result reached 90% due to the introduction of this remote monitoring application (Kunawararak et al. 2011).

5.4.4 Smart Wearable Technologies for Health Monitoring

Wearable technologies can be defined as electronic devices that can be worn or can be implanted. Around 88% of doctors want patients to monitor their vital stats at home (MHealth 2019). These help patients keep track of their health and allow healthcare professionals to make more educated decisions about diagnosis with the help of the information provided by these devices. These technologies are making a huge leap in the healthcare industry by providing a wide range of devices and use cases that they can solve. As per Marketwatch predictions, it is thought that the number of wearable devices will total 51.50 billion by 2022 (Wearable Devices Market 2018). With these high number of wearable devices entering the market, it is essential that these devices are designed with green technology in mind for environmental benefits.

A new wave of innovation is looking toward making even these devices more energy-efficient and environmentally beneficial. The devices are designed nowadays with recycled or compostable material and of lower power consumption. For example, technologies like Bluetooth Low Energy (BLE), and low Wi-Fi power modules, which are the backbone of numerous IoT devices, are specifically designed to consume low energy. A great deal of research is also happening in the development of devices to use human kinetic energy as a power source for the device (10 Human Motion 2012).

Another breakthrough with the potential to revolutionize green healthcare devices is In-Vivo Networking (IVN) technology. IVN is designed to power and communicate with the devices implanted in the human body wirelessly. IVN-powered devices do not require any batteries to be attached to the sensors but instead powers them through radio frequency waves which are designated as safe for humans. Researchers were able to showcase the powering and fetching information from devices that were located 10 cm deep in the tissue from a distance of one meter (In-Vivo Networking 2018).

Wearable devices are also meant for preventive care and making the life of patients more comfortable. Devices or systems that are designed for preventive care can be re-engineered for advancement and reusability purposes. For instance, rather than making separate wearable devices – one for the heartbeat, one for steps and one for glucose level – companies are investing in research projects to provide these capabilities in the same system/device. These devices enable doctors to collect information remotely without the need for patients to travel to hospitals. In some cases, it can even avoid admission to the inpatient facility. For example, if the wearable device can alert a patient about high blood sugar, the patient can immediately take insulin, and the situation gets handled there. In another example, the latest Apple Watch 4 has the capability of not only measuring heartbeat, but can also take an ECG of the patient (ECG App and Irregular 2018). This ECG can then be shared with the healthcare service technologies for faster and easier analysis. This sharing also allows hospitals and doctors to make a better data-driven decision along with reducing carbon footprint and medical waste generated per patient.

5.4.4.1 Innovative Smart Wearables and Their Benefits

While there are continuous breakthroughs happening in the wearable industry, the ones that are more commonly used that help patients enter the triangle of healthy living are listed below:

Smart Sleep Aids

No one can deny the fact that proper sleep is a must for maintaining good health. On this concept, there are a number of devices that can track and monitor sleep patterns of consumers. For example, Apple Watch, Fitbit Ionic, Emfit QS, Jawbone UP3 and Motiv Ring are some of the devices that are trending in the market for sleep tracking. MYIA Labs place sensors under the bed to track heart rate and breathing levels of a sleeping person (How Wearables Will 2019). Further, active feedback sensors are being innovated which can play calm music whenever a sensor detects a patient stirring. The kinetic energy of the patient can power these devices and underlying sensors.

Smart Mobility

Companies like Neofect have developed tools that blend gaming techniques with the rehabilitating muscular and skeletal system. RAPAEL Smart Glove is a high-tech rehab device that follows patients' hand motions, measuring the slightest movements in hand with accelerometer and bending sensors. Clinical results have suggested that the motor recovery rate of clinical rehabilitation therapies can be improved when therapists use wearable sensors and therapeutic games in their routine therapy practice (Jung et al. 2017). Advancements like these considerably reduce the effort and resources required for rehabilitation.

Smart Blood Sugar

While high blood sugar has the potential to damage organs and cause heart attacks, low blood sugar can cause confusion, seizures or loss of consciousness. To avoid any such problems, patients usually go for regular glucose level tests or tend to carry a glucose testing kit. These traditional approaches have a detrimental effect on the environment as patients need to commute, there is increased pressure on healthcare facilities and, of course, the generation of biohazard waste.

In new wearable devices, makers have integrated complete glucose testing in smart watches. On the press of a button, a micro needle from the device creates micrometer sized perforations and can gather all required glucose level information (Khanna et al. 2008) Further, the new devices like Abbott's device are challenging the existing methodologies of measuring glucose and claim to measure glucose level even without pinching/inserting any needle (Revolutionizing CGM With Freestyle Libre 2019). Innovations like these further reduce biohazard waste.

Diabetes Sentry is another company that leverages skin-contact sensors to monitor body chemistry which helps in detecting diabetes. The readings collected by skin-contact sensors with millions of data points provide a pattern over a period of time. Diabetes Sentry then raises an alarm when patients' blood glucose begins to trend downward (Howsmon et al. 2015). This new technology not only reduces the resources needed for blood sugar measurement but also reduces the need for going to the emergency services due to proactive monitoring, thus further reducing the environmental impact of the healthcare sector.

Smart Temperature

These are the devices that monitor the patient's temperature. These devices are used both in hospitals as well as outside. In hospitals, there are patches available that are placed on the patient and the real-time temperature can be seen on the monitors. At the consumer level, there are numerous devices that can capture real-time temperature. Opting for mercury-free temperature measurement devices reduces the environmental impact of cleaning when the thermometer breaks. There are states in the US which have laws not allowing the sale of mercury-based thermometers (Mercury Thermometers 2018). Opting for and legalizing smart environmental options empower the goal of green healthcare.

Smart Pain Management

Wearable technology for pain relief is an intriguing emerging area. For example, the Quell device is a band that patients wear around the upper calf. This wearable device, rather than using any medications, leverages electric signals that stimulate the body to produce naturally occurring substances (endogenous opioids) which then inhibit nerve signals that lead to feeling pain (Gozani 2018). Further, it reduces the environmental footprint pertinent to research, manufacturing, transportation and storage of pain killers

Smart Heart Monitoring

Another company, Kardia Band, uses the track pad attached to a mobile phone to calculate heart rate and atrial fibrillation (Alivecor 2018). The latest version of the Apple watch, series 4, is capable of generating an ECG similar to a single-lead electrocardiogram (ECG App and Irregular 2018). These, when integrated with AI technologies, reduce the need for patients visiting healthcare facilities.

Wearable technologies have a broad spectrum, and in the healthcare industry wearable devices are used by consumers to record and monitor a wide range of data. These technologies, on the one hand, are leading to the development of more personalized healthcare and on the other hand, are saving the planet.

5.4.5 eLearning for Green Healthcare

With ongoing advancements in technology, medical educators are now facing different and new challenges in teaching tomorrow's medical professionals. Improvement in healthcare systems and the increased need for personal healthcare have increased the demands of medical professionals and specialists. This increase is forcing new educational institutions to come up. New institutions need more resources, causing a detrimental impact on the environment. Innovative technologies that can deliver learning experience to students in an environmentally sustainable way are required. eLearning is a concept that has evolved with the vision of sustainable, affordable and approachable learning.

eLearning can be defined as the use of Internet technologies for enhancing the user learning experience. eLearning technologies benefit by providing control over content, personalized learning sequence, individualized pace and schedule of learning. It enables students to access study material related to their field of interest at the ease of a click of a button instead of going through piles of books to learn a new thing. It enriches the education experience by providing the ability to tailor the content according to the personalized learning objective. This way of learning can reduce the carbon footprint and eliminate paper wastage. eLearning has the ability to revolutionize the education system and make it sustainable.

eLearning can also be seen as a tool to improve the efficiency and effectiveness of medical education in an environmentally sustainable way. It can help bring the competency as a focus of education, rather than traditional knowledge. Bringing this competency as the focus of education is critical in the medical field where any mistake in assessment could result in risking the lives of patients.

Many companies are working towards creating and improving eLearning platforms. One of the most famous eLearning platforms, Udacity, offers video lectures, animations and online projects designed in order to provide an immersive education experience. The platform also provides the ability to hold One-2-One discussions with educators. But Udacity currently only provides courses on computer science and related technologies. There are similar platforms for medical science and healthcare. For example, Medskl provides video lectures, whiteboard animations and summary notes specially designed to teach the fundamentals of clinical medicine in a more engaging way. The platform provides access to over 100 lessons to teach topics like cardiology, dermatology, neurology and other medical courses. With improvements in eLearning platforms, the need for traditional schools/colleges will decrease, thus reducing the consumption of resources attached to them.

In the medical industry, medical schools from all over the world have come together to create an (IVIMEDS). IVIMEDS aimed to create a massive online education system where students/medical professionals can get access to specialized programs from across the globe. Centrally managed systems improve efficiency and reduce the environmental impact of developing the same things over and over (Harden et al. 2002).

5.4.5.1 Benefits of eLearning in Green Healthcare

eLearning, along with being an environmentally friendly approach to teach medical professionals, possesses several other benefits.

eLearning is not bound by physical location
eLearning allows healthcare professionals to study in the office, at home or even on the train. This flexibility ends the need to travel acts a big step towards reducing the emission of toxic gases into the environment caused by the burning of fuels used in vehicles. This flexibility also helps in reducing the stress on natural reserves of fuel to meet the ever-increasing energy need.

eLearning is Cost-effective and Resource-efficient

eLearning is cost-effective and resource-efficient when compared to traditional classroom methods. The traditional training methods call for a lot of operational costs. The is an environmental cost for building classrooms. With increasing demand, classrooms need to be created in traditional methods for knowledge transfer. However, with eLearning, these are no longer required. eLearning is very cheap and an environment-friendly way to learn. There is no need for buildings and classrooms for training as professionals can learn from anywhere they want.

eLearning Provides an Effective Educational Resource

Since the learner can access the content of eLearning throughout the learner's lifetime and with access to tools like discussion forums, eLearning provides the ability to refer back to video/audio/text content in case they need/review any information. They can use discussion forums to ask questions and seek clarification. These forums are coming out as alternatives to in-person classroom discussions. As lectures can be accessed throughout the future, the environmental impact of creating new lectures every semester to new students has vanished.

5.4.6 Modeling and Simulation Methodologies

Simulation is the imitation of a situation or process. The act of simulating something first requires that a model is developed. This model represents the key characteristics, functions and behaviors of the selected physical or abstract system or process. The model represents the system itself, whereas the simulation represents the operation of the system over time. With the advancement in computer science and access to higher computing power, simulation has been extensively used in many fields such as education and training, biomechanics and military and more, especially healthcare.

5.4.6.1 Benefits of Simulations and their Environmental Affect

Simulations in healthcare have four primary goals – education, assessment, research and health system integration in facilitating patient safety.

Medical Education

The simulations provide a better and enriched training experience for medical professionals. In healthcare, proper training is of utmost importance where a simple mistake could be fatal for the patients. Simulations help beginners to gain confidence in the medical procedure by performing on them in various simulated conditions without wastage of resources. It allows medical professionals to simulate and train on disease and events which are rare and do not present enough opportunities for practice. These simulations lower both the resourcing needs as well as waste produced when compared with non-simulated learning techniques.

Health Assessment

Medical professionals can perform medical procedures on computerized mannequins that perform dozens of human functions realistically in a healthcare setting such as an operating room or critical care unit that is indistinguishable from the real thing. With simulation techniques, the assessment of ability can be based on demonstrating actual tasks and procedures in a simulated environment rather than relying on traditional methods of written question answers and oral exams. Since no humans/animals are involved in this type of learning, biohazard waste is not produced.

Medical Research

This gives medical professionals the freedom to make mistakes and learn from them, which allows them to gain insights from their mistakes and their effects on patients. Further, with advancement in technologies, standard simulation platforms are coming into the market which further reduce the environmental impact. For example, companies can use AWS backed AR & VR technologies like Amazon Sumerian. Leveraging these publicly offered solutions not only help the companies in reducing the development cost and time-to-market but also reduces the digital and carbon footprint significantly, since public cloud companies manage operations overhead.

It can help the researchers to understand and test the effect of new drugs or medical devices in simulated conditions. It can help validate the new method of providing drugs under various simulated environments. This methodology relieves the resources that were supposed to be consumed for conducting research in real life. The research uses community datasets and simulation techniques to try to predict the community health and onset of the epidemic. It can also help in developing quality of service (QoS) in the medical industry for new procedures and systems. With closer collaboration across research techniques, the datasets can be shared. Sharing of datasets further reduces the effort and environmental impact of creating new datasets.

Integration with Patient Safety

Simulations act as a powerful tool to mitigate common errors and enhance patient safety. Similar to aircrafts where the crew goes through various simulations to learn and understand the behavior, common mistakes, effective communication and team work, simulations in medical institutions can do the same (Moyer n.d.). Further, the simulations for patient safety can be performed without harming the patient along with the risk and resources associated with it.

5.4.7 Intelligent Healthcare Services

Artificial intelligence (AI) is disrupting every sector across the globe. AI in healthcare helps doctors, researchers, and policymakers to make better decisions, and diagnose and provide treatment for chronic and acute illnesses. It uses data science and mathematical algorithms to analyze patterns and data from the human body to make predictions and perform a diagnosis (7 Ways Data Science 2019). AI further allows doctors and other medical professionals to provide more and better-suited treatment options due to early diagnosis, which may not have been possible in case of late diagnosis. With early diagnosis, doctors can provide treatment options at home which was not traditionally possible due to late diagnosis. These tools allow hospitals to reduce medical waste per patient along with lowering the pressure on healthcare centers.

5.4.7.1 Benefits of Artificial Intelligence for Green Healthcare

AI has numerous benefits in the field of healthcare ranging from personalized learning to assisting medical professionals in making decisions. Some of the key benefits are listed as below.

Personalized Medical Information

AI possesses the potential to deliver personalized information to doctors based upon relevance. The relevance can be calculated from search queries, specialization of a doctor, blog posts, major media publications and so on.

Clinical Documentation

AI can also reduce the clinical documentation process. With natural language processing (NLP) capabilities embedded in healthcare AI, doctors can just dictate the medical notes, which can then be saved automatedly in electronic healthcare records. This NLP-powered AI can save significant man hours for medical professionals (3 Ways Medical 2019).

Patient Safety

AI also helps in preventing fatalities due to medical errors by providing extra data and prediction during the process of diagnosis. During research, AI was able to diagnose cancer more accurately, even in cases where doctors were not able to detect cancer and misdiagnosed the case as something temporary (Laura 2018). A lot of these tools are available through intelligent decision support system.

Intelligent Decision Support System

Intelligent decision support systems (IDSS) can be defined as a system that helps in decision-making by the use of AI and machine-learning technologies (Chang et al. 1994). These systems generally use a hybrid approach, utilizing simple rule-based engines, knowledge-based engines and neural network engines. It performs an interpretive analysis of large-scale data with intelligent and knowledge-based methods.

IDSS in healthcare works by analyzing the data from different sources such as:

- real-time data from monitoring devices
- patient and family history
- demographic data, such as ethnic background
- genealogical data
- medical test reports.

It then analyzes this data and tries to find the common characteristics and trends between the data and other medical records to generate predictions about the diagnosis.

IDSS can generally be classified into three main categories:

1. knowledge-based IDSS
2. machine-learning techniques-based IDSS
3. hybrid IDSS.

a. Knowledge-based IDSS This kind of IDSS is generally based around a set of rules which are a compiled set of information around the subject. In medicine, it will capture information about the disease along with symptoms and patient data. The rule engine contains the logic to match the rules in the system with the patient data to provide outcomes such as diagnosis, confidence percentage and possible treatments (Smith 2007).

b. Machine-learning Techniques-based IDSS These types of systems enable the computer to learn and gain knowledge from past experiences and through identifying patterns in the clinical data. Here, the system employs machine-learning and artificial intelligence techniques to process the historical data and convert it into a knowledge base. The patient data is then matched against this knowledge base to provide diagnosis and treatments. The common techniques employed in this kind of system are artificial neural network, genetic algorithms, and decision trees (Smith 2007).

c. Hybrid IDSS Some systems use a combination of knowledge-based techniques and machine-learning techniques to provide outcomes. These systems are known as hybrid IDSS. The primary purpose behind hybrid IDSS is to combine the most powerful features from machine-learning techniques-based IDSS and knowledge-based IDSS to provide optimal output (Smith 2007).

IDSS aids doctors and clinicians in diagnosing and exploring treatments by converting the raw medical data into actionable insights. The system can use configurable rules and patterns for the patients, based on factors such as medical history, patient symptoms and clinical parameters and raise warning alarms to doctors when such rules/conditions are not under the threshold. These flags/alarms can help healthcare providers to intervene and save the lives of patients. These systems can help to diagnose diseases in the early stages by combining the insights from various data sources and comparing them against the historical data. For example, IDSS can help to diagnose cancer early by analyzing and comparing complete blood counts, X-rays and CT scans with historical data to point out cancerous tumors which otherwise might look malignant to the physician. Also, the system can be trained to spot incompatibilities between prescribed medications for patients and raise alarms to the doctors. It can be connected to epidemiological databases to help predict the future spread of an epidemic.

Hence, the IDSS system can not only help in supporting healthcare workers and doctors with everyday decisions, but also proves especially useful when they have to take a decision where there is ambiguity or contradiction. It provides them with a confidence level for each option, with calculation and reasoning used to reach that confidence level. It helps them to solve the problem and helps them improve the justification and explanation for the solution being pursued.

Conclusion

Healthcare is a major sector that needs to undergo significant changes to strengthen its commitment to environmental sustainability. The iron triangle of green healthcare describes how a healthy environment will be beneficial to both patients and healthcare facilities. The movement toward making the healthcare sector green mandates the need to rethink, innovate and adopt green practices while building and designing healthcare facilities as well as medical devices and processes. Further, harnessing the power of digital healthcare technologies like EHR, smart wearable technologies, telemedicine, eLearning, simulation and modulation, along with AI can significantly reduce the environmental impact compared to traditional means. These digital technologies can also assist the healthcare ecosystem to provide better health services by continuous and easy monitoring, opening a reliable communication channel between doctors and patients as well as assisting medical professionals to make the right decisions. Over the last decade, significant improvements have been seen in the healthcare sector towards the green initiative, but even today we have barely scratched the surface. There are still significant challenges that need to be addressed to achieve the end goal of green healthcare, a carbon neutral healthcare ecosystem, for a better quality of life.

References

3 Ways Medical AI Can Improve Workflow for Physicians. Selecting & Using a Health Information Exchange | AMA. Accessed February 22, 2019: www.ama-assn.org/practice-management/digital/3-ways-medical-ai-can-improve- workflow-physicians.

7 Ways Data Science is Reshaping Health Care. 2019. Altexsoft.Com. Accessed February 24, 2019: www.altexsoft.com/blog/datascience/7-ways-data-science-is-reshaping-healthcare/.

10 Human Motion Energy Harvesting Technologies. 2012. Accessed February 23, 2019: www.element14.com/community/groups/energy-harvesting-solutions/blog/2012/12/20/different-approach--still-energy-harvesting.

ACR, Radiological. 2019. Ultrasound (Sonography). Radiologyinfo.Org. Accessed February 24, 2019: www.radiologyinfo.org/en/info.cfm?pg=genus.

Alivecor. 2018. Alivecor.Com. Accessed February 23, 2019: www.alivecor.com/getkardiamobile/.

Azmal, M., Kalhor, R., Dehcheshmeh, N. F., Goharinezhad, S., Heidari, Z. A. and Farzianpour, F. 2014. Going toward Green Hospital by Sustainable Healthcare Waste Management: Segregation, Treatment and Safe Disposal. *Health* 06(19): 2632–2640. doi:10.4236/health.2014.619302.

Boto, E., Meyer, S. S., Shah, V., Alem, O., Knappe, S., Kruger, P., Fromhold, M. T., Lim, M., Glover, P. M., Morris, P. G., Bowtell, R., Barnes, G. R. and Brookes M. J. 2017. A New Generation of Magnetoencephalography: Room Temperature Measurements Using Optically-pumped Magnetometers. *NeuroImage* 149: 404–414. doi:10.1016/j.neuroimage.2017.01.034.

Campbell, B. A., Wells, G. A., Palmer, W. N. and Martin D . L. Reuse of Disposable Medical Devices in Canadian Hospitals. *American Journal of Infection Control* 15(5) (1987): 196–200. doi:10.1016/0196-6553(87)90095-2.

Cassandra, T. 2013. *Understanding and Improving Healthcare Using Environmental Life Cycle Assessment and Evidence-based Design*. PhD diss., University of Pittsburgh.

Chang, A.-M., Holsapple, C. W. and Whinston A. B. 1994. A Hyperknowledge Framework of Decision Support Systems. *Information Processing & Management* 30(4): 473–498. doi:10.1016/0306-4573(94)90035-3.

Coussens, C. and Frumkin, H. (eds.) 2007. *Green Healthcare Institutions: Health, Environment, and Economics: Workshop Summary*. National Academies Press: Washington, DC.

Department of Energy Announces the Launch of the Hospital Energy Alliance to Increase Energy Efficiency in the Healthcare Sector. 2009. Energy.Gov. Retrieved from: www.energy.gov/articles/department-energy-announces-launch-hospital-energy-alliance-increase-energy-efficiency.

Dhillon, V. S. 2015. Green Hospital and Climate Change: Their Interrelationship and the Way Forward. *Journal of Clinical and Diagnostic Research.* doi:10.7860/jcdr/2015/13693.6942.

Digitized Knowledge Processing Service. IndiaMART.com. Accessed February 24, 2019: www.indiamart.com/proddetail/digitized-knowledge-processing-service-19125596455.html.

Donnelly, L. 2018. Robots Are Better Than Doctors at Diagnosing Some Cancers, Major Study Finds. *The Telegraph.* Accessed February 24, 2019: www.telegraph.co.uk/news/2018/05/28/robots-better-doctors-diagnosing-cancers-major-study-finds/.

ECG App and Irregular Heart Rhythm Notification Available Today on Apple Watch. 2018. Accessed February 23, 2019: www.apple.com/newsroom/2018/12/ecg-app-and-irregular-heart-rhythm-notification-available-today-on-apple-watch/.

Ehospital – Best Hospital Management System | EHR Software. 2019. Adroitinfosystems.Com. Accessed February 24, 2019: www.adroitinfosystems.com/products/ehospital- systems.

Frequently Asked Questions About the Patient Portal. 2019. Healthit.Gov. Accessed February 24, 2019: www.healthit.gov/sites/default/files/measure-tools/nlc-faqs-about-patient-portal.docx.

Going Green: 5 Ways to Build a Sustainable Hospital. The Sustainability Craze Has Reached the Healthcare Industry, as Designers Look to Build New, State-of-the-art 'Green' Facilities and Existing Hospitals Upgrade Their Energy Policies. Jeff Hull, Director of Architecture, Construction, and Real Estate with Idaho Based St. Luke's Health System. Becker's Hospital Review. 2011. Accessed February 24, 2019: www.beckershospitalreview.com/hospital-management-administration/going-green-5-ways-to-build-a-sustainable-hospital.html.

Gozani, S. 2018. Science Behind Quell™ Wearable Pain Relief Technology for Treatment of Chronic Pain. Accessed June 24, 2019: www.quellrelief.com/wp-content/uploads/2018/10/Science-Behind-Quell.pdf.

Gunter, T. D. and Terry, N. P. 2005. The Emergence of National Electronic Health Record Architectures in the United States and Australia: Models, Costs, And Questions. *Journal of Medical Internet Research* 7(1): e3. doi:10.2196/jmir.7.1.e3.

Harden, R. M. and Hart I. R. 2002. An International Virtual Medical School (IVIMEDS): The Future for Medical Education? Current Neurology and Neuroscience Reports. May 2002. Accessed February 24, 2019: www.ncbi.nlm.nih.gov/pubmed/12098412.

He, B., Baird, R., Butera, R. J., Datta, A., George, S., Hecht, B., Hero, A. O., Lazzi, G., Lee, R. C., Liang, J., Neuman, M. R., Peng, G. C. Y., Perreault, E. J., Ramasubramanian, M., Wang, M. D., Wikswo, J., Yang, G.-Z. and Zhang, Y.-T. 2013. Grand Challenges in Interfacing Engineering with Life Sciences and Medicine. *IEEE Transactions on Biomedical Engineering* 60: 589–598.

Healthcare Facilities Using Green Spaces to Help in Healing. *Wisconsin Department of Natural Resources Forestry News*, 2017. Accessed February 23, 2019: https://forestrynews.blogs.govdelivery.com.

HISTree (Hospital Information System) – Reviews, Pricing, Free Demo and Alternatives. HISTree (Hospital Information System) – Reviews, Pricing, Free Demo and Alternatives Histree (Hospital Information System) – Reviews, Pricing, Free Demo and Alternatives. 2019. Softwaresuggest.Com. Accessed February 24, 2019: www.softwaresuggest.com/us/histree-hospital- information-sy.

How Wearables Will Take Health Monitoring to The Next Level – Readwrite. 2019. Readwrite. Retrieved from: https://readwrite.com/2018/02/26/wearables-will-take-health-monitoring/.

Howsmon, D. and Wayne Bequette, B. 2015. Hypo- and Hyperglycemic Alarms: Devices and Algorithms. *Journal of Diabetes Science and Technology* 9(5): 1126–1137. doi:10.1177/1932296815583507.

How Energy-Efficient Light Bulbs Compare with Traditional Incandescents. Department of Energy. Accessed February 24, 2019: www.energy.gov/energysaver/save-electricity-and-fuel/lighting-choices-save-you-money/how-energy-efficient-light.

In-Vivo Networking: Powering and Communicating with Tiny Battery-free Devices inside the Body. 2018. Accessed February 23, 2019: www.media.mit.edu/projects/ivn-in-vivo-networking/overview/.

Intelligent Decision Support in Healthcare – Analytics Magazine. 2012. *Analytics Magazine*. Accessed February 24, 2019: http://analytics-magazine.org/intelligent-decision-support-in-healthcare/.

Jung, H.-T., Kim, H., Jeong, J., Jeon, B., Ryu, T. and Kim, Y. 2017. Feasibility of Using the RAPAEL Smart Glove in Upper Limb Physical Therapy for Patients After Stroke: A Randomized Controlled Trial. *39th Annual International Conference of the IEEE Engineering in Medicine and Biology Society (EMBC)*. doi:10.1109/embc.2017.8037698.

Khanna, P., Strom, J. A., Malone, J. I. and Bhansali, S. 2008. Microneedle-based Automated Therapy for Diabetes Mellitus. *Journal of Diabetes Science and Technology* 2(6): 1122–1129. doi:10.1177/193229680800200621.

Kunawararak, P., Pongpanich, S., Chantawong, S., Pokaew, P., Traisathit, P., Srithanaviboonchai, K. and Plipat, T. 2011. Tuberculosis treatment with mobile-phone medication reminders in northern Thailand. *The Southeast Asian Journal of Tropical Medicine and Public Health* 42(6): 1444–1451.

Lutheran, G. and Zarecki, C. 2010. *Energy Challenges Faced by Hospital Sector*. US Department of Energy, January 28, 2010.

Manual on Healthcare Waste Management. 2011. Wpro.Who.Int. Retrieved from: www.wpro.who.int/philippines/publications/health_care_waste_management_manual_3rd_ed.pdf.

Medical Imaging Equipment Study. 2017. Greenhealthcare.Ca. Retrieved from: http://greenhealthcare.ca/wp-content/uploads/2016/11/Medical-Imaging-Equipment-Energy-Use-CCGHC-2017.pdf.

Medical Imaging. 2019. FDA.Gov. Accessed February 24, 2019: www.fda.gov/Radiation-EmittingProducts/RadiationEmittingProductsandProcedures/MedicalImaging/default.htm.

Medical Isotope Production without Highly Enriched Uranium. Washington, DC: National Academies Press, 2009. Retrieved from: https://books.google.com/books?id=HCQndTBSlGAC&printsec=frontcover&source=gbs_ge_summary_r&cad=0#v=onepage&q&f=false.

Mercury Thermometers. EPA. 2018. Accessed February 23, 2019: www.epa.gov/mercury/mercury-thermometers.

mHealth New Horizons for Health through Mobile Technologies. World Health Organization. 2014. Accessed June 24, 2019: www.who.int/goe/publications/goe_mhealth_web

MHealth. Three Technologies to Prevent Medical Errors and Add Important Data for Analytics – Innovatemedtec Content Library. Accessed February 22, 2019: https://innovatemedtec.com/digital-health/mhealth.

Neurobiology – Science & Justice Research Center. Cmasseng, 2017. Accessed February 23, 2019: https://scijust.ucsc.edu/tag/neurobiology/.

NHE-Fact-Sheet. CMS.gov Centers for Medicare & Medicaid Services. 2019. Accessed February 23, 2019: www.cms.gov/research-statistics-data-and-systems/statistics-trends-and-reports/nationalhealthexpenddata/nhe-fact-sheet.html.

Quinlan, D. 2014. Health Care & Climate Change. Accessed June 24, 2019: https://noharm-uscanada.org/sites/default/files/documents-files/2704/Health Care Climate Change – Opportunity Transformative Leadership.

Radioisotopes in Medicine | Nuclear Medicine – World Nuclear Association. 2019. World-Nuclear.Org. Accessed February 24, 2019: www.world-nuclear.org/information-library/non-power-nuclear-applications/radioisotopes-research/radioisotopes-in-medicine.aspx.

Revolutionizing CGM with FreeStyle Libre. Abbott. Accessed February 23, 2019: www.abbott.com/corpnewsroom/product-and-innovation/revolutionizing-cgm-with-freestyle-libre.html.

Safe Management of Wastes from Health-Care Activities. 2014. Apps.Who.Int. Retrieved from: http://apps.who.int/iris/bitstream/handle/10665/85349/9789241548564_eng.pdf?sequence=1.

Sicurello, F. n.d. Towards Standards for Management and Transmission of Medical Data in Web Technology. In *Proc. Workshop on Standardization in E-Health*.

Simulation and Integration into Patient Safety Systems by Michael Moyer, Ph.D., MS, Program Director, Trihealth Simulation & Education Center. Simulation and Integration into Patient Safety Systems. Accessed February 24, 2019: https://simulation.healthcaretechoutlook.com/cxoinsights/simulation-and-integration-into-patient-safety-systems-nid-793.html.

Smith, A. C., Patterson, V. and Scott, R. E. 2007. How Telemedicine Helps. *Reducing Your Carbon Footprint*, 335–1060. Accessed February 23, 2019: www.ncbi.nlm.nih.gov/pmc/articles/PMC2094190/.

Solutions. Hospital Management System. Accessed February 24, 2019: www.mediwarehms.com/solutions.html.

The Importance of Water Sustainability. Global Environment Facility. 2017. Accessed February 24, 2019: www.thegef.org/news/importance-water- sustainability.

Top 10 Energy-Efficient Hospitals in the World. Top 10 | Energy Digital. December 2, 2015. Accessed February 24, 2019: www.energydigital.com/renewable-energy/top-10-energy-efficient-hospitals-world.

Top Ten Green Medical Devices. 2011. Accessed February 23, 2019: www.medicaldevice-network.com/features/feature128184/.

Types of Telemedicine. 2019. Chiron Health. Retrieved from: https://chironhealth.com/definitive-guide-to-telemedicine/about-telemedicine/types-of-telemedicine/.

Water Management Options. n.d. Ocw.Un-Ihe.Org. Accessed February 24: https://ocw.un-ihe.org/pluginfile.php/745/mod_folder/content/0/6_Water_Management_Options_Fact_sheet.pdf.

Wearable Devices Market is Expected to Exceed US$ 51.50 Billion By 2022. MarketWatch. 2018. Accessed February 23, 2019: www.marketwatch.com/press-release/wearable-devices-market-is-expected-to-exceed-us-5150-billion-by-2022-2018-08-27.

WHO | Epidemiology. 2019. Who.Int. Accessed February 23, 2019: www.who.int/topics/epidemiology/en/.

Wind Power for Hospitals. Inerjy Vertical Axis Turbines. 2015. Accessed February 24, 2019: www.inerjy.com/hospitals/.

6
Green Smart Education System

Aditya Pratap Singh and Pradeep Tomar

CONTENTS

6.1 Introduction 131
6.2 Overview and Evolution of a Smart Education System 133
    6.2.1 Evolution of Smart Education 133
6.3 Framework for Smart Education 136
    6.3.1 Advance Technologies for Smart Education 137
    6.3.2 Smart Learner 139
    6.3.3 Smart Teacher 140
    6.3.4 Smart Pedagogy 140
    6.3.5 Smart Learning Environment 141
6.4 Green Computing 142
6.5 Green Smart Education 143
Conclusion 144
References 145

6.1 Introduction

As per historical records, ancient India had the 'Gurukul' education system where the learner would have to stay at the teacher's (Guru's) place (Gurukul) and help and learn all activities along with a formal education. The Guru taught everything the scholar wanted to learn, from language to the Veda and from mathematics to medicine and even training as a soldier. The ancient education system was highly influenced by religion. Different religions were carrying out education in their own way. For example, in India, the Hindu religion had a deep impact on human life and the field of education was no exception. Similarly, other regions of the world followed different religions and had their own education system which was influenced by the geographical structure of that region. India is known for its great tradition of learning and its strong education system from back in antiquity. The educational system of ancient China had as its principal aim the selection and training of civil servants. Later the communist Government of China reformed it. In

China, the prime difficulty was size and geographical diversity along with population (Price 2017). In other regions of world, the Church primarily carried out education along with other religion-influenced organizations: the education system of Israel is based on the Jewish tradition; the Buddhist education aimed at all round development of humans. This education system has emphasis on formation of his character of students and inculcation of social responsibility. The Buddhist education system had a teacher at the center who was responsible for health, studies and the spiritual progress of his pupils. Athens was based on a very slight foundation of formal education. The young Athenian learned to read and write; he went to the palestra for physical training and to the music master for instruction in lyre playing.

Later the whole world shifted towards a modern education system to meet the growing need for communication and coordination among humans from different parts of the world. The various technological advancements also influenced the education system throughout the world.

The world has entered into a new era of an exponentially growing advanced technological environment. With socio-technical advancements, every aspect of human life is becoming more comfortable. Several efforts are being made to formalize new means for the betterment of human life, which are interconnected and enriched with intelligent design. Education can be one of those means. The education methodology has evolved over the years and many projects have converted to a conventional education system of smart education globally (Kim et al. 2012). Most of the developed and developing countries are adopting new educational practices using new devices with advanced technologies. The education environment and educational means are converting to digital. Digitized study material with smart educational devices are supporting and replacing paper study material and making it eco-friendly. The high-speed network is providing unlimited possibilities to stretch teaching and learning globally to be available to the masses in digital form. The Massive Open Online Courses (MOOC) are a good example of this. In India, where the growing student population is a key issue for education providers, MOOC is becoming a workable solution to cater for all. In India, many universities are now providing credits based on performance in MOOC courses being run by the Indian Institute of Technology (IIT). Now many more universities are also adopting MOOC courses to enable students to earn some credits. The Government of India initiated a program called SWAYAM. The goal behind this initiative is to strengthen three fundamental properties viz. access, equity and quality of education policy. This is an initiative to bring persons who remained untouched by the digital revolution into the mainstream of knowledge. It is a platform for everyone who is eager to learn new concepts from almost all fields of education. In this, all the courses are free for the residents of India, and are prepared by excellent teachers and range from ninth grade to advanced post-graduation courses. An indigenous digital platform was developed to deliver the huge content for these courses, and provide facility to access from anywhere. The course contents are developed in digital format like video lectures, presentations and text-based content, which can be easily downloaded or printed if required. This platform provides a self-assessment facility after completing the course and the facility of an online discussion forum to discuss with an instructor or among closed group of learners any questions or concerns.

Green smart education follows green computing principles that say it will be responsible for designing and developing a system that is energy-efficient and eco-friendly. The combination of environment-friendly means and technology, combined with the intention to mitigate the bad effects of technological advancements on environment, is called green technology. When environment-friendly means are used for educational technological advancements, it can be called a green smart education system. This can be solar-powered green smart centers to provide education in schools/colleges, even in remote areas, and will enable computer facilities without external support constraints like electricity.

The computer-enabled education centers may use a smart board, which is an interactive replacement of the non-interactive traditional green/black board. These smart boards enable the teacher to create interesting notes along with the option to save lessons and add verbal explanations for better understanding. These notes will be saved onto a connected server and can be accessed later in other places. The cloud storage may also be used to store all data related to the education center like progress reports, educator notes and assessment data. These facilities provide great support to educators who can build unlimited education resources with minimal effort.

The green smart education system will work towards preparing twenty-first-century compatible young minds who can participate in the new economy. This education system will be a user-driven and technology-centric education system in which the contents are developed as per the current and future needs of humans and the environment and this content will be made available throughout every corner of the world with no obstacles.

6.2 Overview and Evolution of a Smart Education System

An education system involves educational institutions as an instrument to deliver education governed by a principle of sequence in the form of stages and levels (de Babini 1991). The education system is made up of various areas at federal, state and community levels. These things include law, policies, regulations, funding, resources, facilities and administration at different levels.

6.2.1 Evolution of Smart Education

The Instructor-led Training (ILT) system has been followed for centuries throughout the world, and in India, Lord Thomas Babington Macaulay introduced it in the 1830s. In this system, the subject experts deliver new skills and information in classrooms at a particular location as per a defined schedule. The learner has to be physically present in that classroom at that scheduled time.

The teaching-learning system has also benefited from the recent technological advancements, especially in the IT field. Tantatsanawong et al. (2011) predicted that employing information and communications technology (ICT) will enhance the existing education system. Currently, the education industry is revolutionized by many ICT advancements especially in the last decade. The Moodle ("Moodle – Open-source learning platform") is one such examples: it is an elearning platform intended to facilitate educators, learners and administrators with a secure, robust and integrated environment for developing personalized learning resources. It is a good example of an elearning management system. Moodle is a free download and an open-source-type software system that has a global reach and can be translated into many languages. Moodle helps to make teaching learning well organized at many educational institutions, mainly in universities.

The foundation of smart education was laid down with the help of smart devices and intelligent systems (Lee, Zo and Lee 2014). The use of technology can help learners learn and the use of smart devices makes it more convenient and accessible. Employing technology provides flexibility in the approach to teaching and learning. Modern technology helps in content as media, tools for accessing learning media, ease of enquiry, communication and collaboration, expression and evaluation (Zhu, Yu and Riezebos 2016).

The growth of wireless communication technology and the reach of advanced mobile devices to every part of society has helped m-learning to emerge. The concept of m-learning supports and adds facilities to the concept of elearning. The m-learning concept adds more flexibility in terms of location, cost of hardware, time and ease of use. After this, educational advancement coined a new term, ubiquitous learning (u-Learning) (Lee et al. 2012). Figure 6.1 shows the paradigm shift of advance learning environments. The u-learning educational environment allows learners to learn at their convenience. u-Learning removes constraints of time, place and environment as the content can be accessed at anytime from anywhere through diverse terminals. The digital revolution is providing the opportunity to use powerful handheld digital devices connected with other resources through a wireless network for educators to explore and exploit u-Learning.

The Korean Ministry of Education, Science and Technology (MEST 2011) elucidated the term *smart* in smart education as:

FIGURE 6.1
Shift of learning paradigm.

- S: self-directed
- M: motivated
- A: adaptive
- R: resource-enriched
- T: technology-embedded.

Here, as per the definition, the term self-directed refers to the teaching learning system and strengthens the concept of self-learning in which students change their role from knowledge gainers to knowledge creators and educators are converted to facilitators. This s can also be referred to as student-centeredness.

The term motivated refers to the involvement of students in personalized "learning by doing" method and "M" may refer to learning management which helps students to achieve their goals. The term adaptive refers to the flexibility of the education system with the facility to tailor the learning as per individual preference and industry need. Sometimes this "A" is also referred to as assessment and its importance in the education system.

The term resource-enriched refers to the large volume of content base through cloud services, the storage medium to enable collective intelligence. With rich content and technology enhancement, a responsive environment can be created for the remediation of content.

The term technology-embedded refers to the use of technology for the education system to provide an environment where the teaching learning process can take place without any constraints for place and time. The learning process can be tracked and measured, analyzed and reported upon.

With reference to the education environment, "smart" refers to the process of delivering education in a more effective and efficient way (Spector 2014). A learning environment becomes smart by efficient exploitation of its adaptive technologies, having support and flexibility for innovative features to improve understanding and performance. The smart education environment includes planning and innovation for the learner as well as for the teacher/instructor to achieve the desired outcomes.

The increasing bandwidth and decreasing cost of the Internet, along with revolutionized mobile technology, has changed the way to find, select, consume, interact and store educational content. A big manifestation of this divergence is disappearance of paper books and journals form the libraries. Learners are now accessing e-books and online resources (Johnson et al. 2016).

A smart education system does not refer only to a paradigm shift for the way students access education and the delivery of education, but it is much more than that. In the current scenario, the education system is getting support from intelligent technologies like cloud computing, big data, wearable technology, Internet of things (IoT) and data analytics. The concept of green smart education is not only limited to advance devices, efficient software and a robust network infrastructure, but it also refers to the ease of humans, environment safety and custom-made content for different users.

6.3 Framework for Smart Education

Recent technological advancements have nurtured global education reforms. The introduction of new possibilities like digital content, online learning, online course management, tailored learning and cost-effective mobile devices and applications changed the way of traditional teaching-learning environment. The reforms are worldwide, as every country wants to benefit from the smart education environment. From a pedagogical point of view, new approaches like flipped classrooms and a creative community changed the organization of education delivery and participation of learners (McLaughlin et al. 2014). O'Flaherty and Phillips (2015) review the utilization of the flipped classroom in higher education: they found evidence that student performance and their perspectives are rising, however these effects need to be proven over time. This shift of traditional perspective for the education process needs to have a new framework for smart education to grow with a smarter world. A principal for smart education was given by Zhu and Bin (2012):

> The essence of smarter education is to create intelligent environments by using smart technologies, so that smart pedagogies can be facilitated as to provide personalized learning services and empower learners to develop talents of wisdom that have better value orientation, higher thinking quality, and stronger conduct ability.

Based on existing theories and models, the possible smart education framework for a smarter world may be as shown in Figure 6.2.

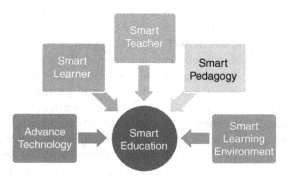

FIGURE 6.2
Smart education framework.

The principle of this framework is to describe the use of the advanced, technology-influenced education environment to achieve enhanced learner-oriented facilities to enable them for innovation and creativity beyond boundaries. The main elements of this framework are advance technology, smart learner, smart teacher, smart pedagogy and smart learning environment.

6.3.1 Advance Technologies for Smart Education

6.3.1.1 *Cloud Computing in Education*

Many cloud-based elearning services (Bora and Ahmed 2013) have been proposed. Many of these related systems emphasize sharing learning material. Cloud computing is one of the suitable technologies to support a smart education framework. Cloud computing has attracted many educational institutions. It provides a reasonable platform for upgrading hardware and software services for smart education. There are several cloud services in our day-to-day lives, such as Gmail, Docs and Calendar from Google, iCloud from Apple, SCloud from Samsung, SkyDrive from Microsoft and Dropbox from Dropbox. Pocatilu, Alecu and Vetrici (2010) discussed cloud computing as a great advantage for elearning establishment. According to the authors, the concept of cloud computing reduces cost and increases data security. It also enhances the opportunity of monitoring data access. The study specifies that elearning benefit from cloud computing service models:

- IaaS (Infrastructure as a Service) – the elearning can be accessed on receiver's/provider's infrastructure;
- PaaS (Platform as a Service) – the elearning system may be implemented as per provider's development interface;
- SaaS (Software as a Service) – the elearning system can incorporate provider-developed software solutions.

Many higher education and research organizations are making use of cloud services to facilitate better infrastructure and efficient software services at low cost. With cloud computing making analysis less expensive and more accessible, it means many more users can set up and customize their own systems and investigators can analyze their data in greater depth (Sultan 2010).

6.3.1.2 *Wireless Technology in Education*

The smart education system is based on the notion of fast connectivity. Smart education depends on a wireless-optical broadband access network, which also demands high speed, reliability and availability. In the last decade, network technologies, wireless communication and mobile devices have drastically advanced. The integration of these technologies also affected the modern education system. The use of wireless communication empowers the educator and learner to learn anytime, anywhere. The educator and learner can both work from diverse physical locations and can participate in collaborative learning and group work on projects without physical availability at a particular location. The smart classes are utilizing wireless handheld devices and smart boards, and can access the learning content from anywhere using wireless communication (Liu et al. 2003). Wireless communication is moving into its fifth generation (5G) and supporting all kinds of digital content with high-speed communication. The key features of 5G includes a packet switched wireless network covering a wide area with high throughput. 5G technology enables speed greater than 100 Mbps at high mobility and higher than 1 Gbps at low mobility. This provides great support to smart education as the large educational content which includes video lectures, presentations, audio and text content can be easily accessed.

6.3.1.3 Learning Analytics

With the use of modern technology and devices, everyone is generating enormous data all the time. There are several new algorithms and technologies evolved to handle this structured and unstructured data and finding useful results from it. This kind of work is termed data analytics (DA). Data analytics refers to the process of getting useful results from data sets, to find some useful conclusions, and to predict information to utilize in enhancing business. If the data is related to the education system, or learner's generated data, then the analysis of such data may be termed learning analytics. The Society for Learning Analytics Research comes up with a definition of learning analytics as the measurement, collection, analysis and reporting of data about learners and their contexts, for purposes of understanding and optimizing learning and the environments in which it occurs (Siemens and Baker 2012). Irrespective of online delivery constraints, the use of various technologies in education has created a new opportunity to get the insight of student learning. In an online mode of learning, the digital traces of student activities are generally marked and stored. The log of these digital traces can be extracted and used for learning analytics. Through these types of analytics, the educators or organization involved in rolling out digital content may understand the pattern of learner learning behavior. Learning analytics include information about learner behavior and psychology and its correlation with the outcome. The outcome from learning analytics may influence course content, access technology, and evaluation schema. It is a tool to empower educators to create and deliver more needed and interesting content. Learning analytics is just the use of analytics in the education field. The technologies used for learning analytics are similar to other analytics, such as Artificial Intelligence (AI), computer science and statistics, and so on.

6.3.1.4 Educational Data Mining

The term educational data mining (EDM) is defined as a new discipline for handling and developing methods for data generated from smart education systems by the International Educational Data Mining Society. EDM helps in understanding learners and evolving new settings suitable for learners (Siemens and Baker 2012). EDM is another application of data mining (DM) and includes machine-learning, statistics, information retrieval, cognitive psychology and recommender systems and techniques to various educational data sets so as to resolve educational issues (Dutt et al. 2017). The goal of this technology is to develop models to store and mine educational data or learner digital traces in a way that can be utilized effectively to improve the learner experience, the effectiveness of content/educator and institutional effectiveness. The process of EDM stores raw data coming from various educational organization systems and converts it into useful information which can have a great impact on the teaching-learning process. With the help of EDM, learner motivation, attitude, behavior, learning style, content access choices and so on can be extracted.

6.3.2 Smart Learner

The learner/student may be defined as a person who studies or investigates to gain knowledge and understanding (Azoury, Lindos and Khoury 2014). In this changing scenario of education, the learner should also be smart and adopt technology to deal with the smart education environment. Since the last decade, learning methodology is shifting, or adjusting, to meet learner demand and results. The learning is transitioning toward inquiry-style techniques and procedures. Learners require a match in their multi-disciplinary system and the availability in current educational settings. The personalized learning needs flexible teaching strategies and learners are freer in their learning based on personal preferences and abilities. To adopt new technological changes the learner must also be technology aware. For efficiently utilizing the smart education environment, the learner becomes a smart learner equipped with the knowledge of utilizing the right technology to achieve learning goals. The smart learners are more digital-native than their predecessors. The learner has to be a competent user capable of exploiting technology efficiently in an integrated environment of learning with strengthened technology. Students should know when, how and for what purpose a given portion of knowledge and competences may be applied, and how the technological means like smart devices and competent applications can be utilized.

6.3.3 Smart Teacher

The role a teacher plays is a very vital role in any kind of education system. In traditional classrooms, the teacher is an essential member of the system. Traditional teaching was based on a passive lecture model. Now the teacher becomes a facilitator.

In the smart world filled with technology advancements, many researchers have discussed the role of the teacher. Berge (1995) proposed the role of a teacher as a moderator (working online), including being a manager for aspects related to the pedagogical, technical and social. Garrison, Cleveland-Innes and Fung (2010) discussed the presence of an educator in the community of inquiry framework. The enquiry framework is a well-accepted framework for computer-aided learning. The role of a teacher can be seen as essential in terms of the design of instructions, organization of a course, facilitating discourse and direct instructions. Garrison and colleagues (2010) point toward the necessity of association of an educator with a technology-integrated learning environment to support higher order thinking.

The use of modern technological means is not limited to online courses, rather many technological supports are being used in classrooms also. Development in wireless communication has removed the constraints of time, space and place. These technology-filled courses can be designed in a way to be conducted beyond computers also. The teacher becomes a smart teacher with the use of technology-facilitated means and social media for conducting classes, creating technology-influenced course structure, new smart pedagogies and teaching models for different online and offline courses. The smart teacher also encourages and supports learners to use technological means. The smart teacher authors assignments in a way that caters to learner requirements. Using author tools enable the teacher to create various media content for the course. The smart teacher becomes a facilitator for providing custom-made learner-centric guidance. The student–teacher interaction is more personalized and based on guidance (participative) in spite of only lecturing (passive) (Lage et al. 2000 and Baker 2000).

6.3.4 Smart Pedagogy

Pedagogy evolved in the Montessori schools of Europe between the 7th and 12th centuries. The term is derived from the Greek words *paid*, meaning "child" and *agogus*, meaning "leader of". Thus pedagogy literally means the art and science of teaching children (Knowles 1990). The teaching-learning strategies tend to be pedagogical nowadays. Currently, education is changing. New approaches like flipped classrooms, creative community, cooperative learning and, above all, advancement in technology have changed the structure of pedagogy. The way of organizing the teaching–learning process and the view of student/learner have changed, and will change further with the adoption of advanced technologies. Zhu et al. (2016) considered transforming pedagogies in their framework on smart education. The changes in conventional pedagogy to smart pedagogy facilitate a personalized learner experience and empower the learner to have better value orientation, better quality of thinking and better conduct. Smart pedagogy uses ICT supported instruction organized in such a way that increases the effectiveness of a student's effort for learning. Smart pedagogies use technological means for the diverse needs of students; they use strategies to support critical and creative thinking. Smart pedagogies create strategies in integration with ICT for managing various learning and other educational activities. Innovative pedagogies may face some resistance from learners as well as from teachers that change the role of teacher from teacher to facilitator.

The change of perspective for education and the invention of new methodologies can lead to the new meaning of smart in smart pedagogy as:

S: sharing

M: mode of delivery

A: assignment

R: revision

T: testing.

The pedagogy should be prepared in a way so that whether learning is conducted in physical classrooms or online, students should perform collaborative learning among students with different performances and achieve desired goals.

6.3.5 Smart Learning Environment

In the traditional education environment, learners' characteristics and differences cannot be met all the time because of the limited resources in view of time and place. With the development of technologies, researchers have developed various learning systems such as a web-based learning system (Chen, Lambert and Guidry 2010) and interactive e-books (Huang et al. 2012) to provide personalized learning paths for students. Other technology like smart mobile devices, advanced wireless communication and sensing technologies support the u-learning environment. The basic need from a smart education environment is that it should provide required educational content to learners at the right time and the right place. The smart education environment allows the learners to access digital resources and exploit the learning system in any place and any time. This environment enables learners to have the needed guidance and supportive tools to achieve their learning goals. In this way, a smart learning environment can be seen as a technology-filled learning system capable of providing correct guidance about learning in real-world scenarios by exploiting available digital resources. Zhu et al. (2016) discussed ten key features to define smart learning:

1. Location-aware: the location of the learner in real-time is treated as significant data required by the education systems in order to get used to the content and situation for the learner.
2. Context-aware: it is important to explore various activity scenarios and information.
3. Socially-aware: the data related to social relationships is also an important feature of smart learning.
4. Interoperable: to provide a platform for setting and collaborating standards for different technology, resources and services.
5. Seamless connection: to provide unbroken service availability when any smart device needs to connect.
6. Adaptable: to be adaptive and deliver needed learning resources according to access, preference and demand.
7. Ubiquitous: to foresee learner demands until evidently expressed, providing visual and transparent access to learning resources and services.
8. Whole record: to log every step the learner takes and to use EDM for analyzing in-depth, then providing reasonable assessment, suggestions and pushing on-demand service.
9. Natural interaction: to transfer the senses of multimodal interaction, like speech, handwriting, gestures and haze, including position and facial expression recognition.
10. High engagement: to engage in multidirectional collaborative learning experiences in technology-enriched environments.

The smart learning environment will be capable of predicting and advising on the learner's need based on the gathered data analysis.

6.4 Green Computing

Since the last decade, the use of advanced computing technology has exploded in several areas, including the education system, to improve all aspects of life. However, these advanced technological means also contribute to several environmental problems, which generally are ignored. The advance-computing infrastructure requires a significant amount of electrical energy and causes greenhouse gas emission. The hardware involved in advanced IT infrastructure also has a point of concern at its production and disposal level. The need of computing practices with minimum environmental impact is already identified. Green computing refers to environmentally sustainable computing or the technological practices which have minimum impact on environment. According to Murugesan (2008), green computing or green IT is the "study and practice of designing, manufacturing, using, and disposing of computers, servers, and associated subsystems such as monitors, printers, storage devices, and networking and communications systems efficiently and effectively with minimal or no impact on the environment". The following approach can be adopted to implement green computing:

- green use of IT Systems
- green design of applications and other IT systems
- green manufacturing of IT hardware
- green disposal of hardware.

The green smart education system is also required to follow green computing approaches. The whole education system may be designed in such a way that it uses less energy, power management at each level, green data centers and other eco-friendly measures. The smart education system is also encouraging green computing in several other ways. The modern education system, especially at higher education levels, needs significant traveling of scholars and teachers for various academic activities. The smart education system reduces this physical relocation for academic activities with the help of technological means like video conferencing, webinars and so on. . In support of green computing, activities like online assessment can reduce use of paper; the concept that learners can learn from anywhere can reduce physical traveling; and the use of mobile devices can reduce the requirement of large computers.

6.5 Green Smart Education

The word green in green computing or in green smart education (GSE) points to energy-efficient technology. Greener IT leads to use of energy-efficient hardware, optimized use of energy and proper disposal measures to minimize bad effects on the environment. Green smart education is a form of education environment that uses new models of learning, supporting smart sharing of resources, the flipped classroom concept, blended learning, digital content, as well as providing flexibility for time and place. The GSE system should have a model to monitor energy performance of the system and corrective actions should take place whenever required as per the monitoring report. The building used in GSE should also be energy-efficient rather than only consume energy: it should help in generating energy like using solar power. GSE encourages minimum travel requirement for education delivery. The concept of GSE is for not only the use of green computing, but it also requires smart building and smart transport to make the overall system greener (more eco-friendly). The concept of GSE has components like smart education with smart learning and sharing, smart campus, smart transport, and all blended with energy-efficient measures. Smart learning refers to the use of digital content through IT in the education system and has a blend of open access contents like MOOCs. Smart education diminishes the requirement of physical presence and gives freedom from time and place constraints, and smart sharing refers to the content using an IT infrastructure like cloud services. The usage of these techniques for sharing data has a positive impact on energy usage. The researchers also use smart sharing by promoting the growing trend of free sharing of research data and results as well: open access (Corrado 2005).

With the use of smart education and smart sharing, the need for physical presence is significantly reduced but physical presence cannot be eliminated. For this reason, the physical infrastructure of educational institutions must be developed in an energy-efficient manner, like maximum use of solar light, use of solar power systems, more green campus and with the necessary implementation of other eco-friendly measures like water harvesting and so on. The advancements in transport facilities must be used to make them smart transport facilities. The use of public transport instead of private vehicles (except bicycles) should be encouraged. The electric vehicle may also reduce greenhouse gases.

All the technology-integrated smart components blended with energy-efficient measures that lead to green smart education. The GSE system supports a sustainability approach towards the education system and should ensure all education institution/university campuses are green and support green technology.

Conclusion

Advanced technology has offered new and exciting opportunities for the education system to move towards a smart education system. Smart education is one of the key parts of a smart world having smart cities. At the same time environment, friendly technological changes may make it a green smart education system. This is a new concept supporting more efficient and resourceful learning environments and the development of existing technology-enhanced learning approaches by incorporating new technologies and new criteria for learning.

This chapter gives an insight to the green smart education system, its participating technologies and its outcome, with a framework for smart education. The state and need of various ingredients for smart education are discussed. The aim of smart education is to bring improvements in the learner's quality of learning experience. Smart education provides a smart environment to promote the learner's ability to learn and become a prospective contributor to fulfill the need of the modern world. The participating technologies will definitely reduce the learning load of the learner and enable learners to focus on sense-making and facilitate ontology construction.

These technological advancements will be continued and it will be tough to predict what the future will be like 20 years from now. The coming years will make more use of virtual reality, augmented reality and artificial intelligence for making the education system more smart. The use of affective computing along with robotics for a smart education system may also be seen in the near future.

References

Azoury, N., Lindos, D. and Khoury, C. E. L. 2014. University Image and Its Relationship to Student Satisfaction: Case of the Middle Eastern Private Business Schools. *International Strategic Management Review* 2(1): 1–8.

Baker, J. W. 2000. The Classroom Flip. Using Web Course Management Tools to Become the Guide by the Side. In *The 11th International Conference on College Teaching and Learning*, 9–17.

Berge, Z. L. 1995. *Facilitating Computer Conferencing: Recommendations from the Field.* Educational Technology: Saddle Brook NJ, 22.

Bora, U. and Ahmed M. 2013. E-learning Using Cloud Computing. *International Journal of Science and Modern Engineering* 1(2): 9–12.

Chen, P.-S., Lambert, A. D. and Guidry K. R. 2010. Engaging Online Learners: The Impact of Web-based Learning Technology on College Student Engagement. *Computers & Education* 54(4): 1222–1232.

Corrado, E. M. 2005. The Importance of Open Access, Open Source, and Open Standards for Libraries. *Issues in Science and Technology Librarianship*, 42.

deBabini, A. M. E. 1991. Convergence and Divergence of Education Systems in Today's World. *Prospects* 21(3): 330–339.

Dutt, A., Ismail, M. A. and Herawan, T. 2017. A Systematic Review on Educational Data Mining. *IEEE Access* 5: 15991–16005.

Garrison, D., R., Cleveland-Innes, M. and Fung, T. 2010. Exploring Causal Relationships among Teaching, Cognitive and Social Presence: Student Perceptions of the Community of Inquiry Framework. *The Internet and Higher Education* 13(1–2); 31–36.

Huang, Y.-M., Liang, T.-H., Su, Y.-N. and Chen N.-S. 2012. Empowering Personalized Learning with an Interactive e-Book Learning System for Elementary School Students. *Educational Technology Research and Development* 60(4): 703–722.

Johnson, L., Becker, S. A., Cummins, M., Estrada, V., Freeman, A. and Hall, C. 2016. *NMC Horizon Report: 2016 Higher Education Edition.* The New Media Consortium.

Kim, T., Cho, J. Y. and Lee B. G. 2012. *Evolution to Smart Learning in Public Education: A Case Study of Korean Public Education.* In IFIP WG 3.4 International Conference on Open and Social Technologies for Networked Learning, Springer: Berlin, Heidelberg, 170–178.

Knowles, M. S. 1990. *The Adult Learner: A Neglected Species (Building Blocks of Human Potential)*. Gulf Publishing: Houston, TX.

Lee, B. G., Kim, S. J., Park, K. C., Kim, S. J. and Jeong E. S. 2012. Empirical Analysis of Learning Effectiveness in u-Learning Environment with Digital Textbook. *KSII Transactions on Internet & Information Systems* 6(3): 869–885.

Lee, J., Zo, H. and Lee, H. 2014. Smart Learning Adoption in Employees and HRD Managers. *British Journal of Educational Technology* 45(6): 1082–1096.

Liu, T. C., Wang, H. Y., Liang, J. K., Chan, T.-W., Ko, H. W. and Yang J. C. 2003. Wireless and Mobile Technologies to Enhance Teaching and Learning. *Journal of Computer Assisted Learning* 19(3): 371–382.

McLaughlin, J. E., Roth, M. T., Glatt, D. M., Gharkholonarehe, N., Davidson, C. A., Griffin, L. M., Esserman, D. A. and Mumper R. J. 2014. The Flipped Classroom: A Course Redesign to Foster Learning and Engagement in a Health Profession's School. *Academic Medicine* 89(2): 236–243.

MEST: Ministry of Education, Science and Technology of the Republic of Korea. 2011. Smart Education Promotion Strategy, President's Council on National ICT Strategies.

Moodle – Open-source learning platform (n.d.). Accessed June 20, 2018: https://moodle.org/.

Murugesan, S. 2008. Harnessing Green IT: Principles and Practices. *IT Professional* 10(1): 24–33.

O'Flaherty, J. and Phillips, C. 2015. The Use of Flipped Classrooms in Higher Education: A Scoping Review. *The Internet and Higher Education* 25: 85–95.

Pocatilu, P., Alecu, F. and Vetrici, M. 2010. Measuring the Efficiency of Cloud Computing for e-Learning Systems. *WSEAS Transactions on Computers* 9(1): 42–51.

Price, R. F. 2017. *Education in Communist China*. Routledge: London.

Siemens, G. and d Baker, R. S. 2012. Learning Analytics and Educational Data Mining: Towards Communication and Collaboration. In *Proceedings of the 2nd International Conference on Learning Analytics and Knowledge*. ACM, 252–254.

Spector, J. M. 2014. Conceptualizing the Emerging Field of Smart Learning Environments. *Smart Learning Environments* 1(1): 2.

Sultan, N. 2010. Cloud Computing for Education: A New Dawn? *International Journal of Information Management* 30(2): 109–116.

SWAYAM, swayam.gov.in/courses/public.

Tantatsanawong, P., Kawtrakul, A. and Lertwipatrakul W. 2011. Enabling Future Education with Smart Services. In *2011 Annual SRII Global Conference*, 550–556.

Zhu, Z. T. and Bin, H. 2012. Smart Education: A New Paradigm in Educational Technology. *Telecommunication Education* 12: 3–15.

Zhu, Z.-T., Yu, M.-H. and Riezebos, P. 2016. A Research Framework of Smart Education. *Smart Learning Environments* 3(1): 4.

7
Green Smart Agriculture System

Garima Singh and Gurjit Kaur

CONTENTS

7.1 Introduction 147
7.2 Features of Green Smart Agriculture System...148
7.3 Designing of Green Smart Agriculture System......................................149
    7.3.1 Smart Monitoring of Climate Conditions 150
    7.3.2 Smart Greenhouse Automation..150
    7.3.3 Smart Crop Management..151
    7.3.4 Smart Cattle Monitoring and Management 151
    7.3.5 End-to-End Green Smart Farm Management Systems...............151
7.4 Forms of Green Smart Agriculture 151
    7.4.1 Precision Farming..152
    7.4.2 Agricultural Drones 152
    7.4.3 Smart Greenhouses 153
    7.4.4 Light-emitting Diode 153
7.5 Integrated Pest Management...154
    7.5.1 Monitor and Identify Pests 155
7.6 Smart Green Water Management Systems 155
7.7 Agriculture Tourism..157
7.8 Precision Farming..158
7.9 Yield Mapping 159
7.10 Role of ICT for Green Smart Agriculture..160
Conclusion 162
References..162

7.1 Introduction

According to a recent report (Drucker 2014), agriculture generally contributes to around 26% of the whole GDP of India. Either directly or indirectly, about 75% of the total population in India rely on agriculture. To feed this population, around 90 million farmers engage in agricultural activities converging over lakhs of villages. Therefore, the field of agriculture needs a revolution to gather wider attention and a much larger market value for its overall development. To serve this purpose, the concept of green smart farming comes which involves computer technology in farming. The agriculture data is therefore transmitted under a virtually ubiquitous computing environment which is sustainable with nature. Smart agriculture in general is an idea that has derived from Farm Management Information Systems (FMIS) (Sørensen et al. 2010) which is used to collect, process, store and disseminate the data in the prescribed format (Kaloxylos et al. 2012).

Green smart agriculture uses IoT applications and Artificial Intelligence (AI) as a solution to farming problems which still need to be consumer friendly. The use of IoT as a solution in agriculture is growing day by day as the market is dynamic in nature. The global market survey states that use of IoT devices in agriculture will increase by 20% annually to 75 million by 2020. By 2025, the market size of global smart agriculture will be tripled, reaching $15.3 billion. The agriculture field needs to enhance its production to meet the requirements using energy-efficient technologies like the use of solar energy for irrigation purposes, making it green smart agriculture (Agriculture and Food 2009).

7.2 Features of Green Smart Agriculture System

By 2050, the global population will increase to 9.6 billion. So, to nourish such a large population, the use of technology must be embraced by the farming industry. Moreover, there will be rising climate change and extreme weather conditions, and impact to the environment due to intensive farming practices. Against these challenges, one needs to ensure that the demand for more food is met. Green smart farming is the outcome of a third green revolution which uses modern information and communication technologies (ICT) in agriculture as precision equipment for genetic revolutions and plant breeding, such as sensors, robotics, the Internet of things (IoT), big data, geo-positioning systems, unmanned aerial vehicles (UAVs, drones) and so on as shown in Figure 7.1.

Green smart farming works more precisely and resource efficiently to deliver a sustainable and productive agriculture. Green smart farming will provide better decision-making and exploitation operations and management efficiency. It is strongly related to three interconnected technology fields.

Management Information Systems: In this, the system is designed for collecting, storing, processing and disseminating the data required to perform a farm's functions and operations (Jiuyan et al. 2013).

Precision Agriculture: This form reduces the environmental impact to improve economic returns. It consists of a decision support system (DSS) for whole farm management enabled by the widespread use of Global Positioning Systems (GPS), Global Navigation Satellite Systems (GNSS), aerial images by drones and the latest generation of hyper spectral images provided by

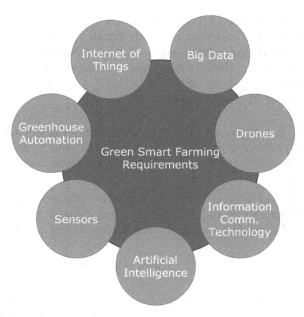

FIGURE 7.1
Requirements of green smart farming.

Sentinel satellites with the aim of optimizing return on input while ensuring resource preservation (Hostiou et al. 2016). This allows the creation of maps of the spatial variability of as many variables that can be measured (e.g. organic matter content, moisture levels, terrain features/topography, crop yield, nitrogen levels, etc.).

Agricultural Automation and Robotics: This includes automation control, robotics and AI techniques for agriculture production including farm bots and farm drones. Green smart farming applications are not only beneficial for conventional farming exploitations, but they can also provide great benefits in environmental issues by boosting other growing trends in agricultural exploitations.

A smart irrigation system is also one important feature of a green smart agriculture system in which sprinklers are used for irrigation that run on renewable energy like wind and solar.

7.3 Designing of Green Smart Agriculture System

Technologies like Internet of things (IoT), information and communications technology (ICT) and AI play a vital role in the implementation of green smart farming at root level (TongKe 2013). The best green smart farming tool is sensors which are used to sense and collect all the information required in farming like rainfall, humidity, temperature, weather information and much more (Singh and Kaur 2019). Technologies in combination with IoT have the potential to change agriculture in many ways. Among them, the five most promising are listed here (Meera et al. 2004):

- Tone of data collected by the sensors about the quality of soil, progress of crop growth and cattle health for smart agriculture can be utilized to update on business state, staff performance and equipment efficiency.
- The control of internal processes is required to reduce production risks and predict the quantity of production. This will help in planning product distribution systems so that none of the products lie around unsold.
- The increase in production control will manage cost and reduce waste and mitigate the risks of losing yield.
- Multiple processes, e.g. fertilizing, irrigation, or pest control, can be performed across the production cycle with the help of automation technologies which increases business efficiency.
- Automation will allow maintaining higher standards for crop quality and yield capacity. This eventually leads to higher revenue.

There are many areas in agriculture where technologies like IoT sensors, big data, AI and ICT are being used and have potential applications like the following:

7.3.1 Smart Monitoring of Climate Conditions

Weather stations, the most popular agriculture device, consist of various smart farming sensors positioned across the field. These sensors are used to collect the environmental data and transmit it to the cloud where it is mapped with climate conditions to choose the right crop to improve capacity: this precision farming (Ishii, Noguchi and Terao 1997) is green smart agriculture.

7.3.2 Smart Greenhouse Automation

Weather stations adjust the conditions to fulfill the given conditions while sourcing environmental data. In particular, the same principle is used by greenhouse automation. There are certain IoT agriculture-based applications like Farmapp and Growlink which offer such capabilities. One other attractive product is GreenIQ which uses smart agriculture sensors to manage the sprinkler used for irrigation and lighting systems remotely. Using renewable energy like solar and wind to power the irrigation pumps, greenhouse lights, drones and robots makes this green and smart in nature.

7.3.3 Smart Crop Management

Crop management is another product type of IoT and an important component of precision farming. Sensors placed in the field are to gather the information specific to the farming of crops like precipitation and temperature and also to monitor overall crop health to improve farming practices. This will prevent crops from disease or infection (Rad et al. 2015).

7.3.4 Smart Cattle Monitoring and Management

Like the above-stated use of IoT sensors in agriculture, they can also be attached to cattle to keep a continuous watch on their health and can help prepare log book. For example, smart agriculture sensors such as collar tags can be utilized to deliver activity, health, temperature and nutrition insights on each individual cow, as well as collective information about the herd (Patil et al. 2015).

7.3.5 End-to-End Green Smart Farm Management Systems

A farm productivity management system is used to represent a complicated approach to IoT products in agriculture. These systems normally consist of agriculture sensors and IoT devices mounted in the premises, plus a strong dashboard having in-built accounting/reporting features and analytical capabilities. This allows us to streamline most business operations and offers us remote farm monitoring abilities. Many countries like China, Japan and the UK are following the green smart agriculture ideas and have developed centers like Agrimetrics, which focus on agri-tech/food by using AI big data; Agri-EPI, which is an innovation center of precision engineering; CHAP, which is a crop health and protection center; and CIEL, which is the innovation center for excellence work in livestock (this is the largest applied animal research group in whole Europe enhancing the green smart agriculture system at root level) (Innovateuk 2018).

7.4 Forms of Green Smart Agriculture

Green smart agriculture allows the use of technology in enhancing farming yield. This can be applied at root level farming in many ways, like precision farming, farming through drones, building automated greenhouses and much more as illustrated in Figure 7.2.

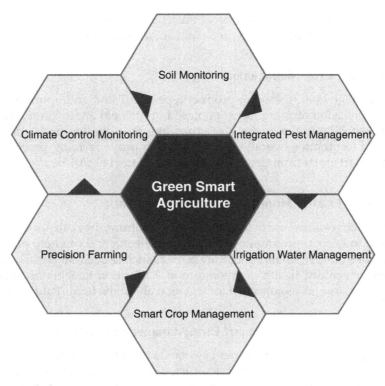

FIGURE 7.2
Features of green smart agriculture.

7.4.1 Precision Farming

Farming which is more controlled and accurate in terms of crop growth and livestock is called precision farming. Components like use of information technology, sensors, robotics, automated hardware and software are required for this approach to farm management (Simon 1994). Low-cost satellites (used for imagery and positioning), high-speed Internet and mobile devices are some of the few key technologies which characterized the agriculture trend: this is the most prominent application of IoT in agriculture.

7.4.2 Agricultural Drones

Agriculture is a good area for the use of technology, such as aerial and ground-based agricultural drones used to enhance various practices like irrigation, crop monitoring, health assessment, planting, spraying crops and field analysis. Drones have the benefit that they are easy to use (Stehr et al. 2015).

Drones are used mainly for integrated GIS mapping and crop health imaging, which is time efficient and has the capability to enhance crops. The technology of drones is based on real-time collection of data and its processing with strategic planning. The use of drones will action high-tech transformation to the agriculture industry.

Drones in agriculture are used by many organizations: one is PrecisionHawk which uses them to collect valuable data through a series of sensors, used for imaging, mapping and surveying of agricultural land. Details like field, altitude and ground resolution are entered by the farmers in order to perform in-flight monitoring. Drones can collect multispectral, thermal, and visual images during their flight and then land at the location from where they took off.

7.4.3 Smart Greenhouses

Greenhouse farming, a tactic that is used to enhance the yield of crops, vegetables, fruits and so on, is where the environmental parameters are controlled by greenhouses via proportional control mechanisms. Smart greenhouses use IoT technology through which climate is intelligently monitored and controlled which eliminates the requirement for manual intervention.

For the process of environmental control in smart greenhouses, different sensors are planted accordingly, to measure the environmental parameters. Then the cloud server can be developed to access the IoT connected system remotely and also can apply the control mechanism. This eliminates the need for constant manual monitoring (Kodali et al. 2016). This design delivers cost-effective and optimal solutions to the farmers with marginal manual intervention.

7.4.4 Light-emitting Diode

Light is important for regulating the growth of plants. The intensity and quality of light along with the photoperiod are very crucial for plant morphogenesis. Crop disease and pests decrease crop growth, and unpredictable climate further worsens the yield. However, the sustainability and feasibility of such systems largely depend upon the power requirements. The electric lamps which drive the light reactions of photosynthesis by providing the actinic light amount to 40% of the recurring cost of plant factories. These lamps are a major restriction for controlled environment agriculture therefore fluorescent lights, high-pressure sodium, metal-halides and incandescent lamps are used as light sources. Among them, the most popular is fluorescent lamps. These lighting systems possess wide wavelengths of 350 to 750 nm, but have a limited lifetime. To achieve the objective of high crop productivity these have limited utilization because of low quality plant development and growth (Bula et al. 1991). Therefore LEDs are the most suitable choice for controlled environment agriculture, making it a green smart agriculture system for smart cities.

7.5 Integrated Pest Management

In general terms the concept of integrated pest management represents an effective environmentally delicate approach that is used for the proactive management of pests. The programs which use integrated pest management (IPM) provide comprehensive and recent information about the life cycles of pests and their behavior with the environment (Kogan 1998). These types of information are quite useful in managing the damage done by pests in the most economical way with negligible hazard to environment, property and people. Thus, IPM makes judicious use of pesticides. While working on IPM, a four-tiered approach is followed by growers to avoid pest infections as shown in Figure 7.3. These four steps are setting an action level to enforce pest control action. Finding a single pest does not always mean that control is needed. The threshold at which pests will turn out to be an economic threat is

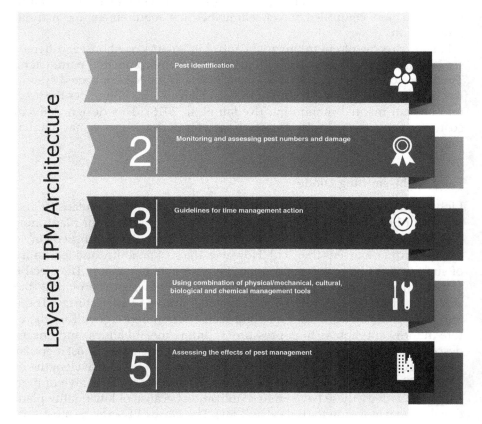

FIGURE 7.3
Layer architecture of IPM.

crucial to guide future pest control decisions (Bottrell 1979). The main motive of IPM program is identifying and monitoring volumes of pest as explained in next section.

7.5.1 Monitor and Identify Pests

Pests are identified and monitored accurately in IPM programs through smart green technologies like sensors for the sake of appropriate control decision-making. This identification and monitoring is useful in deciding when and where pesticides are required and also the type of pesticide. After this, biological and environmental factors of pests are studied to decide the need to control pests or to decide whether they can be tolerated. If control is required, this information helps to select the most effective management method and the most suitable time to use that. Most of the IPM programs are to establish a pest outbreak, and some trends and patterns require a record-keeping system to store the information of every inspection and pest treatment. Pests can never be eradicated but they can be tolerated if at low levels. For this, a certain population size is needed, called an action reply. This is required to define the remedial action for human health, economic or aesthetics. IPM believes in a prevention policy which says that the conditions which attract pests should be removed. This modified use of pests in agriculture has turned it green and smart in nature, which helps in developing a green smart agriculture system for smart cities (Hassanali et al. 2007).

7.6 Smart Green Water Management Systems

The use of water in irrigation is important like gasoline to a vehicle. It is the driving force of irrigation thus its smart management is required. Globally, fresh water used by the agriculture sector amounts to approximately 70% while many countries have below 50% efficiency in water use (Robles et al. 2015). Techniques like isotopic and nuclear gives data about the use of water, including all the losses through soil evaporation which helps the user to optimize the scheduling of irrigation with improved water use efficiency. Smart water management means utilizing water effectively with certain rules and regulations. A decrease in rainfall and its abuse has made it a more valuable commodity. Smart water management requires planning and smart dispensing of water for its optimal use to cater to all competing needs. Although it's quite hard to practice this at a practical stage because 0.08% of world's fresh water is being abused by humans in drinking, sanitation, agriculture and leisure. To solve this global water scarcity problem there is a need to make use of technology like IoT, AI and ICT for green smart water management in agriculture. To ensure food security and sustainable water management, more crops per drop of water is needed to be produced in the agricultural sector. This will ensure that water is used more efficiently without afflicting downstream water quantity and quality (Stewart et al. 2010).

Green smart water management (GSWM) in agriculture is accepting the challenge through integration of ICT products for irrigation and other uses of agriculture. Deployment of IoT connected sensors is required to consistently sense water sources and analyze the quantity of water that is required for agriculture. By using smart sprinklers for irrigation of farms which turns on-off automatically as per the water requirement and availability, use of water can be managed. Therefore, a sustainable green water management plan is achieved by proper operational performance. Green smart water systems offer great potential to improve the water system in agriculture through, first, improving the bill payments system and making it more convenient for customers as well as the revenue collection system. Second, by reducing the administrative cost by utility, and, third, by detecting the non-revenue water or theft of water. At the same time, leaks could be spotted and repaired, thus decreasing pumping and purification costs and maintaining water conservation.

Green smart water management systems can also be achieved by using smart meters which:

1. detects leakage and illegal connections remotely;
2. improve billing accuracy;
3. prevent corrupt practices related to illegal connections and meter hampering.

A soil-water-plant-nutrient management approach is needed to be used in the handling of water resources. This approach provides an optimized irrigation schedule with efficient irrigation like drip irrigation. Improving soil fertility ensures that crop growth isn't stunted by nutrients or due to physical constraints so that every drop of water is utilized for growth. Taking into account prevailing the environmental conditions and growth stages of different crops, water uptake is different. This difference in requirement can be managed by efficient water methods such as demand-based irrigation scheduling. Water use efficiency in agriculture can be enhanced through minimizing the losses from soil evaporation relative to plant transpiration in the field. Soil evaporation and plant transpiration gives quantified information about total water required for irrigation, specific to crop type and stage, used to conserve and manage water use (Ntuli et al. 2016). Water management in agriculture can be improved in green smart ways by:

1. increasing crop yield;
2. avoiding unnecessary water loss caused by evaporation of water from the soil; and

3. increasing storage of soil water within the roots of plants with the help of good water and soil management techniques.

The above-mentioned ideas are very useful in managing the use of water in agriculture, making it a green smart agriculture system.

7.7 Agriculture Tourism

As the name says, in agriculture tourism, agriculture and tourism are combined to offer tourists a knowledgeable experience on various agriculture practices which usually starts from a tour of a farm or ranch and a festival celebration in farms (McGehee 2007). To promote this, farm lands are often converted into a destination by farmers to attract members of the public to experience what real rural life is all about, along with a taste of the local food. Agriculture is a culture which is more than a profession or business. Therefore, these types of activities making additional income for farmers will definitely increase the agriculture contribution in the national GDP. Agriculture tourism is one serious effort to improve employment and poverty in agriculture as it stimulates national integration, and encourages local handicrafts and cultural activities. One solution to implement this is the involvement of the private sector in agriculture that will give good publicity and provide a platform to agriculture tourism. Moreover, this is booming rapidly to become an increasingly popular industry globally and even in most of the states in India. This tourism offers a plethora of unique experiences from picking your own garden-fresh fruit, to a hay ride through a pumpkin farm. These types of a leisure activity often result in the understanding of rural area settings which ultimately helps a person to appreciate the land and the people who live on it. Travelers now not only want to slow down but want to meet locals in their natural environments and become more involved with the land they are visiting whenever they discover a new destination. With this in mind, agriculture tourism was born. At its most basic level, this is a style of travel that entails a meaningful visit to a farm or ranch and usually offers the opportunity of a producer of land-based products and services to help with on-site farming or ranching tasks during the visit. This is not at all about staying in a village and savoring the food. It is also a chance to be close to where 75% of Indians live. The whole idea behind this big concept is to provide tourists with the chance to experience the life of a villager, right from ploughing the field to milking a cow, or from bathing in a well to climbing a tree and plucking fruit.

With commercialism and mass production, agriculture tourism has given people who work in the agricultural and horticultural sectors an opportunity to share their work with the masses (McGehee 2007). Some agriculture tourism experiences allow visitors to buy organic food products or handcrafted products. Purchasing these goods provide farmers with another source of income. Education has given way to technology courses in middle and high schools and many children are growing up without ever actually knowing what the countryside is or what it is like to interact with live farm animals. Agriculture tourism, therefore, gives parents the chance to introduce their children to something different from city life. For farmers the potentials are the following:

- expansion of farm operations;
- innovative ways of utilizing farm products;
- improving farm income streams;
- development of a new consumer market;
- increasing awareness of local agricultural products;
- increasing appreciation of the importance of sustaining agricultural land;
- channelizing additional on-farm income directly to family members;
- improving farm living conditions, working areas and amp; farm recreation opportunities;
- development of entrepreneurial spirit and managerial skill; and
- increasing long-term sustainability of farm businesses.

7.8 Precision Farming

Precision farming is an emerging paradigm used to amend the prevailing techniques for adding new ones in order to create a new set of tools. A significant amount of computation and electronics is integrated with a higher level of control for a systems approach. Precision farming, therefore, not only requires the application of treatments at the local level, but also needs the monitoring and assessment of an agricultural enterprise at a local and farm level with sufficient understanding of the procedures involved to apply the inputs to achieve the particular goal. The detailed pictorial representation of precision farming is given in Figure 7.4. It isn't necessarily maximum yield; it may be maximum financial gain for given environmental constraints (Auernhammer 2001). Furthermore, precision farming techniques should be viewed as inseparable from the concept of sustainable land management. In sustainable development, the requirement is for non-negative changes in both the stock of natural resources and waste capacity of the receiving environment. This classification is helpful in precision farming as it implies the concept of stewardship of the land for future generations, conservation of the

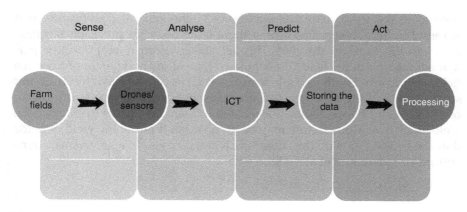

FIGURE 7.4
Precision farming process.

potential for multiple uses and the significance of off-site as well as on-site effects (Grisso et al. 2005). It is well-known fact that agricultural activities like planting crops, maintaining them and their harvestation at individual level takes energy and labor. The use of AI in agriculture is the best solution to reduce these efforts. Robots are available in markets which help in harvesting the crops and packaging them and also monitors the soil quality using deep and machine-learning techniques. Another important digital technology used in precision farming is ICT which is helping organizations to make green smart agriculture available to common men, like "Bosch" is doing "Precision Livestock Farming" with the motive where "connected agriculture means more efficient agriculture" thus making it green smart agriculture for smart cities.

7.9 Yield Mapping

Yield mapping and soil sampling tend to be the first stage in the implementation of precision farming where the produced yield is mapped by a vehicle positioning system and then integrated with a yield recording system. Massey Ferguson was the first company to produce a commercial yield mapping system. This system had a differential global positioning system (DGPS) installed to it that can be recognized by the GPS receiver on the roof of the cab and the differential aerial above the engine. This gives the data file as output which records the position of combined longitude and latitude for every 1.2 seconds and matches with the yield at that point. A data set is prepared which can be processed by a variety of geostatistical techniques to portray the map of field to represent the yield in tons per hectare. After producing the yield maps, its variability is checked for significance. If it is found to be non-significant then there is no need for precision farming, although this is never the case. Through this geostatical mapping the permanent features of the field, like topology, type of soil and height of trees in a field, are recorded with variable features, such as type of seed, fertilizer and spray. After getting this data various statistical methods are used to make out correlations between the above-stated influencing factors to enhance the yield smartly. Various factors may be involved in limiting the yield of fields that can be identified by this, therefore it can give rise to a green smart agriculture system (Kiura et al. 2000).

7.10 Role of ICT for Green Smart Agriculture

In recent times technological developments like the use of electronic systems and data transmission in agriculture has introduced major changes in the agricultural working environment. These changes demand updated information from production systems and from markets and agents involved in production to provide decision-making information for production as well as for the strategic and managerial issues involved. The use of ICT in agriculture is a new field which needs to be systematically introduced. Green smart agriculture based on the integration of information and communication technologies into machinery, equipment and sensors in an agricultural production system allows a large volume of data and information to be generated with progressive supplementation of automation into the process. Through this, farmers will get relevant scientific and technology-based information in a timely way which help them in raising their income by using their resources effectively in a competitive manner. Private consulting services will also be needed for commercialized farms to get their technological advice in terms of marketing strategies and farm management (Sugahara 2007). ICT will also play a key role in other agricultural activities like online training and educating farmers with new advancements, transaction and processing of their subsidies, and also for monitoring farms and consulting the farmers. Green smart agriculture mainly relies on transmission of data and its storage in online databases for the purpose of analysis to make decisions regarding the farming. Integrating ICT in agriculture will keep farmers up-to-date in information like management skills, packages of production, market information, techniques/practices, weather forecasting, agricultural statistics, policy and programs for greater efficiency. Online databases can provide farmers with current market prices for a wide range of agricultural commodities. It helps them to decide what crops to plant. This helps them to decide where and when to market their produce. With the help of ICT, farmers will get early warning signs about crop problems, natural disasters, livestock disease and pest problems so that they can take precautionary measurements (Lomas, Milford and Mukhala 2000). There is a need to develop such ICT tools that enable information sharing relevant to local national and international commerce. For example, the local and regional collation and dissemination of agricultural information will help the development of both internal and external markets. This technology will also set a direct link between producer, retailer and supplier through e-commerce and solve the problem of food crises, and can contribute to the attainment of sustainable agricultural development progressing to be a green smart agriculture system (Chumjai 2006 and Majid 2006). Some farmers have created their own website on an individual basis for marketing purposes. All these benefits progress towards a green smart agriculture system for smart cities.

ICT has a bright future in agriculture with some issues:

- availability and knowledge of the Internet for farmers at the same speed as it is in cities.
- only very few farmers have computer skills because of their low literacy rate which needs to be improved to make them computer-friendly.
- the basic info regarding food safety and security and labeling should be updated on the website in a timely way so that the farmers can utilize the knowledge for their own benefit.

In order to efficiently utilize the benefits of ICT in agriculture, the following green smart agriculture system practices are needed to be enforced for smart cities of the future:

- establishing an efficient network that provides information on different farm products;
- creating an e-commerce portal to provide an independent platform to farmers to sell their produce directly to the customers;
- The need to establish workshops for training of the marketing staff and farmers to handle e-commerce-based websites, and being trained about new farming technologies for enhanced yield;
- The need to establish various community network centers that can provide technical information on existing farm production technologies, along with the skillset from agricultural various research-based institutes and marketing groups; and
- associating with groups of farmers engaged in dealing with electronic instruments and tools in agricultural-related tasks.

Conclusion

Digital technologies have people and things the opportunity to connect at any time. They have provided a good platform for the agriculture industry also. The use of green technologies such as sensors, IoT, big data, geopositioning systems, unmanned aerial vehicles (UAVs, drones), AI, and satellite which helps in providing information about remote areas in agriculture are explained in detail in this chapter. With the help of these technologies one usually gets up-to-date data collected from the sensors in real-time, which ultimately helps in determining season crop growth and development, moisture of the soil and other dynamic variables that make it a green smart agriculture system for smart cities. The main objective of this chapter is to enforce the usage of green smart technologies in agriculture at root level.

References

Auernhammer, H. 2001. Precision farming – The Environmental Challenge. *Computers and Electronics in Agriculture* 30(1–3): 31–43.

Blackmore, S. 1994. Precision Farming: An Introduction. *Outlook on Agriculture* 23(4): 275–280.

Bottrell, D. R. 1979. *Integrated Pest Management*. Supt. of Docs., U.S. G.P.O: Washington, DC.

Bula R. J., Morrow R. C., Tibbitts T. W., Ignatius R. W., Martin T. S. and Barta D. J. 1991. Light-emitting Diodes as a Radiation Source of Plants. *HorticSci* 26: 203.

Chumjai, P. 2006. Farmers' Organization Approach: An Alternative to Effective Extension in Thailand, International Workshop on Effective Methods of Disseminating New Technology Considering the Viewpoint of Farmers, Taiwan.

Drucker, V. 2014. Agriculture springs into the digital age. Fund strategy. Retrieved from: www.fundstrategy.co.uk/ issues/fund-strategy-sept-2014/agriculture-springs-into-the digital-age.

Food and Agriculture Organization of the United Nations. 2009. *Global Agriculture towards 2050*. Rome: Food and Agriculture Organization of the United Nations.

Gondchawar, N. and Kawitkar R. S. 2016. IoT Based Smart Agriculture. *International Journal of Advanced Research in Computer and Communication Engineering* 5(6).

Grisso, R. D., Alley, M. M., Holshouser, D. L. and Thomason, W. E. 2005. Precision Farming Tools. *Soil Electrical Conductivity*. Retrieved from: https://vtechworks.lib.vt.edu/handle/10919/51377

Hassanali, A., Herren, H., Khan, Z. R., Pickett, J. A. and Woodcock, C. M. 2007. Integrated Pest Management: The Push–Pull Approach for Controlling Insect Pests and Weeds of Cereals, and its Potential for Other Agricultural Systems Including Animal Husbandry. *Philosophical Transactions of the Royal Society B: Biological Sciences* 363(1491): 611–621.

Hostiou, N., Fagon, J., Chauvat, S., Turlot, A., Kling-Eveillard, F., Boivin, X. and Allain C. 2016. Impact of Precision Livestock Farming on Work and Human–Animal Interactions on Dairy Farms. A Review. *Biotechnologie, Agronomie, Sociétéet Environnement/Biotechnology, Agronomy, Society and Environment* 21(4): 268–275.

https://innovateuk.blog.gov.uk/2018/02/15/smart-farm-a-new-approach-to-farming/.

Kaloxylos A., Eigemann R., Teye F., Politopoulou Z., Wolfert S., Shrank C. S. 2012. Farm Management Systems and the Future Internet Era. *Computers Electronics in Agriculture* 89: 130–144.

Kodali, R. K., Jain, V. and Karagwal S. 2016. IoT-based Smart Greenhouse. In *IEEE Region 10 Humanitarian Technology Conference (R10-HTC)*, 1–6. IEEE: Pennsylvania.

Kogan, M. 1998. Integrated Pest Management: Historical Perspectives and Contemporary Developments. *Annual Review of Entomology* 43(1): 243–270.

Laurenson, M., Kiura, T. and Ninomiya, S. 2000. Accessing On-line Weather Databases from Java, Internet Workshop, *Journal of the Japanese Society of Agricultural Machinery* 63(4): 4–11.

Lomas, J., Milford, J. R. and Mukhala, E. 2000. Education and Training in Agricultural Meteorology: Current Status and Future Needs, Agric. *Forest Meteor* 103: 197–208.

Majid, M. N. A. 2006. Effective Methods of Disseminating New Technology Considering the Viewpoint of Farmers in Malaysia, International Workshop on Effective Methods of Disseminating New Technology Considering the Viewpoint of Farmers, Taiwan.

McGehee, N. G. 2007. An Agritourism Systems Model: A Weberian Perspective. *Journal of Sustainable Tourism* 15(2): 111–124.

McGehee, N. G., Kim, K. and Jennings G. R. 2007. Gender and Motivation for Agritourism Entrepreneurship. *Tourism Management* 28(1): 280–289.

Meera, S. N., Jhamtani, A. and Rao D. U. M. 2004. Information and Communication Technology in Agricultural Development: A Comparative Analysis of Three Projects from India. *Network Paper No.* 135 (report).

Nanseki, T. 1994. Agricultural Marketing Information Systems in Asian and Pacific Countries The agricultural marketing information system in Japan: An overview of it as a public service, Navigating the Marketing Information Systems of Asia workshop Seoul, Korea.

Noguchi, N., Ishii, K. and Terao H. 1997. Development of an Agricultural Mobile Robot Using a Geomagnetic Direction Sensor and Image Sensors. *Journal of Agricultural Engineering Research* 67(1): 1–15.

Ntuli, N. and Abu-Mahfouz A. 2016. A Simple Security Architecture for Smart Water Management System. *Procedia Computer Science* 83: 1164–1169.

Patil, A., Pawar, C., Patil, N. and Tambe R. 2015 Smart Health Monitoring System for Animals. In *International Conference on Green Computing and Internet of Things (ICGCIoT)*, pp. 1560–1564. IEEE.

Rad, C.-R., Hancu, O., Takacs, I.-A. and Olteanu G. 2015. Smart Monitoring of Potato Crop: A Cyber-physical System Architecture Model in the Field of Precision Agriculture. *Agriculture and Agricultural Science Procedia* 6: 73–79.

Robles, T., Alcarria, R., Andrés, D. M. de, Cruz, M. N. de la, Calero, R., Iglesias, S. and López M. 2015. An IoT-based Reference Architecture for Smart Water Management Processes. *JoWUA* 6(1): 4–23.

Sørensen C. G., Fountas S., Nash E., Pesonen L., Bochtis D., Pedersen S. M. 2010. Conceptual Model of a Future Farm Management Information System. *Comput Electron Agric* 72(1): 37–47.

Singh, G. and Kaur G. 2019. Role of Communication Technologies for Smart Applications in IoT. In their *Handbook of Research on Big Data and the IoT*, pp. 300–313. IGI Global: Pennsylvania.

Stehr, N. J. 2015. Drones: The Newest Technology for Precision Agriculture. *Natural Sciences Education* 44(1): 89–91.

Stewart, R. A., Willis, R., Giurco, D., Panuwatwanich, K. and Guillermo C. 2010. Web-based Knowledge Management System: Linking Smart Metering to the Future of Urban Water Planning. *Australian Planner* 47(2): 66–74.

Sugahara, K. 2007. Traceability System for Agricultural Products Based on RFID and Mobile Technology, *National Agriculture Research Center*, Tsukuba: Japan (conference paper).

TongKe, F. 2013. Smart Agriculture Based on Cloud Computing and IoT. *Journal of Convergence Information Technology* 8(2): 210–216.

Tsouvalis, J., Seymour, S. and Watkins C. 2000. Exploring Knowledge-cultures: Precision Farming, Yield Mapping, and the Expert-farmer Interface. *Environment and Planning A* 32(5): 909–924.

Ye, J., Chen, B., Liu, Q. and Fang Y. 2013. A Precision Agriculture Management System Based on Internet of Things and WebGIS. In *21st International Conference on Geoinformatics*, 1–5. IEEE: Kaifeng, China.

8

Green Smart Security System

Yaman Parasher, Prabhjot Singh and Gurjit Kaur

CONTENTS

8.1 Introduction to the Green Smart Security System..................................165
8.2 Features of the Green Smart Security System...166
8.3 Technologies to Design the Green Smart Security System 168
    8.3.1 Green RFID 168
    8.3.2 Green Wireless Sensor Networks 170
    8.3.3 Green Cloud Computing-based Security Systems.......................171
    8.3.4 Green Machine-to-machine-based Security Systems...................172
8.4 Security of Mobile Devices, Applications and Transactions.................172
8.5 Smart Embedded Security for Smart Cities 175
8.6 Digital Identity and Access Management for Cloud-based Services...176
    8.6.1 Advantages of Using IAM Systems 177
8.7 Challenges to Designing the Green Smart Security System 179
    8.7.1 Factors that Hinder its Growth 179
Conclusion 183
References..183

8.1 Introduction to the Green Smart Security System

A smart city can be defined as an urban landscape that comprises highly advanced integrated technologies combined together to improve the infrastructure, communication, transportation and many other essential utilities needed for the upliftment of living standards. However futuristic this concept may appear, it is unimaginably closer to be practically implemented in most parts of the world. The interconnection of these small, self-reliant, low power embedded units is something that makes it very favorable for its widespread adoption in various smart cities around the world. This world of smart utilities can be envisioned through a collection of smart things like mobile phones, watches, sensors, cars and so on that are capable of serving a ton of people intelligently and automatically in a cooperative way.

Preparing the growth of the world in a smarter direction, the Internet of things (IoT) is standing firmly to represent itself as the backbone in various sorts of applications like medical, security, monitoring, supply chain and so on (Zhu et al. 2015). Amongst all of the critical aspects of these systems, the most important feature that can lead to its scalability to further adoption lies in its security. To build the same, one needs to create a web of solutions that must be able to integrate monitoring and surveillance of the premises, alarm and fire detection, detection of intrusions, entrance and access control solutions with screening and scanning. However, despite its numerous benefits, it has been observed that these security mechanisms often result in the increment of the computation and energy demand of these conventional security systems to many significant levels. It is therefore not surprising to say that the characterization of these security systems in terms of energy consumption still remain in the infancy stage. However, with the rapid technological advancement of the twenty-first century, the proliferation of new energy-efficient technologies is coming into the picture in various applications related to security in the smart cities. A number of researchers around the world are working on designing smart security solutions that can take into consideration the critical aspect of sustainable development with minimal energy consumption. This chapter, therefore, discusses various issues associated with the existing security infrastructure and provides exposure to the new green technologies that can be used to build a new set of smart security systems that can abide by sustainable development of the present world.

In the beginning, a brief overview of the smart security system is presented. Then, the critical features of the green smart security systems for smart cities along with the technologies needed to build it are summarized (Casini 2017). Also, the latest developments with the future vision of integrating the smart embedding solution in smart cities along with the focus on the development of security measure for mobile devices, applications and transactions have been put into perspective. Finally, open problems and future research directions related to the introduction of cloud-based services in the security system are presented through a well-known framework known as identity and access management (IAM) altogether with a brief discussion on the challenges that may intervene while developing the whole smart green security system for a smarter world. The chapter thus attempts to enlighten the most recent direction for innovative research and development of the green smart security system for a smart and sustainable city.

8.2 Features of the Green Smart Security System

Energy consumption associated with every security system that exists today depends mainly on the electrical power used for its operation. Generally, whenever a security system is visioned in the past, the power utilization factor does not seem to stand as an essential criterion. With each add-on to the integrated security environment, the pressure on demand for a large energy supply has begun to surge to enormous levels.

To cope with the same, one most adopt energy-efficient policies and technologies to take the whole setup towards a more sustainable eco-friendly and cost-savvy security system. One of the initial phases in building a green security system is by finding a set of ideal optimal solutions that can help in the merging of the green technologies into the pre-existing ones. Every smart security system that presents around us, like the intrusion detection systems with panels and sensors and the swing gate automation depending upon access, always has options for the integration of green technologies that holds the potential to reduce electrical consumption to much lower levels.

There are many instances where appropriate energy-efficient green technological advancement can bring many effective solutions to conventional integrated security systems. For example, choosing an Internet protocol-based surveillance system over the simpler one holds the potential to produce more benefits as its single piece can cover what three of such old pre-existing surveillance cameras cannot cover altogether with much lower energy consumptions. Similarly, in smart video analytics, the amount of video storage dedicated to each user in the cloud storage is reduced to much lower levels by taking into consideration only the important events like the footage during intrusion detection, line crossing, object removal and so on. Thus, by doing so, a large amount of energy consumption and storage space can be saved. Another way to achieve the same benefits for the same security measure designed for a smart home can be through automatically turning the hard disk into sleep mode when the security system is not recording anything.

It has been observed that for a 16-channel surveillance security system, the total amount of storage space requirement can scale up to 10 TB per month which ultimately means a 625 GB share of space by each camera into operation. However, with the recent upgradation in the video compression standard, H.265 seems to become the prime choice of the surveillance systems manufactures. On comparing it with its old version H.264, H.265 holds the ability to send ultra-high definition streams at the much lower bit rate. It therefore, represents approximately a 50% bandwidth reduction which ultimately enables the system to make more room for the power requirements and recordings.

Solar energy is another player that represents another energy-efficient approach for the video surveillance system. Using such methods in the integrated security environment, a lot of money can be saved on the electricity bills. It has been successfully deployed in remote areas of most countries where surveillance monitoring was a mandate due to any local electric chain not being available and was proved to be a vital solution to the existing smart security systems.

Apart from this, another significant feature of these systems is the false alarm shutdown function that is capable of reducing the number of false alarms from an intrusion detector during a period of false triggering. This is also a smart add-on feature to the green-based smart security system that can help in reducing power consumption to significant levels. Now, taking into consideration the hardware aspect, there are a number of green certifications needed to build devices for such smart security systems: one is RoHS under which a some security equipment and devices are produced. In order to secure the pre-existing systems, all the security devices involved in a smart security system need to be developed under the set of instructions that comes under such certification. Due to their hazardous impact on the environment, the RoHS certification discussed above provides a directive to prohibit a number of substances like cadmium (Cd), mercury (Hg), lead (Pb), hexavalent chromium (CrVI), polybrominated diphenyl ethers (PBDE), along with four different phthalates in the process of manufacturing. Further, another factor that prohibits their usage in the industry is their inability to recycle products made from the listed compounds more safely. Thus, by using RoHS-certified security products, one can efficiently manufacture a comprehensive green smart security system for the desired units wherever it needs to be deployed in a smart city.

8.3 Technologies to Design the Green Smart Security System

In spite of the existence of a number of technologies in today's world, summing up the most vital technologies which have an enormous impact on the day-to-day lives of people which constitute the core foundation of almost every green smart security system for smart cities is explored. This section seeks a green approach associated with every single one of these technologies, and each is discussed in an extremely brief manner to give a standpoint of the probability of their integration with the present or existing frameworks. The technologies that play an essential role in building of green smart security system can be seen in Figure 8.1.

8.3.1 Green RFID

RFID is an important innovation that has been deployed as an essential part of smart integrated security systems for a long time. These cards, in general, incorporate a few ID labels and a little subset of label readers that serve a ton of applications related to security infrastructure. The RFID tag in simple terms is a small microchip connected to a radio wrapped in a glue sticker, with a unique identity. The primary function of these tags is to store data concerning the entities attached to them. The fundamental process involves triggering data stored on these tag readers through transmitting an inquiry signal, followed with the feedback of nearby similar tags. These

Green Smart Security System

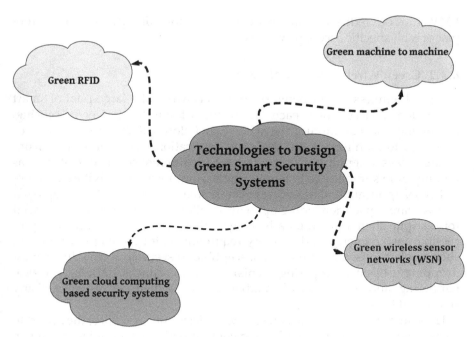

FIGURE 8.1
Technologies to design green smart security systems.

systems operate on low range transmissions, generally a few meters, and are considered beneficial in various sorts of identity-related security protocols.

These RFID tags are generally available in two types. One is the tags with batteries empowering an increased range of communication, termed active RFID tags, whereas the other tags utilize energy from the receiving signal through the process of induction due to the absence of onboard batteries and are called passive RFID tags.

The following points highlight key ideas to utilize RFID in a green manner in a number of security-related applications:

1. the amount of non-biodegradable substances used in the production of RFID tags can be depreciated significantly with minimizing the size of RFID (e.g., printable RFID tags, biodegradable RFID tags, paper-based RFID tags), as tags themselves are quite inefficient for recycling.
2. the optimization of tag estimation, dynamic adjustment of transmission power parameter and avoiding tag collision along with overhearing, and so on, can be done with the help of energy-efficient algorithms and protocols.

All these solutions will be quite useful in various sort of applications related to electronic article surveillance, secure payment through these RFID-based

EMV chips on credit cards and in identity cards for gaining access in different places with specific access privileges.

8.3.2 Green Wireless Sensor Networks

A typical wireless sensor network (WSNs) consists of a large pool of smart embedded devices where each one of them holds the power of sensing, processing and communicating with their fellow collaborative devices or gateways to monitor various real-time applications. This kind of network usually plays a significantly important role in different security-related areas ranging from surveillance to monitoring various intruder activities. In a typical WSN application, these small portable smart devices are often deployed across a large space depending upon the application type there are employed for. Despite their numerous advantages, they usually lack sufficient processing power, energy and memory requirements for large applications. In such applications, they are often susceptible to security attacks like eavesdropping, physical tampering, denial of service attacks (DDOS), physical tempering, and so on, that often adversely affect their performance to any significant level.

Thus, in order to cope with these new policies and strategies, they need to be implemented at grass root level to counter with the loopholes. However, apart from all this, one significant concern that is inhibiting its widespread adoption is energy requirement. All these WSN nodes usually employ energy restrained devices that often hold the pressure of handling multiple operations at the same time. All this typically results in the draining of energy and thus makes it futile for further usage. Therefore, in order to counter this problem for various applications in smart cities, the concept of energy efficiency needs to be taken into consideration with the utmost consideration

A shift towards green-based WSN will indeed be the ideal concept that needs to be put into practice in the existing WSN infrastructure as this will not only improve the reliability of security applications, but also will increase its ability to handle a number of processes altogether. Thus, to put it into mainstream use, the following methods can be implemented concerning green WSN in various security-related applications:

1. operating sensor nodes only, and only when their work is necessary and enabling sleep mode for the rest of the time to reduce power consumption;
2. adopting radio optimization procedures like energy-efficient cognitive radio, transmission power control, and so on;
3. incorporating a number of energy harvesting techniques like solar radiation, vibrational energy from piezo, and so on, for the more extended low power demand-type security applications;

4. deploying energy-efficient routing techniques; and
5. employing data-saving mechanisms like adaptive sampling, network coding, compression, and so on.

Incorporating all these solutions will help in improving the performance in many operations related to security, like military surveillance and sensing. These methods, if adopted with proper care, can adapt to the pre-existing smart city infrastructure and will provide a way to enrich more secure applications for different case scenarios.

8.3.3 Green Cloud Computing-based Security Systems

Cloud computing is generally characterized into three basic sets of services. The first IaaS (Infrastructure as a Service), the second, PaaS (Platform as a Service) and the third, SaaS (Software as a Service). The concept of incorporating cloud computing services is often considered one of the most significant aspects needed to shape a smart security system. The main reason behind this is its capability to deliver and maintain large infrastructure facilities for the data which is associated with the security systems.

Apart from this, it also offers a flexible set of resources (e.g., high-performance computing resources) to users based on their requirements. As discussed earlier in this section, these arrangements provide the users with hassle-free access to a massive organized pool of resources within a fraction of a second which ultimately enables them to secure their respective premises on a real-time basis.

Even though it provides many merits, such flexibility often involves the shifting of a number of concerned applications to a cloud platform which ultimately leads to more energy consumption for a given set of services. Adoption of such practices brings much more harm to the environment as a whole by increasing CO_2 emissions to much higher levels. Thus, to move towards a greener approach in cloud computing for security-based solutions, a number of methods can be adopted:

1. adapting low-power consuming hardware and software. In such contexts, equipment arrangements should focus on machines which devour less energy. Software improvizations should utilize approaches that can help us in achieving maximum efficiency utilization from the resource available to us;
2. incorporating energy-efficient virtual machine practices for a number of tasks;
3. deployment of numerous low-power consuming task-scheduling mechanisms and associated resource-distribution techniques;

4. implementation of precise and powerful models and assessment methodologies in coordination with energy-saving policies.

The methods that are discussed above provide a unique approach to handling massive security-based services effectively and regularly. Since big security enterprises depend hugely on these high computational-intensive, large storage systems for their services, these systems often tend to play a massive role in handling the needs of their customers on a regular basis. Therefore, it is quite essential for such security organizations to scale down the unnecessary cost incurred during the operation by incorporating unique green technology-based solutions into their concerned security infrastructure.

8.3.4 Green Machine-to-machine-based Security Systems

These type of systems generally involve installation of a considerable number of M2M nodes employed for the collation of monitoring data collected from a number of secure systems (Lu et al. 2011). This data is further escalated down to the base station via a number of wired and wireless technologies. The base station that has been deployed for the purpose generally involves many M2M applications via the cloud platform. Involvement of such a large number of M2M nodes generally results in the consumption of a lot of power. Since the primary goal is to achieve a green M2M environment that can help us to enhance energy efficiency in a sustainable way, a number of prospective approaches can be adopted. Some of which are listed below:

1. adjustment in the power of transmission (e.g., to the minimally acceptable levels);
2. using competitive strategies for energy-efficient communication protocols like the routing protocol, with the involvement of distributed and algorithmic computing procedures;
3. using proper scheduling schemes to put the necessary tasks in the queue takes into consideration the vitality of the whole security system.
4. deployment of energy-savvy mechanisms;
5. employing energy harvesting techniques for cognitive radio networks like spectrum management, power optimization, interference mitigation, spectrum sensing, so on.

8.4 Security of Mobile Devices, Applications and Transactions

Mobile devices in today's world have already enabled us to do almost everything on the web on the basis of demand, irrespective of the time and location. At this very moment, it is not easy to envision a future without the immense assortment of cell phones that are presently part of our everyday lives.

In the present time, they have transformed themselves from just a means to serve the purpose of voice communication to a multifunctional device that is capable of allowing users to engage in a ton of different activities like health tracking, news and information sharing, financial services and so on. The main success behind them is the multitude of applications that allow many services by connecting to APIs and cloud servers around the world to deliver information, value and services to users.

Alongside developing a reliance on these mobile devices, new technological advancement has been witnessed both in equipment and programming designs. These advancements can be seen through the research that is being carried out by various researchers around the world for improving the security of mobile phones. An examination of dealing with physical or logical access, the authentication of digital identities, the platform for software tokens or even for the employment of mobile devices as tools for verifying transactions in desktop personal computers show that mobile devices, by default, possess a preferred security threat over normal PCs. If properly configured and protected, mobile devices present themselves as a viable platform for securing online transactions and digital identities.

With the colossal reliance on mobile devices to execute business activities, security system organizations are provisioning arrangements that address security needs. Authentication of the mobile device while connecting to a network, embedding identity protection inside mobile applications and monitoring sensitive transactions to detect unauthorized or fraudulent activities are some of the measures that need to be incorporated into mobile devices to enhance the security level to acceptable levels. Thus, in order to realize these measures, the following points should be taken into consideration to achieve sufficient security of mobile devices, applications and transactions.

1. Enabling authentication of user: mobile devices must be protected with PINs or passwords. Additionally, the password tab must be censored to keep it from being watched, and to protect it from unauthorized access, and the devices should set a minimum stipulated time to lock the screen again, in case of an idle condition.
2. Authenticity verification of applications: appropriate measures must be employed for evaluation of the applications using digital footprints or signatures, to verify that it is the same as that of the original.
3. Introduction of firewall: an individual firewall is capable of ensuring protection against unauthorized access by examining outgoing and incoming attempts of connections and blocking or allowing them in view of a set of guidelines for gaining access as laid down by the system administrator.
4. Introducing antimalware ability: secured protection from malware must be introduced to ensure protection against threats such as malicious

applications, spyware, viruses and infected digital cards. Moreover, such abilities also ensure protection against undesirable (spam) text messages, voice messages and email attachments.

5. Introducing security updates: vital software updates need to be consequently exchanged from the manufacturer to mobile devices on a regular basis. To ensure that these security updates are transmitted promptly to the destined mobile device, a definite set of procedures must be implemented.
6. Remote disabling of stolen or lost devices: this feature is handy for a stolen or lost device as it takes command over the mobile devices despite the location of operation and deletes the user's data according to his/her wish to maintain integrity and protection against potential data theft.
7. Empowering data encryption: file encryption secures confidential data stored on cell phones and memory cards. These devices must be able to work in encryption-based capacities or should be able to utilize industrially accessible encryption tools.
8. Establishment of security policies for a mobile device: there are basic fundamental security guidelines that define the rules and protocols which determine standard operation procedures for the organizations and individuals to use mobile devices in a regulated manner.

    Security policies of a security organization characterize the principles, rules and practices to treat mobile devices. This set of policies must cover responsibilities and roles, devices and infrastructure security, together with the assessment of overall security. By setting up such policies, a security organization can create a structure that can apply tools, training and practices to help support the security of the whole system.
9. Two-factor authentication for secure transactions: the Two-factor authentication should be employed when one needs to perform critical secure transactions on mobile devices. This process is capable of providing a higher level of security in comparison to the conventional password types. The whole process refers to kind of a verification protocol in which users are authenticated through a minimum of two unique parameters. Such two-factor authentication is very much necessary when there is a need for secure transactions, like for mobile banking or during other various online monetary transactions.
10. Providing training for the security of mobile devices: preparing skilled employees in a smart security system company can guarantee that mobile devices are assembled, configured and operated in an appropriate way.
11. Performing assessments of risk: performing risk analysis helps us to locate the threats and vulnerabilities, list potential attacks and evaluate

their success probability. It also assesses the potential damage caused by successful attacks on mobile devices.
12. Performing management and control of configuration: managing configuration in mobile devices ensures protection against the introduction of any type of modifications after, during and before deployment.

8.5 Smart Embedded Security for Smart Cities

A typical smart city generally comprises of infrastructures that are usually integrated with advanced embedded technologies to serve a variety of operations related to security on a daily basis.

Access control, intrusion alarm, fire alarms and protection, video surveillance along with efficient and effective emergency evacuation protocol frameworks are some crucial embedded security solutions, which require precise planning and vigilant monitoring from the concerned authorities to ensure smooth operation whenever needed. However, despite their ability to provide a useful solution for enhancing the security features, it often results in the dissipation of useful energy that often leads to a hindrance in its operational efficiency. Therefore, there is a need for adaptation of sustainable policies, and strategies that can ensure the safety and security requirements of the residents without much capital expenditure on the energy aspect.

Implementation of these practices would be much more beneficial if we introduce the green concept into conventional pre-existing security solutions. These green alternatives are a way to provide a sustainable integrated approach to the safety and security of smart city infrastructure. Advanced sensor technology, AI, smart analytical tools and platforms generally can serve as a useful tool for both old conventional and the latest unconventional integrated surveillance infrastructures which can ultimately lead to assist in controlling traffic, spotting criminals and managing crowds and so on for a number of different scenarios. These security measures also additionally include integration of digital surveillance and the physical safekeeping along with perimeter securing frameworks to a central database hub of the concerned police department.

Embedding such smart security solutions will reduce criminal activities by regularly updating and notifying the law enforcement organizations to take necessary or preventive actions against the offender. Another example is that of smart alarm systems, which in addition to triggering high-pitched sharp, loud, noisy alarms during an invasion, are also capable of texting an alert message to the mobile device of the owner of the premises due to the integration with other technologies. All such embedded solutions and tools are becoming smarter day by day with the rapid evolution of technology. Clearly, technology advancement of the twenty-first century has a crucial role in making cities smart, empowering its solutions for effective administration and incorporation of a city's various technical ecosystems. Most of the existing solutions in the smart cities today are indeed very advanced and have some sort of intelligence embedded in them. However, besides the potential advantages of these systems, there is also a concern for security.

Due to technological advancement in recent times, there has been a rapid increase in ownership of smart devices which subsequently has led to an increase in cybercrime, making society ultimately more unsafe. The current need is to objectify the much-needed embedded smart security measures to tackle such intervention into the integrated system of smart cities. One way to deal with this problem is through the process of authentication where devices should be able to validate themselves to the network and each other by authenticating any messages they exchange. This basically confirms that data remains unaltered during its course of transmission, and would not respond to the commands from any unauthorized source.

Another way to deal with this problem is by introducing cryptographic technologies to the existing embedded security devices which will ultimately ensure a much higher level of security without any computational or energy overheads. Many devices used for smart cities are left without any sort of resiliency and upgradability. It is therefore required to update these devices with recent security updates and incorporate in them a much flexible cryptographic solution that makes them last for extended periods in comparison to the existing ones.

8.6 Digital Identity and Access Management for Cloud-based Services

IAM presents a systematic framework to provide management of electric and digital identities in the cloud. It is like a security layer to your security system. This layer will reduce the complexity of your user network also reducing the risk by limiting their access. In the context of the green smart security system, such frameworks in cloud computing are equipped with the required technologies and policies needed to support significant user base identities that are cumbersome and impossible to allocate manually. Such type of structure is often developed to provide each user with a single identity. After allotment of these digital identities, the IAM framework generally has a task to maintain, monitor and modify the identities that have been allotted to each user depending upon the privileges. In the given service framework the administrator holds the privilege to make alterations in the user's role depending upon the role, work, limitations and so on. Apart from this, each user activity can be tracked by administration through a report. The need for such a framework is felt when most of the integrated smart security companies face identity problems in the cloud storage of each user's data.

These changes in the system and security have been a very long, busy and devastating practice that usually shatter the whole security. Due to this cumbersome process, they are more focused on managing data resolving identity problems. Their focus is moved from securing the clouds first, and that makes it vulnerable. This creates a blind spot and increases the security risk. It is like an essential and significant member of the security family.

It is a critical process to provide perfect information to the user without any form of delay and problems. Let's think about it as an example. There are many organizations who have lots of information and users. Most of the time users can access the information that they should have access to. So, here IAM works as a filter layer. It will eliminate these people who shouldn't have access to that information.

This framework provides services like password security, provision software, and security policy enforcement. This is the work of Identity Access. Then on the other hand identity management works with software like Microsoft SharePoint, or cloud-based Office 360.

The solutions of the digital identity and access management (IAM) cloud-based services solutions are depicted in Figure 8.2.

8.6.1 Advantages of Using IAM Systems

Nowadays to increase the revenue and efficiency of service delivery to the end-users, most of the security companies are providing access privileges to the internal systems. These end-users generally comprise customers, contractors, suppliers, partners, and so on. Opening these systems to clients, suppliers, partners and employees can increase the efficiency of management with lower working expenses.

From the framework, identity management can allow an organization to stretch out access to its data over an assortment of internal and external server-based apps, SaaS-based tools and mobile applications without risking the essential security standards that a company needs to abide by. By giving more access to outsiders, it drives a coordinated effort all through the organization, improving productivity, worker fulfillment, innovation and, at last, increasing revenue.

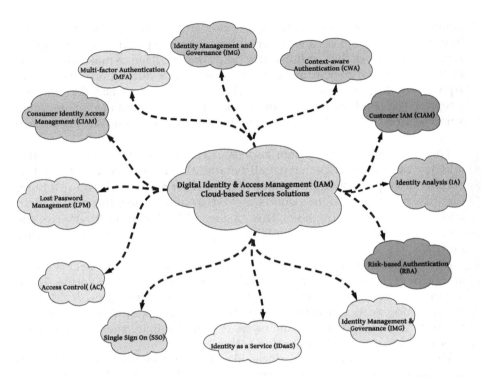

FIGURE 8.2
Solutions of the digital identity and access management (IAM) cloud-based services solutions.

Identity management can also reduce the number of assistance calls to IT support for password resets by cost-effectively automating all these trivial tasks. Quite often it is considered as the foundation stone for a secure network due to its ability to manage access control dedicated to a large user identity base without any hassle. It works on the access policies defined by security system companies which outline the access to specific data resources according to a particular condition associated with it.

Due to the proper management of digital identities associated with each client, the smart security system generally encounters reduced risk of external as well as internal breaches. This feature is vital because with ever-increasing dangers of external attacks, internal attacks are rising at a similar pace. According to a report by IBM cybersecurity intelligence index, approximately 60% of all breaches of information involves employees of an organization. Of that 60%, 75% was found to be malicious in intent while the other 25% was encountered accidentally.

As said already, the IAM framework holds great potential to support the associated security system organization by giving them sufficient resources for the implementation of multidimensional security, access, as well as review strategies to achieve safer cloud storage.

8.7 Challenges to Designing the Green Smart Security System

Though the adaptation of green smart security systems holds the potential to reduce the problem of energy consumption, for example, there are also several issues related to the integration of these new advent technologies into the pre-existing conventional type systems. It is, therefore, necessary to keep a check on the vital loopholes that usually cost the user a significant amount of energy and money if not properly taken care of. These challenges will eventually prove fruitful in designing more advanced security solutions for the smart cities that are not only efficient, but which also follow the mandate for an eco-friendly sustainable environment.

Therefore in this section, an attempt has been made to make the user aware of the critical challenges that one faces while working with conventional security systems. However, despite the methods there still exist many ways in which security of the system can be compromised.

8.7.1 Factors that Hinder its Growth

8.7.1.1 Capital and Availability of Resources

Each new method or security framework needs a significant amount of resources whose usage frequently prompts an augmentation in cost. It, therefore, seems vital for industrial organizations to address this issue, as energy consumption rate is multiplying day by day. Along these lines, the enhancement of energy efficiency in effective green smart security frameworks has become a burning matter in the recent security system domain. Each of these security systems is simply made up of smart embedded devices that together are responsible for forming a massive IoT network. However, there are specific issues related to the security of these smart devices that actually prove to be a hindrance in their establishment. Some of the leading challenges in securing the relevant IoT devices for green smart security systems are mentioned below.

8.7.1.2 Securing Underpowered Embedded Devices

Most of the conventional smart security embedded devices lack memory, processing power and storage, and are required to operate on minimum power to save the limited amount of battery power available. Deployment of various encryption-based security methods is often not considered a solid match here because of these devices mainly because of their incapability to carry out the complex process of encryption and decryption fast enough to serve real-time applications.

Such gadgets are generally defenseless against side-channel attacks, for example, power analysis attacks, that can prove to be handy in reverse engineering their firmware and installed algorithms. Instead, small, underpowered embedded devices usually just apply fast, low-end encrypting algorithms. These systems can deploy numerous strata of protection: as an example, isolation of these devices onto the separate environment and installing firewalls, to catch up for restrictions imposed on these devices.

8.7.1.3 Authentication

Considering the vast numbers of devices vulnerable to probable chances of failure while operating in a security framework, genuine gadget approval, verification and authentication are quite elementary for anchoring such green systems.

Devices must authenticate their unique credentials before they can access gateways and over-the-air facilities and applications. There are numerous devices that fall short with regards device validation, for instance, by utilizing weak passwords or utilizing passwords which are kept unchanged from their initial default values.

Deploying a smart security protocol which provides security by default cuts down these issues, like in the case of empowering two-factor authentication (2FA) and upholding the use of strong and highly unpredictable passwords or signatures.

8.7.1.4 Manage Device Updates

Updating the software and firmware running on these smart devices is also a challenge. You are required to monitor the availability of update packages and security patches continuously, and applying them across the diverse operational environments and heterogeneous devices that demand a variety of networking protocols to communicate.

Availability of updates to all the devices is a challenge, specifically older devices. Moreover, there are devices which cannot be updated over-the-air, so they need to be physically retrieved or pulled back from the system temporarily. The owner of a smart device may refuse to apply an update even after its availability. Device management requires keeping an eye on the versions that are installed on every smart device and which of them are eligible for withdrawal from the system after the updates discontinue. Device managing frameworks regularly release updates automatically to the suitable devices along with handling rollbacks if the process of updating miscarries abruptly. Additionally, they can confirm that genuine updates are installed only via methods like the digital signing.

8.7.1.5 Protecting Communication

After ensuring the security of the devices, an upcoming hurdle in the path of security is to safeguard the network which enables communication among devices, their applications and web-based services, such as the cloud. A considerable number of these gadgets omit encryption of data before transmitting them to the system. Be that as it may, the optimum solution is to deploy transport encryption, and to embrace norms like TLS. Configuring discrete systems to disengage devices additionally assists with building up safe and protected communication, so communicated data stays confidential.

8.7.1.6 Securing Data Privacy

It is quite critical that when the data is transferred successfully to the desired location in the system, it is processed and stored safely. Data which would not be required in future must be handled and disposed of safely without compromising its integrity, and if it is needed to store the information, following up consistency with a legal and administrative framework is also an essential task. Safeguarding data integrity, which may include applying digital signatures or checksums to guarantee information, is unaltered. Blockchain – as a dispersed appropriated record for IoT information – provides a versatile and robust path for ensuring the integrity of IoT information.

8.7.1.7 Protection of Applications

Applications and services based on mobile, web and cloud are deployed to monitor, operate and secure IoT devices and information, so these applications also need to be protected in order to approach the security of IoT systems in a hierarchal manner. While developing green smart framework-based applications, make sure to engage secure building practices to sustain a tactical space away from vulnerabilities and malfunctioning. Similar to the devices, applications also require to be equipped with secure authentication protocols, both for the end-user of the application and the use within itself, through efficient alternatives such as protected password recovery, 2FA and so on.

8.7.1.8 Ensuring Adequate Availability

With the increase in dependence on these coordinated smart security frameworks inside the day-to-day lives of people, system engineers need to consider the convenience in accessing the data, Internet and mobile-based applications that rely on that data along with the access of people to the physical amenities managed and coordinated by these smart green secure systems. The probability of disorder triggered by connectivity issues or device malfunctioning, or developing due to attacks like denial of service attacks, is apparently way beyond mere inconvenience. There are applications where the impact of the absence of accessibility would result in damaged hardware, money loss, harm to gear, or in extreme cases even loss of limb or human life. For example, in smart cities, these types of infrastructure control important security facilities in traffic control and further monitoring tasks. To guarantee high availability, these smart devices shall be sheltered against physical damage and cybersecurity threats. These systems should be able to incorporate redundancy to avoid particularly one reason of failure and must be manufactured to be robust and capable of tolerating errors, designed in a way to adjust and revive quickly when abnormalities occur.

8.7.1.9 Identify Susceptibilities

Even after the adoption of excellent techniques in the incorporation of the green smart security system, there is still a high possibility of intervention of safety susceptibilities and weak loopholes. Identifying vulnerability and occurrence of any failure is a difficult task in large-scale integrated smart security systems, considering the complexity of the framework due to the number of diverse devices connected operating on varied applications, network protocols and services. Constant monitoring of the activities within the network framework is required to detect any possible security breach. More of these strategies can come in handy in identification of system susceptibilities; for example, indulging in ethical hacking and penetration testing to uncover hidden loopholes. Moreover, making use of analytical, intelligent security algorithms to determine the occurrence of any security breakthrough and notifying it well within time needs to be adopted to realize a reliable, secure smart environment.

8.7.1.10 Managing Loopholes

Endless complications of integrated smart security frameworks make it quite tough to examine the consequences of susceptibility and the depth of a security breakdown, to analyze its impact. Hindrances include determining which of the security modules were influenced, how much and what information was leaked and the number of users affected, and, afterwards, implementing the required appropriate measures to avoid similar breaches in future. To facilitate the realization of these counteractions, device management services keep track of devices, which when needed can be deployed to disable affected or infected modules, till they are patched and become fit to operate again into the central system. This feature, in particular, is crucial for critical devices such as gateways. This entire process can be automated very effectively and efficiently with the help algorithms based on susceptibility management guidelines.

8.7.1.11 Anticipate Security Breaches

Aiming towards long-term goals, the primary challenge regarding security is not just the detection and resolution of breaches whenever they happen, but to be able to predict potential threats and take action in advance to avoid compromising the security of green smart systems in every scenario. Threat modeling is one such approach that can be used to predict security issues well in advance. Further solutions include the application of AI and deep learning which would analyze and correlate the past threats and security breaches, providing precise real-time predictions protecting against future issues, while intelligently updating the methodologies to approach such problems, through learning based on the effectiveness of previous algorithms and protocols, hence deploying a fully functional and evolving green smart security system.

Conclusion

With the rapid agglomeration of different technologies for security in the smart cities, there is an inevitable need for the adoption of sustainable measures and practices that can help us to alleviate the global problem of depletion of resources and energy. In this chapter, an attempt has been made to accustom the potential reader with the vital issues and solutions that can help in bringing the concept of a green smart security system into existence in the current world. With the introduction of existing smart security infrastructure, approaches to embedded green solutions into the current pre-existing security infrastructure were also discussed thoroughly in detail with the focus to provide a vision for a secure sustainable smarter world.

References

Arshad, R., Zahoor, S., Shah, M. A., Wahid, A. and Yu, H. 2017. Green IoT: An Investigation on Energy Saving Practices for 2020 and Beyond. *IEEE Access* 5: 15667–15681.

Bastos, D., Shackleton, M, and El-Moussa, F. 2018. Internet of Things: A Survey of Technologies and Security Risks in Smart Home and City Environments. In *Living in the Internet of Things: Cybersecurity of the IoT*. IET/IEEE: London.

Casini, M. 2017, August. Green Technology for Smart Cities. In *IOP Conference Series: Earth and Environmental Science* 83(1): 012014. IOP Publishing (conference research article).

Casini, M. 2017. *Green Technology for Smart Cities. IOP Conference Series: Earth and Environmental Science*, 83.

Gou, Q., Yan, L., Liu, Y. and Li, Y. 2013, August. Construction and Strategies in IoT Security System. In *IEEE International Conference on Green Computing and*

Communications and IEEE Internet of Things and IEEE Cyber, Physical and Social Computing, 1129–1132. IEEE.

Latif, S. and Zafar, N. A. 2017, November. A Survey of Security and Privacy Issues in IoT for Smart Cities. In *Fifth International Conference on Aerospace Science and Engineering ICASE*, 1–5. IEEE (conference research article).

Lu, R., Li, X., Liang, X., Shen, X. and Lin, X. 2011. GRS: The Green, Reliability, and Security of Emerging Machine to Machine Communications. *IEEE Communications Magazine*, 49(4): 28–35.

Matharu, G. S., Upadhyay, P. and Chaudhary, L. 2014. The Internet of Things: Challenges and Security Issues. In *International Conference on Emerging Technologies ICET*, 54–59. IEEE.

Number of Mobile Phone Users Worldwide from 2015 to 2020 in Billions. Accessed July 20, 2018: www.statista.com/statistics/274774/forecast-of-mobile-phone-users-worldwide/.

Robles, R. J., Kim, T. H., Cook, D. and Das, S. 2010. A Review of Security in Smart Home Development. *International Journal of Advanced Science and Technology*, 15.

Sivaraman, V., Gharakheili, H. H., Fernandes, C., Clark, N. and Karliychuk, T. 2018. Smart IoT Devices in the Home: Security and Privacy Implications. *IEEE Technology and Society Magazine*, 37(2): 71–79.

Solangi, Z. A., Solangi, Y. A., Chandio, S., bin Hamzah, M. S. and Shah, A. 2018, May. The Future of Data Privacy and Security Concerns in the Internet of Things. In *IEEE International Conference on Innovative Research and Development ICIRD*, 1–4. IEEE.

Szymanski, T. H. 2017. Security and Privacy for a Green Internet of Things. *IT Professional* 195: 34–41.

Talwana, J. C. and Hua, H. J. 2016. December. Smart World of the Internet of Things IoT and Its Security Concerns. In *IEEE International Conference on Internet of Things iThings and IEEE Green Computing and Communications GreenCom and IEEE Cyber, Physical and Social Computing CPSCom and IEEE Smart Data*, 240–245. IEEE.

Wang, M., Zhang, G., Zhang, C., Zhang, J. and Li, C. 2013. An IoT-based Appliance Control System for Smart Homes. In *Fourth International Conference on Intelligent Control and Information Processing ICICIP*, 744–747. IEEE.

Zhu, C., Leung, V. C., Shu, L. and Ngai, E. C. H. 2015. Green Internet of Things for the Smart World. *IEEE Access*, 3: 2151–2162.

9
Green Smart Transport Systems

Arsh Javed Rehman, Shweta Yadav and Pradeep Tomar

CONTENTS

9.1 Introduction 186
    9.1.1 Identification of Problem 186
    9.1.2 Need for Smart Green Transportation 187
9.2 Green Transport Systems...188
    9.2.1 Technologies for Green Transportation 188
    9.2.2 General Methods for Improving Transport Efficiency 189
9.3 Projects and Initiatives 190
    9.3.1 Horizon 2020...190
    9.3.2 Climate Smart Initiative 190
    9.3.3 Smart Mass Transit Rail (SMT Rail) 191
    9.3.4 Green Transport in Developing Countries..................................191
    9.3.5 Solar-powered Transport 191
    9.3.6 BuyZET Project...191
    9.3.7 Infrastructure for Green Vehicles..192
    9.3.8 Green Vehicles 193
    9.3.9 Standards for Transportation 193
9.4 Implementations of Green Transport Systems 194
    9.4.1 Green Smart City, Dubai 195
    9.4.2 Greenest City Draft Action Plan, City of Vancouver 196
    9.4.3 Intelligent Transport Systems, Korea 196
9.5 Future Work...198
    9.5.1 Sustainable Transport Development in Smart Cities..................199
    9.5.2 AI Machine-learning Algorithms for Smart Green
            Transportation 199
    9.5.3 Hyperloop Technology...200
Conclusion 200
References..200

9.1 Introduction

The concept of "green" is associated with multiple objectives like sustainability, meeting the principles of the circular economy (recycle, reuse, reduce) and the protection, preservation and recovery of the environment and its resources. Green technologies can be defined as a set of techniques or practices which reduce the technological impact on the environment and also take into consideration the recycling of many of the parts used in these processes, allowing people to benefit from these technologies.

Green transportation is that means of transportation system which has no harm on the ecosystem or the environment. It makes life easier, reduces congestion and decreases the dependence on cars and foreign oil. It is safer and less costly, and would overall support the Earth. Björklund (Björklund 2011) defines "green transportation" as: "Transportation service that has a lesser or reduced negative impact on human health and the natural environment when compared with competing transportation services that serve the same purpose." Most of the people drive their cars, and one of the reasons for this is that it is the most efficient way to get around, which is true. The transportation system around us has been built for cars. There are not a lot of sidewalks or bike paths if one looks at the streets around. The buses are sometimes unreliable and everything is far away, so it is more convenient for us to drive.

It really matters when someone chooses to drive or walk, because it's all about sustainability. It is known that cars are not good for the environment as they contaminate the air with their emissions and use a lot of fossil fuels. The issue with fossil fuels is that it is a finite resource which means that one has a limited amount of it here on earth and according to research already half of all the supply has been used. If we continue to use it in this way then it will be a big issue (Guglielmo 2012). So, one has to be careful with the usage of such resources.

One way to do this would be to use more fuel-efficient cars but the problem with fuel-efficient cars is that if one uses less fuel then it might encourage people to drive longer distances and then in the long run it'll end up using the same amount of fuel just because one can afford it. The best alternative would be to build sustainable and livable communities, and use green and smart transportation by moving towards a medium of transportation like electric or hybrid vehicles which are green, smart and sustainable. Smart transportation is green and the research in the field of green transportation has expanded a lot in recent years mainly due to the increase in the price of petrol and gas all around the world.

9.1.1 Identification of Problem

Transportation is always associated with the growth of modern society. One of the things which has satisfied the everyday lives of human needs is transportation. Today, the transportation system is IC engine-based transportation, with emissions of carbon dioxide (CO_2), unburned hydrocarbons (HCs), carbon monoxide (CO), nitrogen oxides (NOx), in large amount the main cause of environmental pollution (Panday, 2014). The rise in population exponentially and the increase in personal vehicles have upped the volume of automobiles around the world. It is a continuous cause of severe environmental problems and risk to human life.

The emissions from transportation are made up of approximately 23% of CO_2 and overall 15% of greenhouse gases, and CO_2 is the most widespread of all greenhouse gases. The emissions of CO_2 from transportation have grown by 45% in a span of 17 years (from 1990 to 2007), and the environment is expected to degrade more as the greenhouse gases will increase by 40% from 2007 to 2030 (Leipzig 2010).

9.1.2 Need for Smart Green Transportation

For reasons of degradation of the environment, depletion of natural resources, air pollution, emissions of greenhouse gases, rapid depletion and continuous increase in the oil prices, people around the globe should move towards smart and greener transportation. Some of the important factors are as follows:

9.1.2.1 Degradation from Air Pollution

The chemical reaction between the atmospheric air and fuel produces heat. Due to the heat produced from the IC engine, it emits many harmful gases like CO_2 and hydrocarbons (unburned). Air pollution is a serious environmental threat to health and has caused 2 million premature deaths worldwide per year. According to the World Health Organization, 9 out of 10 people inhale toxic air, that means only one in ten people breathe clean air. Every year around 7 million people die due to diseases related to air pollution (WHO 2016).

9.1.2.2 Rise in Oil Prices

Oil is the major constituent that works as a fuel for several types of vehicles used in transportation. Also, the burning of oil contributes to pollution of the environment. As stated by OPEC, the main consumer of oil is the transportation sector and most of the demand for oil is used for transportation purposes such as aviation, road, rail, domestic waterways and marine (World Oil Outlook 2040, 2017). The *Business Line* (Kilian 2009) also stated that an increase of $10 in oil price will eventually reduce GDP by 1.5% in the developing countries.

9.1.2.3 Depletion and Extraction of Natural Resources

The extraction of crude oil around the world is increasing day by day as the consumption of liquid fuel is increasing exponentially. According to EIA, the US has increased its oil production to 15.6 million barrels per day, Saudi Arabia to 12.1 million barrels per day and Russia to 11.2 million barrels per day (EIA, 2017). Developing countries like India are producing crude oil like never before and it ranks 24th in the total production of crude oil around world (Kilian 2009).

9.1.2.4 Relying on Energy from Oil

As per the environmental impact assessment (EIA) (Kilian 2009), the usage of petroleum is increasing day by day, and has increased from 84,918.2 thousand barrels per day (2009) to 89,429.8 thousand barrels per day (2012).

9.2 Green Transport Systems

Today people are working to design various green transport systems, which will have low environmental impact by combining sustainable transport with green technology.

9.2.1 Technologies for Green Transportation

Technologies for smart and green transportation are as follows:

9.2.1.1 Electric-powered Vehicles

On a tank-to-wheel basis, fuel-powered cars are not as efficient as electric cars. Electric-powered vehicles are better in terms of fuel economy than the conventional IC engine-powered vehicles but in case of long distances they lack range as they tend to discharge before reaching your destination. The leading cost in the electric vehicles is the battery but this type of systems leads to a reduction in emission of CO_2, that is 0% to 99% as compared to conventional IC engines, but the reduction of CO_2 depends on where electricity is sourced.

9.2.1.2 Hybrid Electric Vehicles

Hybrid electric vehicles are a combination of fossil fuel, which can be run on bio-fuel and hydrogen or electric. These hybrid transport systems are partly electric powered and partly fossil fuel-powered. In most cases, an electric engine is combined with an IC engine but other combinations can also be used. The IC engine are costly to purchase and are more expensive and are mainly a diesel or gasoline engine. These engines provide a better economy of fuel and the cost of IC engine is redeemed in around five years.

9.2.1.3 Stirling-powered and Compressed Air Vehicles

Stirling-powered, liquid nitrogen and compressed air vehicles produce less pollution as compared to electric cars, and it makes them environmentally eco-friendlier. To promote green technology, solar-powered car races are held to make people aware and to show the significance of these vehicles. The electricity generated is instantaneous from the sun and these cars can travel at high speed for a longer distance.

9.2.1.4 Pedal-powered and Electric Motor Vehicles

Developing two-, three- and four-wheel vehicles is possible by combining electric motors with the characteristics of the bicycle. It provides an eco-friendlier solution to the problem of pollution emitted from the transportation system.

9.2.2 General Methods for Improving Transport Efficiency

Several techniques are becoming popular in countries with the sole purpose of reforming the transport sector and to improve fuel efficiency. This would reduce carbon and other greenhouse gas emissions in the ecosystem and would lead to a sustainable environment. Some of them are briefly explained in the following section.

9.2.2.1 Mechanical Design Improvement

Energy loss (waste heat that is released from the cooling systems and exhaust) from engines leads to large fuel consumption, thereby reducing efficiency. Drivetrain engine-based technologies can decrease such losses and improve total efficacy. This technology includes electric (based on electric motors) and hybrid (multiple power sources used in the same vehicle) drivetrains. Such drivetrains lead to low emissions and lesser fuel consumption.

9.2.2.2 Reducing the Rolling-resistance

Tires have a high rolling-resistance that leads to high fuel energy loss. Innovative design of tire-treads, casings and rubber components have improved the rolling-resistance. Experiments have indicated up to 13% of fuel savings for heavy means of transport.

9.2.2.3 Highly Effective Wheel/ Tire Lubrication

The friction of train wheels amounts to large energy consumption in rail transportation. The usage of lubricants like oil, water or grease to decrease friction levels not only reduces the energy consumption, but also excessive noise and wear and tear. Lubricants can be applied from onboard systems or wayside (systems set up on the side of the tracks). The latter is most common while onboard lubrication has started to become popular around the world.

9.2.2.4 Usage of Lightweight Materials

The usage of high integrity, low-density metals like aluminum has allowed the production of more low weight air, water and sea transport. Such materials, first, decrease the rolling-resistance and, second, advocate the usage of such highly fuel-efficient vehicles as they are lighter and need lesser power to travel. A combination of such materials with improved design of the engine enables higher fuel savings. This method has found its application in cars, trains carrying freight and passengers, aircrafts and so on.

9.3 Projects and Initiatives

There are many projects and initiatives for green and integrated transportation. Some of them are listed in the following sections:

9.3.1 Horizon 2020

Horizon 2020 is the biggest research EU program: it is the implementation of the Innovation Union, EU flagship program. It is the financial instrument focused on securing global competitiveness for Europe (Kugleta 2017). The main focus of Horizon 2020 is to develop smart, green and integrated transportation. The focus of this program is to achieve a transportation system that is climate and environment-friendly, resource-efficient, safe and should be able to give benefits to all the citizens and societies as a whole (Besseso 2018).

9.3.2 Climate Smart Initiative

The transportation sector is responsible for 23% of greenhouse gas (GHG) emissions and emissions are on the increase. This statistic makes it clear that for significant progress on climate, greener and more sustainable mobility is needed. Transportation is vulnerable to the climate changes such as increased precipitations, flooding and higher temperatures. The World Bank has taken the initiative to address these issues on three major areas (Cervigni 2016):

1. climate-resilient transport
2. clean, safe and efficient mass transit
3. efficient and multimodal transport systems.

9.3.3 Smart Mass Transit Rail (SMT Rail)

SMT Rail is aimed at providing the greenest, fastest and safest means of transportation at the most economical, quickest travel time and shortest distance to all the countries around the globe. To protect the ecosystem, the project pledges to reduce the human carbon footprint by reducing travel distance and using smart green transportation. SMT Rail will provide 24/7, 365 days a year green and smart transportation. Municipalities and communities will be provided with on-demand, comfortable and safest means of transportation system. This will reduce transportation cost by smart green transportation.

9.3.4 Green Transport in Developing Countries

In developing countries, the critical issue is how to combine economic growth with green transport. Air pollution and congestion make economic development difficult and so sustainable transport is necessary. Green Projects will also foster employment.

Some of the ongoing projects for sustainable development (transport sector) in developing countries are as follows:

- the Bus Rapid Transit (BRT) in Johannesburg;
- the Sustainable Urban Transport Improvement Project, Indonesia;
- the Dedicated Freight Corridor Corporation, India.

9.3.5 Solar-powered Transport

Solar-charged transport or solar transport is powered by solar energy. Solar power usage has increased almost ten times. The entire public transport should be solar-powered. Before departing, solar transport is charged at the starting station. Solar transport also has regenerative braking systems, in which the energy lost in breaking is reused and harvested. This feature helps to gain another 30% in efficiency and making fares economical at the same time. For example, solar buses have saved more than 14,000 literss of fuel in the first six months. The technology is being used in Australia, China, Europe, UK, Uganda and the United States. Solar transport is part of sustainable transport schemes, used for the purpose of public transport.

9.3.6 BuyZET Project

The European Commission granted 225,000 euros to the city of Oslo to fund a project on green and smart transportation. The cities participate with their goal of zero emission by implementing the BuyZET project. Approximately 200,000 people live in the city of Oslo that has a harmful air pollution level.

Around 63% of nitrogen dioxide, sulfur dioxide and suspended particles emissions derive from transport.

9.3.7 Infrastructure for Green Vehicles

Infrastructure will be required in the future, for a sustainable and green mobility. It will be a great challenge in the future to secure mobility and to supply goods in conurbations and cities. Infrastructure is necessary for friendly mobility and "green" good transport. The required infrastructure for green vehicles must go ahead and overcome future obstacles. "Green vehicles" are still in their developing stage as well as infrastructure, so not yet in proper shape to serve the market.

"Green vehicles" are vehicles which use new technologies that are environment-friendly and are also referred to as "green". Some green technologies for vehicles are as follows:

- hydrogen-driven vehicles known as fuel cell electric vehicles (FCEV);
- full electric vehicles (EV) or hybrid electric vehicles (HEV) with or without plug-in devices;
- vehicles powered by biofuels or with advanced internal combustion engines (ICE) which have a reduced impact on environment.

The development of infrastructure and the introduction of green vehicles to the market must take place concurrently. The sooner the coordination between legislation and road transport stakeholder takes place, the better for the future to secure "green mobility".

9.3.7.1 Benefits

The benefits of infrastructure for green vehicles are as follows:

- an adapted infrastructure for green vehicles will benefit not only public mobility and individual, but also the environment as a whole;
- secure and better roads will provide flexible traffic management, which will ensure sustainable goods transport and a safe and comfortable journey; and
- it will support the needs of the society for green mobility and a sustainable and decarbonized road transport system.

9.3.7.2 Challenges

The challenges in establishing an integrated sustainable infrastructure are:

- implementing the coordination between infrastructure and green vehicles is needed. All road transport stakeholders, e.g. operators/

Green Smart Transport Systems

builders, automotive industry, government/states, have to take concurrent actions;

- the interaction of different technologies such as business models, players, ICT systems with the system integration will be a big challenge;
- the return on investment will take a long time and investment in road infrastructure is very high;
- the absence of standards between the different regions, cities and the differences in the culture needs to be addressed;
- different industries have different cycles.

9.3.8 Green Vehicles

For the development of vehicles in the future, there will be larger diversification in vehicle design, application and propulsion systems.

The need for green vehicles is rising as the traffic in cities increases rapidly and long-distance transport and travel require fuel for a longer period of time. Green vehicles can be implemented in the following categories:

9.3.8.1 Light Commercial Vehicles and Passenger Cars

The concept of the electric car is not new and the electric city car (ECC) will be developed and designed so that on a single charge it can used for most of the day. To complete a full day of driving, if the battery is not sufficient one can still use the fuel-based range extender, until more advanced and better battery technology is available. For the delivery of goods, environment-friendly vehicles will be used with a new design.

9.3.8.2 Passenger Cars and Heavy Duty Trucks for Long-distance Travel

To travel medium to long distances, passenger cars are used which are powered by gasoline and diesel fuels. These can be optimized to reduce emission and consumption. Plug-in hybrids, hydrogen-driven vehicles and range extenders can be used. Eco-trucks are green transport aimed at reducing carbon emissions and have the following capabilities:

1. reduce greenhouse gases by 40%; and
2. are the most sustainable means of transport.

For a longer transport journey cargo or freight eco-trucks are the most sustainable means of transport from an ecological perspective.

9.3.9 Standards for Transportation

Transport standard organizations are involved in maintaining and producing standards that are relevant to the global transport technology. For efficient and safe transportation system, robust and modern standards are necessary. According to the international agreement, the standard development organizations organize the formal development of international standards in three tiers:

- world
- regional
- national.

9.3.9.1 World

The international standard-setting bodies are:

- the International Organisation for Standardisation (ISO) and
- the International Electrotechnical Commission (IEC).

9.3.9.2 Regional

Politically or geographically connected regions coordinate and synchronize through the regional standards bodies with a need to harmonize practices and products, for example in Europe, CEN or the European Committee for Standardisation.

9.3.9.3 National

The National Standard Development Organisation (SDO) are a coordinating body for publishing CEN and ISO standards. They organize and participate in the activities of CEN and ISO. The National SDO are responsible for implementing standards within the country, and implement changes in technology that evolve over time. For example, in the UK, the National SDO is BSI (British Standards Institution).

Standards have been developed for the various modes of transportation. Smart green transportation must follow the standards for intelligent transport systems (ITS), interoperability, transport of dangerous goods and intermodal transport.

9.4 Implementations of Green Transport Systems

A rise in the economy of a country leads to prosperity which in turn causes an increase in the number of cars and other polluting motorized vehicles. This would slowly lead to air pollution in the future and hence degrade the quality of living standards. The current trend is to make or transform the existing transportation systems to greener, eco-friendly and efficient transportation systems for bridging the gap between different locations (Li 2016). Several countries are facing difficulties in combining economic growth with green transport. But at the same time developed nations have taken a step further towards green transport. Some implementations are listed in the following section.

9.4.1 Green Smart City, Dubai

Green and smart technology has found its way in many developed nations. The "sustainable city", a smart and efficient city, has been built in Dubai, United Arab Emirates, and is home to 3,500 people and is still a work in progress. This city can be pointed out as one of the best examples of green and smart efficient technology. The primary goal in construction of the "sustainable city" was to save energy and reduce pollution. The main design element of the city had its focus on eco-friendly means of transportation and to decrease the energy consumption of the city houses.

1. The distance between residential and shopping areas is reduced comparably in order to avoid the need to travel. The city is provided with a safe cycling and walking infrastructure in order to popularize "green traveling".
2. Green traveling should be considered as an environment-friendly and effective mode of traveling with less pollution, low-energy consumption and lesser emissions. In order to promote green traveling, several policies are set up for the residents to follow such as usage of non-motorized vehicles and public transport for traveling.
3. The city constantly advocates the concept of "green traveling" in order to promote green traveling awareness among citizens, communities and schools. Also, people are guided to use cars in a reasonable manner in order to reduce pollution and traffic pressure.
4. The car parking shades (roof tops) are fixed with solar panels. All the cars are parked at one place under the solar panels. The car parking cluster is away from the residential villas so as to provide children with free space and at the same time solar energy is collected for utilization.
5. Fuel standards are intensively regulated in the city and the citizens are motivated to shift to alternative fuels and smart vehicles.
6. The bus networks outside the city have been scientifically optimized and intelligent information systems have been set up to control the flow of traffic.
7. Mobility inside, as well as outside, the city is transformed by the construction of roads and bicycle tracks, but apart from these, two very

innovative steps have been taken by the development authorities. First, charging stations have been set up at different places in the city. Second, an incentive has been set up to provide subsidy to the owners of the villas who would like to purchase electric vehicles.

8. Apart from public transport, rail transport, cycling and walking, there are other vehicles that come under green transportation such as electric vehicles, natural gas vehicles and solar energy-based vehicles. Electric vehicles also include tram cars, the subway, light rail and trolley buses. People are encouraged to use them as a means of transport extensively.

9.4.2 Greenest City Draft Action Plan, City of Vancouver

The city has a goal to be the greenest city by the year 2020. Ten long-term goals have been set for this 2020 target. A draft action plan (DAP) has been set up to meet the proposed targets. Out of the ten goals, number four is green transportation. This goal emphasizes walking, cycling and use of public transport as the preferred choice of transport. This will not only handle the excessive vehicle use but also reduce the heavy release of greenhouse gas emissions. It would also improve the quality of air, thus making living healthier.

1. every year the city will try to reduce emissions by 20% from the benchmark set in the year 2007;
2. the first strategy to reach such targets is the updating of the current transport plan and to create a new active transportation plan for the city which would include everything "without engines";
3. the DAP aims to create people-friendly streets and public spaces that make it convenient and enjoyable to witness the city on foot, with the help from the government. This would promote fast, accessible, reliable and comfortable transport options;
4. other features include the bike chain program, electric vehicle infrastructure, parking policy changes and so on. Also, facilities like childcare, hospitals, shops and so on would be closer to residential areas of the city.

9.4.3 Intelligent Transport Systems, Korea

The roads have become more crowded and the number of cars has increased as society develops. There are many accident-prone areas on the roads, but the ITS, a smart transportation system developed in Korea, helps to tackle the problems or dangers which arise. ITS is a traffic system that combines many transport systems and traffic movement facilities with high technology and

Green Smart Transport Systems

information to increase the efficiency and safety of the roads in Korea. ITS manages and operates everything scientifically. Some key features of ITS are:

1. ITS was introduced with the sole purpose to increase the traffic movement efficiency;
2. Korea has applied ITS in 100% of the expressways, 21% of the highways and 11% of the local government roads;
3. observations after ITS application have been very motivating. The average speed has increased by 15%–20%, a 20% reduction in traffic jams and the cost of 1% of road construction expenses;
4. fuel consumption and hazardous gas emissions have also decreased;
5. various industries have been started based on the concept of sharing of information. ITS applies to several areas such as monitoring of traffic flow, payments, speed control, enforcement and parking;
6. the bus information system (BIS) collects the information related to the location of buses and transmits that information to the traffic information center and hence makes it possible for people to check this information on their mobiles at bus stops in real-time. Information includes bus routes, arrival time, bus intervals, last arrival time and so on. This increases user satisfaction and utilization of public transport.
7. the automatic fare collection (AFC) is also a main part of the ITS. All the payment systems of the public vehicles are integrated and managed. The money collected is sent to a traffic integration card company. The profits are divided into respective public transport companies. The public enjoys an efficient and convenient transport service. Korea runs a "one card all pass" system that can be used for all public transport systems which includes buses, subways, taxis and so on. A single card is required for using any of these systems. All this leads to efficient profit management.
8. the Freeway Traffic Management System (FTMS) is again a very important part of the ITS. It collects various information from the expressways. Various observation devices like closed-circuit television (CCTV), dedicated short range communication (DSRC) and loop-type vehicle detection systems (VDS) are used for collecting and processing information. The traffic information center processes all the information. And then this information is distributed through electronic displays, smartphones, broadcasting, Internet and so on.
9. the electronic toll collection system (ETCS) is a system in which the terminal presents inside the car and the transponders present on the roadside use radio communications in order to pay the toll money without the vehicle having to stop. This not only accelerates the payment process, but also reduces the jams at the toll gates. Usage of the Hipass System (Korea's ETCS) is about 71% in South Korea and the reduction in CO_2 is equivalent to planting 2,000,000 trees per year.

10. the automatic traffic enforcement system (ATES) uses sensors and cameras on the road to sense cars breaking the law and then send the related information to the traffic information center. The information includes reserved lanes regulation data, speed violation regulation data, signal violation regulation data and illegal parking regulation data. This is followed by sending the information to the police and then automatic dispatching of bills for cars breaking the law.
11. the parking information system (PIS) uses various sensors to seek out parking situations in real-time and send the required information to the information center. The center then reprocesses the data received and distributes the traffic information to users.
12. the National Traffic Information Center (NTIC) provides all road-related data. Korea has 66 traffic information centers which find all the information related to roads. Then, all the information is collected and operated through integrated management. It is reprocessed in order to assist government organizations, private enterprises and people to use the roads more efficiently.
13. the Cooperative ITS (C-ITS) is a future initiative in Korea. The usage of the C-ITS system would update drivers in real-time about the dangers on the road such as unexpected stops, fallen objects and so on. This would lead to a huge decrease in traffic accidents and also a decrease in social cost. Apart from these, autonomous vehicles can be linked to the infrastructure of the road and a Cooperative Automated Driving Highway System (C-AHS) project which aids safe and easy driving.
14. Korea has become a leading country with good transport systems due to ITS implementation. Korea shares its ITS establishment with several other nations. ITS technology has been exported to 37 nations since 2006. ITS traffic management was successfully adopted by Baku, the capital of Azerbaijan, in 2008. The bus information and traffic card system in Columbia's Bogotá city was established in 2011.

9.5 Future Work

Green transport is a major sector of a green economy. Green smart transport supports the sustainability of the environment which includes protection of natural resources, ecosystems and global climate. Hence technological advancements in this field will reach new heights in the near future. Green smart transportation will dismantle the current existing transportation systems and will lead to a safer future. Some future prospects are mentioned in the following sections.

9.5.1 Sustainable Transport Development in Smart Cities

Sustainable transport development is an amalgamation of several tasks and techniques adopted to develop green smart transport vehicles which are eco-friendly, sustainable and efficient. The three basic terms followed for sustainable transport development are Avoid, Shift and Improve.

1. *Avoiding* the need to travel by reducing distances between the residential and working offices. This is done by integration of land use planning and transport with intelligent logistics concepts.
2. *Shifting* to eco-friendlier means of transport by providing cycling and walking infrastructure along with good quality public transport. Parking policies which are part of Transport Demand Management should be taken care of.
3. *Improving* the fuel standards and the flow of transport systems in order to improve the overall efficacy of transportation systems must be mandated.

9.5.1.1 Smart Flexibility

There is a need for changes in the behavior of people in cities and urban areas for a sustainable change in the transportation system. An efficient system should be developed that monitors different modes and organize transport that is, use alternate and eco-friendlier fuels. New opportunity should be created for collective mobility. These solutions help to decrease the impact on the environment. The most intellectual systems are the transport system in the smart city.

To maintain safety and sustainability, there are three ways:

- smart users
- smart vehicles
- smart road infrastructure (Makarova 2017).

9.5.2 AI Machine-learning Algorithms for Smart Green Transportation

The decision system for smart green transportation should be developed for the traveler to select the best means of transport from their journey. The route is calculated to reach the desired destination at a given time. Various significant parameters for the selection of route and transport can be made like CO_2 emissions, ticket price, duration of travel, waiting time, connection time to catch transport, time taken by different transportation to reach the destination, and reviews, comfortability feedback. Machine-learning algorithms such as reward-based reinforcement learning, Deep Learning's Q-Learning algorithm, can be applied for validation and then the prediction of the transport is done by using learning techniques such as Support Vector Machine (SVM) (Said 2017).

9.5.3 Hyperloop Technology

The concept of hyperloop technology was invented by Elon Musk. Their work (Musk, 2013) shows a system which can transport cargo and people at a speed not seen before. It is a system which uses electricity generated from solar power and air. The air cushion consumes energy which is approximately 21 megawatts (MW). Solar-powered batteries generate this power which is located on the outer tube's surface. The total energy produced by the cells is 57 MW, that is, it covers all the energy needed (Dudnikov 2017).

Conclusion

The twenty-first century has witnessed progress in almost every field of science from modern marvels to greener technologies. In this due course of development, 50% of forests have been cleared, several species killed and a huge amount of stored fossil energy has been burnt. The greatest impact of this is a steady increase in the temperatures of the earth due to the greenhouse gas emissions. This is the biggest challenge that is to be dealt with by the usage of green smart technologies now. For social and economic development, transportation is one of the most important sector. It is one of the biggest consumers of fossil fuel products. It is one of the main reasons for the existing harmful emissions including CO_2 and other greenhouse gases. People around the globe have realized the urgency to reduce these emissions and pollutants. Communities around the globe have been activated and motivated in a common desire for a "green transportation" paradigm, to protect and secure not only the creatures, but also the planet as a whole. Engineers, doctors, law professionals, researchers and many other fields of knowledge are working together for a better, greener and smarter transportation system and other green smart techniques.

References

Air Pollution Killing 620,000 Indians Every Year: Global Burden of Disease Report. (n.d.). Accessed February 20, 2019: www.downtoearth.org.in/news/air-pollution-killing-620000-indians-every-year-global-burden-of-disease-report--40316.

Besseso. (2018, July 25). Smart, Green and Integrated Transport. Retrieved from: http://ec.europa.eu/programmes/horizon2020/en/h2020-section/smart-green-and-integrated-transport.

Björklund, M. 2011. Influence from the Business Environment on Environmental Purchasing – Drivers and Hinders of Purchasing Green Transportation Services. *Journal of Purchasing and Supply Management* 17(1): 11–22. Retrieved from: https://doi.org/10.1016/j.pursup.2010.04.002.

Cervigni, R., Losos, A. M., Neumann, J. L. and Chinowsky, P. 2016. *Enhancing the Climate Resilience of Africa's Infrastructure: The Roads and Bridges Sector English.* Washington, DC: World Bank Group.

Consumer Reports Revises Financial Analysis in Report on Ownership Costs for Hybrid Cars. Consumer Reports. Consumers Union. 2006. Accessed February 22, 2019: www.consumerreports.org/cro/2012/01/hybrids-diesels-do-they-save-money/index.htm.

Dudnikov, E. E. 2017. Advantages of a New Hyperloop Transport Technology. *Tenth International Conference Management of Large-Scale System Development (MLSD)*. Moscow, 1–4. doi: 10.1109/MLSD.2017.8109613.

Elon, M. 2013. Hyperloop Alpha, *SpaceX*. Accessed February 22, 2018: www.spacex.com/sites/spacex/files/hyperloop_alpha.pdf.

Europa Analytics. (March 15) 2017. What is Horizon 2020? Retrieved from: http://ec.europa.eu/programmes/horizon2020/en/what-horizon-2020.

Kilian, L. 2009. Not All Oil Price Shocks Are Alike: Disentangling Demand and Supply Shocks in the Crude Oil Market. *American Economic Review* 99(3): 1053–1069. doi:10.1257/aer.99.3.1053.

Kumar, N. M., Goel, S. and Mallick, P. K. 2018. Smart Cities in India: Features, Policies, Current Status, and Challenges. Proceeding of *IEEE International Conference on Technologies for Smart-City Energy Security and Power ICSESP*.

Leipzig, International Transport Forum (ITF). 2010. Reducing Transport Greenhouse Gas Emissions: Trends & Data., Germany: Background for the 2010 International Transport Forum.

Li, H. 2016. Study on Green Transportation System of International Metropolises. *Procedia Engineering* 137: 762–771.

Muda, N. and Jin, T. 2012. On Prediction of Depreciation Time of Fossil Fuel in Malaysia. *Journal of Mathematics and Statistics* 8(1): 136–143. Retrieved from: https://doi.org/10.3844/jmssp.2012.136.143.

Organization of the Petroleum Exporting Countries. 2017. OPEC World Oil Outlook.

Panday, A. and Bansal, H. 2014. Green Transportation: Need, Technology and Challenges. *International Journal of Global Energy* 37(5/6): 304.

Said, A. M., Abd-Elrahman, E. and Afifi, H. 2017. A Comparative Study on Machine Learning Algorithms for Green Context-aware Intelligent Transportation Systems. *International Conference on Electrical and Computing Technologies and Applications ICECTA*.

Su, K., Li, J. and Fu, H. 2011. Smart City and the Applications. *International Conference on Electronics, Communications and Control ICECC*. Retrieved from: https://doi.org/10.1109/icecc.2011.6066743.

US Energy Information Administration - EIA - Independent Statistics and Analysis. n.d. Retrieved from: www.eia.gov/.

World Health Organization 2016. Ambient Air Pollution: A global Assessment of Exposure and Burden of Disease.

10
Green Smart Energy Management System

Garima Singh and Gurjit Kaur

CONTENTS

10.1 Introduction 204
10.2 Features of the Green Smart Energy Management System 206
10.3 Technologies to Design Green Smart Energy Management
 System 207
    10.3.1 Sensing Infra 208
    10.3.2 Information Extractor 208
    10.3.3 Data Mining 208
10.4 Green Smart Metering for the Green Smart Energy Management
 System 208
    10.4.1 Embedded Solutions 211
    10.4.2 Networking Solutions, Routers and Gateways 211
    10.4.3 Connectivity and Cloud Services 211
10.5 Supply-side Solutions for Green Smart Energy Management
 System 211
    10.5.1 Upgrading the Gas-fired Combined Cycle Power Plants 211
    10.5.2 Efficient Distribution and Transmission of Electricity 212
    10.5.3 Capture Waste Heat 212
    10.5.4 Dynamic Price for All Consumers 212
10.6 Cloud-based Demand-side Solutions for a Green Smart Energy
 Management System 213
10.7 Energy Efficiency in Buildings and Districts for Green Smart
 Cities 214
    10.7.1 Increase in Production of Renewable Energy and
            Self-Consumption 214
    10.7.2 Including Building Stock with Demand Response
            Capacity 215
    10.7.3 Smart Interconnected Districts 215
    10.7.4 Zero-energy Smart Buildings 216
10.8 Renewable and Distributed Energies for Green Smart Cities 216
10.9 ICT as Enabling Technology for Green Smart Cities 217
Conclusion 217
References 219

10.1 Introduction

With the present exceptional increment in energy demand, it is essential to design systems which require less energy. Solar, hydroelectricity and wind are well-accepted choices, as they are abundantly available and cause low pollution. However, one major challenge comes from the handling requirements of upcoming systems within the existing traditional electric grids.

Green smart energy management is the best paradigm supporting sustainable development. Through monitoring and controlling energy production plants and the energy consumption, smart management of energy flow can be achieved. To make an energy-efficient energy management system, the consumers, production centers and electrical devices must communicate with each other. Thus, in this way the consumption values can be adjusted effectively with the energy production available. For the smart management of energy, not only consumers' energy needs to be reduced, but optimization of all energy sources (renewable and power plants) is also required. It is therefore required that energy should be managed at every single moment while fulfilling the tasks scheduled on request by the user. The "green smart energy management system" in general has a software module which is used to perform energy management algorithms to optimize the energy and hardware modules. The whole system controls the physical implementation of the management policies determined by the software module. The software module is an online simulation-based management system which makes decisions on the basis of different tasks assigned to it. It also possesses the ability to handle the energy resources in order to attain optimized system performance and efficiency. A hardware module is used to implement the physical management required for the optimization decisions of software modules. To give optimized performance and solutions, the software module is required to have access to the user load defining environmental constraints such as the energy cost of loads, the maximum green energy available, for example solar power, and the power limit available from the grid, execution sequences and other relevant information.

There are many open challenges to be explored to achieve the aim of "green smart energy management". Regardless of whether world pioneers make any global climate policies, many agencies have already started their preliminary work which set the rules for the regulation of energy in a smart way. The energy sector can give sound answers to take care of serious global energy wastage problems. These proactive actions taken by industrial consumers and power providers will help improve corporate reputation capital gain significantly. In order to estimate the possible environmental effects of the upcoming smart cities, there is a need to define the statistics of energy requirement. Further, each geographical region has different resources for electricity generation and different rates of CO_2 emissions which are operated

Green Smart Energy Management System

at different times of the day. As a result, the nature and timing of electricity consumption in the smart cities will have a considerable influence on carbon emission reduction. Annual CO_2 emissions in the power sector can be reduced by 5% through conservative approaches in smart cities by 2030, which can be extended further up to 16% through a broader range of technologies by 2030 (Hledik 2009). In the same view, most countries around the world are working continuously to adopt these advanced energy management technologies. Countries like Italy, a pioneer in this field, have already achieved complete deployment in this area. Further, the US and Canada also have the highest rate of deployment in progress right now. In India, Gujarat is the only state which has adopted the concept of smart cities that a require green energy management system. As shown in Figure 10.1, green energy

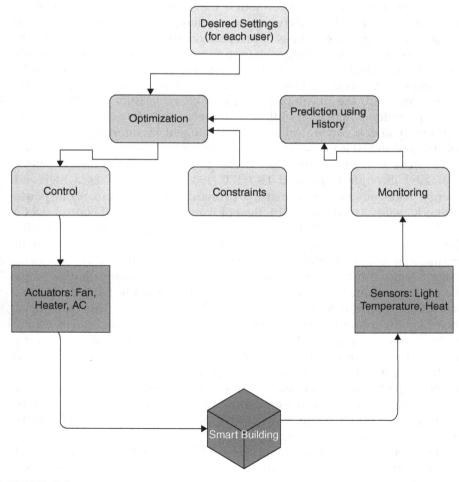

FIGURE 10.1
Smart energy management system.

management is an essential perspective for the interactions and handling of energy resources, and the smart grid is one of the best examples of that.

10.2 Features of the Green Smart Energy Management System

The key objective of this chapter is to list and explain the efficient use of energy in smart cities and offer friendly atmospheres in carrying out the day-to-day tasks of the inhabitants. The exceptional growth in information and communications technologies (ICTs) has turned the page for green smart energy management in smart cities. And it also has been made ready for the promising advanced smart infrastructures, especially for monitoring real-time energy usage. Consequently, the service suppliers have to maintain two-way communication among end-users and need to evaluate the aspects of real-time data consumption, though they will encourage customers to adjust their consumption behavior by controlling different demand response programs. Through these types of program minimization of energy cost per user can be done by providing each user with an energy consumption scheduling device which will automatically control their loads. A household's consumption behavior depends on many parameters, for example, s the appliances used, the developing envelope characteristics and the service utility provider. Calculation of these parameters therefore needs to be done thoroughly and accurately, along with intelligent monitoring to achieve a reliable and robust green smart energy management system.

The work flow of the green smart energy management system is briefly explained through a standard ISO/IEC 15067-3 which gives a high-end energy management model focusing primarily on three demand response methods real-time pricing (RTP), time of use and direct load control (DLC) (Mell and Grance 2011). A direct load control program is mainly planned for the consumers who generally have low energy consumption, such as small commercial and residential users who have appliances like air-conditioners, water heaters and pool pumps which can be further remotely shut down by the service provider. Next, RTP is one of the most efficient DRPs which sets cost according to the devices used which help the smart appliances to adjust their load according to energy price signals. In the RTP program, extensive energy market changes will be indicated along with the prices of energy that fluctuate on an hourly basis or a day ahead. For instance, to adjust the load, the manufacturer will insert a program in the appliances on the basis of energy price, such as air-conditioners in which energy price is controlled through their operation and temperature set point. In this case, the mode of communication is either through smart appliances directly or a wide utility area network to home area networks.

The green energy management agent (EMA) is another system defined by the ISO/IEC 15067-3 standard which is vital for the smart energy management system as shown in Figure 10.1. Specialized computing functions are executed by EMA on the basis of the electricity data rate received from

Green Smart Energy Management System 207

the software algorithms which are useful in checking when and which distributed energy resources (DERs) and appliances can be operated.

The main characteristics are given below:

1. by what means and what time an appliance will operate can be determined using this agent;
2. the energy cost is acknowledged by it;
3. EMA operation considers distributed energy resources and inputs given by consumer;
4. EMA controls the increase or decrease in power consumption of appliances and switches them on or off accordingly;
5. EMA should have the capacity to obtain every hour or every day's data price in advance:
6. EMA has a facility through which consumers can override its scheduling anytime;
7. the signals to appliances are sent through home network in EMA;
8. the energy signal is received from the utility in EMA and usage data is sent back to the utility;
9. to maintain consumer privacy and security, the data stream sent from consumer to utility is encrypted on the gateway side, while the data stream which is published publicly does not need encryption;
10. EMA software is bit difficult for matching comfort level and budget;
11. the switching on and off of the many circuits is controlled by EMA to smartly control the demand of appliances like air-conditioners and fridge;
12. the appliances should have an indicator like a lamp, which will be used to alert the consumer about an appliance which has issues in operation;
13. a display on the appliances or in the home is required which show the alert messages for the cost consequences of overriding consumptions.

The development of ICTs in wired networks, home area networks (HANs) and wireless sensor networks (WSNs) have made HEMS able to gather information related to those parameters which can enhance energy management processing.

10.3 Technologies to Design Green Smart Energy Management System

Many types of technologies are required on the basis of which green smart energy management system is designed. All these technologies require gathering of data from the sensors which are then stored in the cloud for access on requirement. After that different types of technologies need to be designed to process and execute the sensed data.

10.3.1 Sensing Infra

Smart energy management system for smart homes generally requires a sensing infrastructure to request and receive the physically sensed data from the installed sensors associated with the power-consuming devices.

This module handles the sensor nodes of the active sensor network. Sensing data collectors sum up the information concerning each and every sensed data like location or activity. This sensing infrastructure is used to give uniform interface to modules which use this sensed data, used to hide the details of the underlying context-sensing mechanisms and to treat implicit and explicit input in the same method. One needs a sensor bed to sense the complete information because in a dynamic and open atmosphere, no single agent is capable of having complete knowledge.

10.3.2 Information Extractor

After sensing all the data, an information extractor is needed to get the useful informative projections out of the required gathered data. Such as, location context is provided in latitude and longitude form by the context aggregator, which generally needs to be extracted in the form of a street name for an application. For this, a feature extractor will be required to do this abstraction, which will display the information in a user-friendly way to make the application friendly system.

10.3.3 Data Mining

Data mining is required to provide contextual databases for the data generated by the sensors. In order to communicate, a common communication language and ontology must be shared by the independently developed agents. However, the use of a shared database repository has been suggested by some approaches to facilitate knowledge sharing. As the different agents are used to update the database, it will eventually become a shared knowledge base for these agents.

10.4 Green Smart Metering for the Green Smart Energy Management System

To improve the smart energy management in smart cities information technology works as the backbone that facilitates extensive invasion of

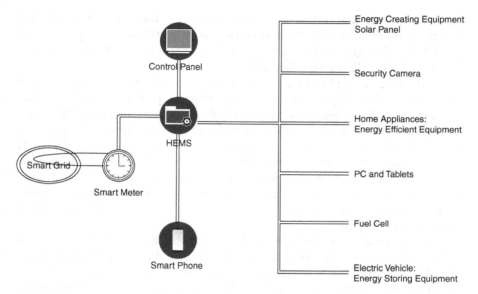

FIGURE 10.2
Green smart metering for the green smart energy management system.

new technologies. These new emerging technologies consist of advanced improvements in transmission, metering, electricity storage and distribution, in addition to providing flexibility and new information to both providers and consumers of electricity as shown in Figure 10.2.

Eventually, the services and products that are offered to consumers can be enhanced by getting this required information, which leads to more proficient consumption of resources and energy, which then leads to electricity management by an advanced metering infrastructure (AMI). To have a mutual communication between electricity providers and meter, the smart meters are used to accompany communications setup. This will enable the providers to have the real-time access of electricity consumption information for every customer. To better understand and control the energy system, end-users need to be enabled with smart meters to measure the energy consumption accurately and provide real-time energy data to demand response services and this will reduce the energy consumption between the buildings and their occupants). Smart control of household appliances will allow building users to regulate their energy use according to their preferences, requirement of load and price signals which results in more efficient use of energy.

To make buildings energy-efficient, building owners and tenants should have permission to install smart meters and regulate the whole system with a universal communication protocol which will allow interoperability between appliances and systems.

Implementing new technologies should also be done in commercial and tertiary buildings to effectively save the energy: 80% of electricity meters need to be changed with smart meters by 2020 for cost-effective energy-saving (UNEP 2015; Project Zero 2016).

Smart metering technology is the basis of every green energy-efficient smart city because it provides the energy-efficient platform to offer digital-era services to the consumers. With the help of this detailed real-time information the customers can be offered new, innovative rate forms by the service providers, commonly known as dynamic pricing. To reduce the electricity cost, dynamic pricing is a much better option than the equal price for each unit scheme. Dynamic pricing will allow customers to pay less during off-peak hours and higher during peak hours. This in turn helps the customers to reduce their electricity bills by shifting from higher-priced consumption hours to lower ones.

To widespread green and energy-efficient applications the smart metering and communication methods play a key role in reducing energy consumption. To proficiently carry out demand-side management and control services the home energy management is the best platform for users. It has home appliances connected all together with a grid of an interconnected network of devices placing user convenience as the first priority. Energy consumption of types of community houses can be managed through this platform. Many utilities manufacturers like water and more are exploring the use of smart metering because through AMI one can detect leaks and inefficiencies which will lower operating costs, and will improve water conservation. Cellular technology has an important position in the development of the smart home management system because governments and utilities want reliable and durable solutions which can be implemented cost-effectively and quickly.

Consumers' energy usage can be controlled through smart metering which will reduce carbon emission and make it a smart green energy management system which also saves money and reduces utility costs, thus making customer satisfaction better. Because of the visibility in power consumption, utilities can shift demand loads which will optimize energy distribution and make it green energy-friendly.

Smart metering helps utilities to:

- manage manual operations remotely which reduces operating expenses;
- enhance streamline power consumption and forecasting;
- segmentation and profiling to make customer service better;
- trim down the theft of energy;
- follow renewable energy power and simplify micro-generation monitoring.

There are a number of technologies which are used in smart metering-based applications in smart cities (see the following examples):.

10.4.1 Embedded Solutions

Embedded SIM (Subscriber Identification Module) has low power consumption which is highly reliable and cost-effective. They are simple to program, easy to integrate, highly secure and can be easily lifted up from 2G to 4G. Legato is their open-source platform which is used to speed up IoT applications development for smart metering and other applications of the smart city (Hledik 2009; Kedia et al. 2018).

10.4.2 Networking Solutions, Routers and Gateways

Routers and gateways are used for rugged outdoor environments where security and reliability are important. They are used to provide constant connectivity and services based on location. They are used in remote monitoring also to make the most demanding smart metering applications simple.

10.4.3 Connectivity and Cloud Services

Cloud services are the best solution to securely integrate retrieved data to develop new cloud services in the application of smart metering. They are used to lower maintenance and deployment costs through continuous monitoring. Cloud services are also used to accelerate the time-to-market.

So, by designing smart meters one can manage the use of energy which in turn will save on energy, making it a green smart management system.

10.5 Supply-side Solutions for Green Smart Energy Management System

By concentrating on users' energy consumption habits and choices, many substantial claims and benefits can be made. There are lots of supply-side opportunities which are a big concern for power plant managers that need to be implemented readily. Because the number of consumers is higher than power providers, solutions need to be found, settled and deployed in an effective manner. Some supply-side solutions that can be used are listed here:

10.5.1 Upgrading the Gas-fired Combined Cycle Power Plants

Existing gas-fired combined cycle power plants can be upgraded to achieve considerable benefits because they are the major power sources in Asia. There exist many retrofit cost-effective solutions for these turbines but regulators and governments, backed by consumer demand, have to produce environmentally as well as economically friendly technology. Retrofitting of these gas-fired plants in small numbers can create great results in terms of operational efficiency. To gain similar effects on the demand side, nearly 5 million households need to be encouraged to take the most out of the efficient and commercially available green technologies, regardless of cost and expenditure.

10.5.2 Efficient Distribution and Transmission of Electricity

To enhance supply-side efficiency, additional opportunities are offered by electric grids. There is always a considerable loss associated with the distribution and transmission of power. Smart grid applications play a major role in such situations by providing solutions that can stop unnecessary energy loss and wastage of money. So, actual energy monitoring technologies are required which in turn helps in minimizing the overcapacity of power generators to significant levels.

10.5.3 Capture Waste Heat

Heat is naturally created in power generation. Whenever an electrical appliance is plugged, it usually gets warm. An overall efficiency of 70% can be achieved by using the heat produced in generating the electricity combined as heat and power. Through encapsulating the byproduct of heat obtained from household and industrial applications, one can achieve strategies and goals for green smart energy management.

10.5.4 Dynamic Price for All Consumers

To encourage customer-friendly smart energy, new technologies are required to give time-varying rates. Though a flat electricity rate is always considered easy to implement, customers are expected to adapt to this dynamic pricing scheme easily as they will pay less if they use power in off-peak hours. Offering cheaper rates during off-peak hours will encourage customers to cut down unnecessary energy demand during peak hours. This in turn will reduce the demand during more expensive periods which will prove advantageous for the users and power suppliers as well.

To manage energy requirement it is important to upgrade the supply-side infrastructure with the latest and upcoming technologies. This will reduce waste and increase the efficiency of the systems, pushing it towards the goal of achieving a unified green energy management system.

10.6 Cloud-based Demand-side Solutions for a Green Smart Energy Management System

Demand-side management (DSM) is an important concept that in a typical smart city. DSM programs enable consumers to partake in electric grid operations by decreasing or moving their peak hour power usage which ultimately helps in achieving proficient utilization of energy. To create IoE (Internet of Everything) one needs to combine cloud computing and IoT to enable ubiquitous sensing services which will gather the sensed data to draw useful projections for smart monitoring. Further, an increase in the number of IoE devices will increase the response time and latency in cloud computing. Fog computing is the solution to conquer the issues related to cloud computing. This will allow the data to be preprocessed and leads to an increase in interoperability, scalability, consistency and connection between the smart devices.

The essential characteristics and advantages of using a cloud computing model lie particularly in the elastic nature of storage and memory devices of cloud computing (Kim et al. 2011, Yang et al. 2012; Li et al. 2012; Tomar et al. 2017). One of its peculiar characteristics comes from its flexibility to reduce and expand itself in accordance with demand from the users. It also offers a metering infrastructure to customers through which cost optimization mechanisms are made available for users where they will pay for consumed resources only. Cloud computing is therefore generally considered as an emerging green technology computation model that is used to share resources over the Internet and gives on-demand computing facilities (Fang, Yang and Xue 2013; Krishna et al. 2013). There are numerous cloud-based two-tier DSMs to manage the residential load of customers, outfitted with local power generation and storage facilities as supplementary sources of energy. To find out the possible schedule for power consumption for customers an edge cloud is used. A typical edge cloud is facilitated with a bi-level optimization approach which is used by local storage to schedule the power consumption levels. It is also used to optimize the amount of power being demanded from both the power grid and local storage facilities. Due to the same reasons, different regional demand from consumers is gathered in the core cloud. Through reducing the overall demand of electricity, demand-side management can be implemented with proper network scheduling mechanisms.

However, in traditional networks, programs such as load shedding are often resolved through predefined factors which change for some defined operational periods.

10.7 Energy Efficiency in Buildings and Districts for Green Smart Cities

Demand flexibility measures and energy efficiency are fully complementary. Compared to previous years, energy demand will reduce in the coming years through energy renovation of the existing building stocks (BPIE 2011). This brings numerous benefits to demand-side problems that will substantially increase the mixing of volatile renewable energy on the supply-side (Kylili and Fokaides 2015). The end-user thus will be guided by an energy-efficient building which will fulfill the heating or cooling demand of the user as a well-designed and proficient building keeps the required indoor temperature better for a longer period. All this will eventually enable them to shift the energy consumption to another time.

Therefore, with proper compliance, implementation and control mechanisms the concept of nZEB (nearly-zero energy buildings) can be achieved. This will also require major renovations with (nearly-zero) energy performance. The construction sector therefore needs to be transformed to accomplish an energy-proficient building stock which results in new service-driven business models concentrating on the needs of the consumers.

10.7.1 Increase in Production of Renewable Energy and Self-Consumption

With rapid technological growth, the EU (European Union) is incorporating small-scale renewable energy systems which are driving new buildings towards a nearly-zero energy level. Technologies installed on buildings like biomass boilers, heat pumps, solar and photovoltaic thermal panels are therefore finding their place in the mainstream.

In spite of provisions on grid parity, the energy performance of buildings and the renewable energy directives have stimulated on-site renewable energy systems (Recast 2010). Self-consumption and on-site renewable energy production are still not used to their maximum potential. However, despite their useful merits green energy which is produced naturally is still not accepted or encouraged by all member states. Due to this, renewable electricity is generally seen to injected to the public network instead of using it at local levels.

To mitigate the stress of the system and for the end-user, empowerment policymakers have to promote production and self-consumption of renewable energy to most areas within a smart city. Therefore, it is essential for the consumers to generate renewable energy sources in their respective household and business and others have to generate their own energy in spite of buying it from the grid: this will save them time and money too. Instead of limiting the energy consumption of end-users, it encourages them to better control and generate their own energy systems, which in turn increases much-needed grid security levels. Empowering self-consumption will make energy suppliers more innovative, agile and faster in reaching a balanced level of centralized and decentralized production. Additionally, it will also promote societal gains due to reduction in the overall associated energy consumption.

Therefore, it is essential that user-friendly administrative procedures should be there to allow self-consumption and regulations and taxes. For the sake of encouragement, energy performance certificates (EPCs) should be given to buildings which consume their self-produced renewable energy. This in turn will result in utilization of green energy which will lead to the adoption of smart green energy management systems.

10.7.2 Including Building Stock with Demand Response Capacity

Demand response (DR) is the capability to shift energy demand by decreasing peak hour consumption and avoiding grid imbalance. In spite of expanding the infrastructure of a smart city one should try to apply the concept of demand response due to the cost-effective solutions it has to offer. Some steps are therefore needed in this regard which promote the development of new applications giving flexibility to end-users to keep a check on their home appliance status and thermostats and to take control of some applications which enable relevant technologies for demand response through smartphones.

Demand response for the commercial and residential market could thus further be facilitated by adopting green energy management systems (GEMS) and other new technologies like smart thermostats, smart meters and other load control technologies with focus on the end-users' smart devices.

10.7.3 Smart Interconnected Districts

The main construction components of urban areas are buildings which need to be empowered with smart technologies. Interconnecting the buildings within a particular region of smart cities will give real value to the whole society and will boost the economy by enhancing green energy efficiency and will bring elasticity in demand and supply.

The spatial planning of smart cities is also required for a sustainable energy system. Using renewable energy in conventional old systems will prove to be a cost-effective solution of heat and cooling demands in urban areas. Waste heat recovered from industrial processes will be used by smart cities to provide heating to offices and residential buildings. Since modern-day urban areas are complex heterogeneous entities, there is a need to develop integrated and holistic energy plans that will help to realize city-wide energy-related solutions involving, assessing and coordinating the various stakeholders necessary for the implementation. It is therefore important that policymakers should steer green energy management systems towards lowering the effects of urbanization in the whole surrounding environment.

10.7.4 Zero-energy Smart Buildings

According to Deng, Wang and Dai, the consumption of energy in buildings accounts for about 40% of the total energy used worldwide (Deng, Wang and Dai 2013). As a result, future buildings have to produce their daily energy demands themselves to complete their requirement of full sustainability, that is, become zero or nearly-zero energy buildings (ZEBs) (Kolokotsa et al. 2011, Pyloudi, Papantoniou and Kolokotsa 2014). This in turn will help in the reduction of the additional stress load on the pre-existing power infrastructure (Li et al. 2013). Achieving a ZEB generally requires well-balanced and optimized operations between production and consumption (Carlisle, Van Geet and Pless 2013). Information and computer-enabled technologies (ICT) and smart grids therefore represent themselves as the main requirements that can help the whole modern-day smart city to achieve the above-mentioned goal of zero energy (Zhang, Lee and Bhatt 2000). Using ICT for the purpose of green smart energy management in buildings leads to a better understanding and penetration of the term "smart and zero buildings" (Nikolaou et al. 2012).

10.8 Renewable and Distributed Energies for Green Smart Cities

Nowadays, the world is facing an important issue of using fossil fuel to meet increased energy demand. Therefore, there is a vital need to use renewable energy sources to address electricity generation. However, the use of renewable energy sources generally comes up with various modifications and serious implementation in the existing electric grid (Chen et al. 2009; Dispersed Generation and Storgae, 2011). With the use of a smart grid, smart home technologies and time-varying energy pricing models, the concept of smart energy management systems therefore needs to be adopted (Siano, Rigatos and Chen 2012; Garrity 2008). But for distributed widespread renewable energy sources adoption there needs to be a strong enabling environment that can regulate and enforce relevant goals and strategies in a much more effective manner. The smart city therefore represents an efficient and sustainable urban center that gives a high quality of life to its inhabitants with the best possible management of resources. In this scenario, a distributed generation will definitely prove to be an adequate tool to deal with renewable sources demand and provide energy reliability with cost-effective solutions (Varaiya, Wu and Bialek 2011). Solar and wind energy for electricity are the best and most powerful source of energy for smart cities. Smart cities will utilize the renewable form of energy like solar geysers, water purifiers and so on which moves towards a much greener technological advancement as shown in Figure 10.3.

10.9 ICT as Enabling Technology for Green Smart Cities

The expansion in technology has significantly raised energy consumption levels in the last century. In 2012 alone, ICT (information and communications technology) consumed approximately 4.7% of the world's electrical energy, releasing into the atmosphere approximately 1.7% of the total CO_2 emissions (Gelenbe and Caseau 2015). Reducing the energy consumed by software systems via designing and carefully selecting green software is considered a vital step towards the green IT goal. With the introduction of new technologies in ICT the concept of Green Smart Energy is expected to reach the application level. The use of sensors and actuators can efficiently control the whole energy chain in smart cities and will provide application-based solutions for them.

Apart from this, ICT also plays a vital role in the smart metering process that is used to reduce the overall power consumption of a particular system. With ICT utilities helping in connecting meters easily and inexpensively with low latency using 4G LTE, a whole bunch of smart power management functions and requirements are met. ICT provides industry-leading cellular machine-to-machine (M2M) technologies including industrial-grade embedded modules with cloud platforms, long life spans, expert application development assistance and much more.

Using Internet of things (IoT) in ICT provides solutions to energy-saving by motivating the behavioral change of the buildings' occupants as shown in Figure 10.4. These IoT-based systems help end-users rate their total energy consumption together with personalized recommendations for the actions required for energy conservation and load shifting (Tomar et al. 2018).

Due to all these reasons, energy systems are often seen equipped with data sensors installed in the building structures. Moreover, it can also be used by the city authorities for the purpose of monitoring and managing the energy status of city's buildings.

Conclusion

This chapter provides a comprehensive literature survey on renewable energy integration and smart management systems and challenges and

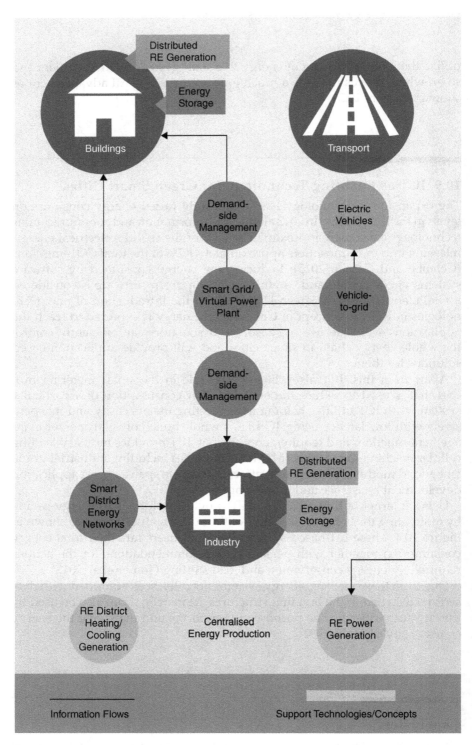

FIGURE 10.3
Renewable and distributed energies for the green smart energy management system.

Green Smart Energy Management System

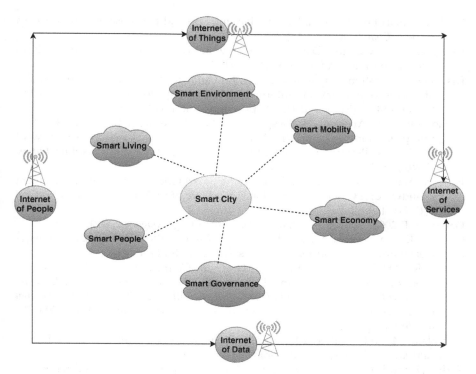

FIGURE 10.4
Components of smart cities.

looks at future research directions in this regard. The proper management of energy is a crucial issue that needs to be addressed to support the economy towards a higher growth trajectory. A holistic approach addressing the issues of energy supply, demand and pricing needs to be undertaken with utmost care in this regard.

References

Carlisle, N., Van Geet, O., and Pless, S. 2009. Definition of a "Zero Net Energy" Community (No. NREL/TP-7A2-46065). National Renewable Energy Lab. (NREL), Golden, CO.

Chen, S. Y., Song, S. F., Li, L. X. and Shen, J. (2009). Survey on Smart Grid Technology. *Power System Technology* 8: 1–7.

Cities, U. D. E. I. 2015. *Unlocking the Potential of Energy Efficiency and Renewable Energy*. United Nations Environment Programme: Washington, DC.

Coalition Smart Energy Deman. 2014. Mapping Demand Response in Europe Today. *Tracking Compliance with Article*, 15.

Deng, S., Wang, R. Z. and Dai, Y. J. 2014. How to Evaluate Performance of Net Zero Energy Building – A Literature Research. *Energy* 71: 1–16.

Dispersed Generation and Storage 2011. IEEE Guide for Smart Grid Interoperability of Energy Technology and Information Technology Operation with the Electric Power System (EPS), End-Use Applications, and Loads (report).

Economidou, M., Atanasiu, B., Despret, C., Maio, J., Nolte, I. and Rapf, O. 2011. Europe's Buildings under the Microscope. A Country-by-country Review of the Energy Performance of Buildings. *Buildings Performance Institute Europe (BPIE)*, 35–36.

Fang, X., Yang, D. and Xue, G. 2013. Evolving Smart Grid Information Management Cloudward: A Cloud Optimization Perspective. *IEEE Transactions on Smart Grid* 4(1): 111–119.

Garrity, T. F. 2008. Getting Smart. *IEEE Power and Energy Magazine*, 6(2): 38–45.

Gelenbe, E. and Caseau, Y. 2015. The Impact of Information Technology on Energy Consumption and Carbon Emissions. *Ubiquity*, (June), 1.

Hledik, R. 2009. How Green is the Smart Grid? *The Electricity Journal*, 22(3): 29–41.

Hongseok, Kim, Y. J., Yang, K. and Thottan, M. 2011. *Cloud-based Demand Response for Smart Grid: Architecture and Distributed Algorithms.* In 2011 IEEE International Conference on Smart Grid Communications (SmartGridComm) . IEEE, 398–403 (conference paper).

Kapsalaki, M. and Leal, V. 2011. Recent Progress on Net Zero Energy Buildings. *Advances in Building Energy Research* 5(1): 129–162.

Kedia, D. and Kaur, G. 2018. Examining Different Applications of Cloud-Based IoT. In Examining Cloud Computing Technologies Through the Internet of Things, 125–146. IGI Global: Pennsylvania.

Kolokotsa, D. E. K. D., Rovas, D., Kosmatopoulos, E. and Kalaitzakis, K. 2011. A Roadmap towards Intelligent Net-zero and Positive-energy Buildings. *Solar Energy* 85(12): 3067–3084.

Krishna, P. V., Misra, S., Joshi, D. and Obaidat, M. S. 2013. Learning Automata Based Sentiment Analysis for Recommender System on Cloud. In *2013 International Conference on Computer, Information and Telecommunication Systems (CITS)* (pp. 1–5). IEEE (conference paper).

Kylili, A. and Fokaides, P. A. 2015. European Smart Cities: The Role of Zero Energy Buildings. *Sustainable Cities and Society* 15: 86–95.

Li, D. H., Yang, L. and Lam, J. C. 2013. Zero Energy Buildings and Sustainable Development Implications – A Review. *Energy* 54: 1–10.

Li, X. and Lo, J. C. 2012. Pricing and Peak Aware Scheduling Algorithm for Cloud Computing. In *2012 IEEE PES Innovative Smart Grid Technologies* (ISGT) (pp. 1–7). IEEE (conference paper).

Mell, P. and Grance, T. 2011. *The NIST Definition of Cloud Computing (Draft), in NIST Special Publication.* Gaithersburg, MD: National Institute of Standards and Technology. Retrieved from: http://ojs.jecr.org/jecr/sites/default/files/12_4_p01.pdf.

Nikolaou, T. G., Kolokotsa, D. S., Stavrakakis, G. S. and Skias, I. D. 2012. On the Application of Clustering Techniques for Office Buildings' Energy and Thermal Comfort Classification. *IEEE Transactions on Smart Grid* 3(4): 2196–2210.

Privat, G. 2013. How Information and Communication Technologies Will Shape SmartGrids. *Smart Grids* 263–280.
Project Zero. Retrieved from: www.projectzero.dk/.
Pyloudi, E., Papantoniou, S. and Kolokotsa, D. 2015. Retrofitting an Office Building towards a Net Zero Energy Building. *Advances in Building Energy Research* 9(1): 20–33.
Recast, E. P. B. D. (2010). Directive 2010/31/EU of the European Parliament and of the Council of 19 May 2010 on the Energy Performance of Buildings (recast). *Official Journal of the European Union* 18(06): 943–952.
Santamouris, M. and Kolokotsa, D. 2013. Passive Cooling Dissipation Techniques for Buildings and Other Structures: The State of the Art. *Energy and Buildings* 57: 74–94.
Siano, P., Rigatos, G. and Chen, P. 2012. Strategic Placement of Wind Turbines in Smart Grids. *International Journal of Emerging Electric Power Systems*, 13(2): 1–23.
Tomar, P. and Kaur, G. (eds.) 2017. *Examining Cloud Computing Technologies through the Internet of Things*. IGI Global: Pennsylvania.
Tomar, P., Kaur, G. and Singh, P. 2018. A Prototype of IoT-based Real Time Smart Street Parking System for Smart Cities. In *Internet of Things and Big Data Analytics Toward Next-Generation Intelligence*, 243–263. Springer, Cham.
UNEP. 2015. District Energy in Cities (report).
Varaiya, P. P., Wu, F. F. and Bialek, J. W. 2011. Smart Operation of Smart Grid: Risk-limiting Dispatch. *Proceedings of the IEEE*, 99(1): 40–57.
Yang, C. T., Chen, W. S., Huang, K. L., Liu, J. C., Hsu, W. H. and Hsu, C. H. 2012. Implementation of Smart Power Management and Service System on Cloud Computing. In *9th International Conference on Ubiquitous Intelligence and Computing and 9th International Conference on Autonomic and Trusted Computing*. IEEE, 924–929.
Zhang, P., Li, F. and Bhatt, N. 2010. Next-generation Monitoring, Analysis, and Control for the Future Smart Control Center. *IEEE Transactions on Smart Grid* 1(2): 186–192.

11

Green Smart Waste Management System

Dimpal Tomar, Pradeep Tomar and Gurjit Kaur

CONTENTS

11.1 Introduction: 223
11.2 Features of Smart Waste Management...225
11.3 Technologies to Design Smart Waste Management 226
    11.3.1 Global System for Mobile (GSM) -based Green Smart Waste Management System...226
    11.3.2 Wi-Fi Module-based Green Smart Waste Management System..227
    11.3.3 Radio Frequency Module-based Smart Waste Management System..227
    11.3.4 ZigBee Module-based Smart Waste Management System 227
    11.3.5 IoT-based Smart Waste Management System...........................228
11.4 Smart Bin 228
    11.4.1 Features of Smart Waste Bin/Container......................................229
    11.4.2 Benefits of Using Smart Bins 230
11.5 Technologies for Disposal, Treatment, and Recycling of Solid Waste 230
    11.5.1 Disposal Method 231
    11.5.2 Treatment Methods..232
    11.5.3 Recycling Methods 235
    11.5.4 Challenges...236
Conclusion 237
References...237

11.1 Introduction

Nowadays, with the ever-increasing development rate of smart cities around the world, the world's population has reached up to 50% in urban areas, and it has been expected that by 2050 it likely to stretch up to 70% (Constantino 2014). This inclination of people towards moving into modern-day smart cities leads to the rise of challenging obstacles in the area of health sector, traffic, pollution, resource, and waste management (Chourabi et al., 2014). In today's world, the problem related to management of waste is one that it isn't able to address properly in comparison to the different domains of the smart city, described earlier.

Based on these statistics, the advancement in sustainable solutions for urban life, dealing with waste is one of the important concerns regarding health (Enevo 2016). Besides that, the upsurge of "smart cities" in recent years is responsible for tackling these huge challenges (Buchanan 1992) and (Weber et al. 2008) using several solutions, driven by technology. There are a few examples of the demands of future cities, such as vehicle exhaust emissions, energy-saving, efficient waste management and air pollution and so on. Due to this, productive utilization and management of resources is more essential. Also, urban areas are also facing poor waste management that signifies immense inconvenience for the residents.

In most cities around the world, one usually encounters dustbins that are excessively overflowing. These kinds of situations are neither good for the environment nor for advancement as a whole. In general, these kinds of problem often lead to the spreading of a considerable number of diseases in through insects and mosquitoes that usually breed on waste. A few years ago, the municipality office of the city of Copenhagen received a number of complaints from their residents regarding overflowing dustbins (Lundin and Ozkil 2017).

Urban cities are more vulnerable to this problem as they face exponential and rapid growth in population which leads to more waste disposal in most of their areas. At present, the modern-day smart metro cities in particular are facing significant problems of waste management. One of the leading causes is the rapid urbanization of these metro cities due to a large number of migrants going there for employment, education and so on. This problem is getting bigger with each passing day due to improper and inefficient practices of waste collection and management. This situation has become more deplorable with large accumulation of the unmanaged dumping and landfill sites.

So, the efficient and proper waste disposal is an essential part of the urban infrastructure as well as the emerging smart cities due to the significant impact on environmental protection and public health as a whole. The process of waste management generally includes various activities and tasks like the door-to-door collection of waste, proper transportation, segregation, treatment and proper disposal of the waste material on a regular basis. To ensure proper waste removal the whole process must be carefully monitored and regulated by the relevant civic agencies responsible for the development of the area. In view to the same, all over the globe, the concept of curbside pickup is considered the most popular way of waste disposal where the door-to-door collection of household waste is done by the people appointed for the task with the help of specially designed vehicles regularly. After which the collected waste is carried to a disposal site where it is disposed of properly with machines. In general, there are various factors which are directly responsible for the large accumulation of waste generated throughout the year and its potential impacts on the environment and society as a whole. These factors are usually discovered due to the skew rate of social and industrial development alongside the existing deplorable state of the civic bodies involved in waste management, and associated waste disposal practices, and so on. The main reason behind this is the absence of effective techniques, methods, strategies and policies.

Due to the following reasons, there is a need for the deployment of effective measures and systems that can deracinate or at least minimize this problem to some extent. To be precise, advancement in the technical sector can be used as a means to curb this problem of improper waste management in much more effective way. The green and smart dustbin can be considered a single solution to the specific and peculiar issues in waste management. Therefore, this chapter discusses different ways that can help the authorities to tackle the menace that has been caused by the improper management of waste.

It is well known the important role that IoT will play in the future smart cities technologies (Cambi 2016), where involvement with other auxiliary service like cloud computing and wireless sensor networks will definitely pave the way to tackling the whole problem of improper waste management in much more effective way in the future. Further, the concept of green smart waste management should advance urban areas and satisfy its natives, a perfect and economic condition along with the utilization of smart arrangements (Shyamalan et al. 2016).

It is due to the following reasons that green smart waste management provides a number of advantages over the conventional systems which have been deployed so far for addressing the whole problem. This whole concept generally requires minimal resource consumption in terms of time and fuel which sets it apart from other competitive methods in the long run.

In simple terms, the concept of green waste management in a typical smart city involves the effective routing of waste collecting trucks to different locations of smart bins spread across the city network. The real-time data about these smart bins can provide overflow information to the concerned authorities by using sensors which are fitted to them. Apart from this, as discussed earlier, it will also help in determining and prioritizing the collected waste location which results in less pollution, is cost-saving and provides a better green environment to the local citizens.

11.2 Features of Smart Waste Management

Many emerging developing countries like India have also taken a similar initiative to solve the problem of waste management. India's biggest cleanliness campaign, "Swach Bharat Abhiyaan", aims to achieve the revelation "Clean

India" where a smart waste management system is an essential focus of the whole movement (Buchanan 1992). The main features of this smart waste management are as follows:

- reduction in smart bin filling and cleaning times, letting people make use of clean and hygienic dustbins;
- based on the routing algorithms, the shortest route will be detected that will help in the reduction of the number of vehicles used for the collection of waste by optimizing routes;
- use of such algorithms in smart waste management system which results in a reduction of fuel consumption, thus economically beneficial as well;
- design of smart dustbins which can judge the level of waste inside it and can send the messages to the concerned authority using sensors that are fitted to them.

11.3 Technologies to Design Smart Waste Management

There are many technologies which can help to design a smart waste management system for a typical green smart city. Some of them are listed as follows:

11.3.1 Global System for Mobile (GSM) -based Green Smart Waste Management System

Many researchers around the world are working on developing effective GSM-based smart waste management systems where users can send an alert message to the concerned authority about status of a particular waste bin. In the whole process a microcontroller along with a suitable sensor will be employed to monitor highlight when there is a certain level of waste inside the smart bin. When the sensor detects that the waste has crossed a certain threshold level, a message will be delivered to the nearest waste collecting driver through the microcontroller to empty the trash inside it within a stipulated time frame. Apart from this the location coordinates of the smart bin will be delivered to the truck driver via GPS, which is again done by the same microcontroller fitted with the bin structure. It is only the combination of microcontroller and sensors that makes the whole bin smart enough to raise call for action when needed the most (Ganesh 2017).

11.3.2 Wi-Fi Module-based Green Smart Waste Management System

The ESP8266 is a low-cost wireless fidelity (Wi-Fi) microchip with full transmission control protocol/Internet protocol (TCP/IP) stack along with microcontroller capability (ESP8266). This small module allows microcontrollers to connect to a Wi-Fi network and make simple TCP/IP connections using Hayes-style commands. Through this, one can send the status of waste's bin to the concerned authority system.

11.3.3 Radio Frequency Module-based Smart Waste Management System

A tiny sized electronic device typically used for transmitting and receiving radio-formed signals holds great potential to be used as a part of the whole waste management process. Such embedded systems are often considered quite capable to convey their information wirelessly with the other devices that lie within a particular operational range. There is generally no power dissipation in RF transmitters during the transmission of logic zero that completely suppresses the carrier signal frequency and enables them to operate on low power supply through batteries. But during the transmission of logic "ones" the RF module demands the full power supply. And the transmission of data from the transmitter to the receiver is done serially. The applications of RF modules mainly involve wireless waste management alarm systems, waste remote controls, waste door openers and various smart sensor waste management applications.

11.3.4 ZigBee Module-based Smart Waste Management System

ZigBee-based waste management systems can be considered as one of the most effective methods for monitoring and controlling waste management applications. The main peculiarity behind its usage come from a couple of reasons which encompass low power consumption at much lower bandwidth. Another vital aspect that makes it stand out among its peers is its support to different network configurations for the master-to-slave or master-to-master communications.

To build the wide area network (WAN), ZigBee provides the facility to extend the network and allowing multi-node interconnections with the help of routers. ZigBee architecture in general consists of three essential devices, which can categorized as ZigBee coordinator, router and end device. For smart waste management a few models have been developed that essentially consist three of modules, namely smart waste bins, trucks and a centralized controlling corporation office. The two sensors which are employed for the purpose consist of IR and gas sensors that are fitted within the bin. Here the IR sensor senses the level of waste and the gas sensor is responsible for sensing the presence of toxic gases that may generate with time. The ZigBee placed inside the waste will communicate the overflow to the centralized control system in the municipal corporation office which is located within the relevant proximity. In order to receive the message by the ZigBee, there must be another ZigBee placed at the controlling center in the municipal office which will be connected with the office's system through a serial interface. Only after proper acknowledgment of the waste collection request from the smart bin to the centralized system will the message be forwarded to the respective area truck for proper further action. Further, the ZigBee placed at the truck informs its driver about the exact location where the waste is placed. Further, if the waste is not removed from that location within the stipulated time, the microcontroller placed at the waste bin will again send a second reminder to the municipal office after a certain time interval.

11.3.5 IoT-based Smart Waste Management System

As the development of smart cities is in fashion, IoT technology can be used further for the development of intelligent transportation systems (ITS) and Surveillance systems that are essentially required for efficient and effective collection of waste throughout the city network. The whole IoT is usually composed of cameras, sensors, radio frequency identification (RFIDs) and actuators which are assimilated into surveillance systems and ITS for efficient collection of waste. The whole system based on IoT creates a model that collects sensory data from smart bins, analyzes it at a central monitoring station and shares the request for action with the drivers of waste collection vehicles roaming in the city for waste collection and transportation. The whole system plays an important role in optimizing the waste collection process by utilizing the resources efficiently and results in reduction of processing and operational cost to much lower levels.

11.4 Smart Bin

With the emergence of new technologies such as IoTs, smart sensors, global positioning system (GPS) tracking systems, ultra-high-frequency RFID and mobile web applications, the task of handling waste management processes has become much cleaner and greener, all over the globe. All these technologies have resulted in improvising waste management operations like the collection, processing, and recycling of waste in a much more cost-effective and time-savvy manner. In view of the same, Pune Municipal Corporation recently implemented these technologies in waste management practices for efficient waste collection around the whole city. The main objective behind enforcement of this initiative is to provide the people of the city with a clean and healthy environment that promotes the concept of sustainable development in much more comprehensive way.

There are number of researchers and companies around the world who are continuously working to design smart bin structures. France has started using these smart bins in their smart waste management processes. About 24,000 bins equipped with chips have been tested in the 14 different districts in Paris. All these smart bins contain facilities that can inform the residents by sending a short message service (SMS) alert on their smartphone. A similar concept of smart bins has also been adopted in the city of Brussels, Belgium where bins were equipped with solar-powered batteries along with facility to compress waste on their own.

With the compression techniques within these smart bin structures the capacity of standard bins is enhanced to significant levels. According to *The Baltimore Sun*, the city of Baltimore has allocated $15 million to install 4,500 similar smart bins. Also, in the United States, Pittsburgh spent $854,000 on installing sensors in recycling bins located on the streets. These smart sensors are responsible for transmitting data that informs workers when these smart bins are fully occupied with waste. This ultimately reduces the unnecessary time spent checking empty containers on a regular basis.

11.4.1 Features of Smart Waste Bin/Container

Technological advancement in sensors plays a vital role in the development of smart bins and their real-time remote monitoring. There are a wide variety of sensors that are used specifically for these smart bins, some of which are listed down below:

- ultrasonic sensor-based smart bins: these sensors are used as fill-level sensors. They are mainly responsible for monitoring the fill level of solid or liquid waste in the bin and send real-time data to the central monitoring system for analysis.
- geo tracking sensor-based smart bins provide real-time location of waste bins to waste collection vehicles.
- IR sensor-based smart bins can be used in waste bins to open its lid automatically if the presence of a person is sensed near the bin.
- weight sensor-based smart bins are placed below the waste bin that alerts the system about the fill level of the waste inside the bin structure.
- optical sensor-based smart bins are widely used in automated waste collection because of their ability to be readily equipped in small places.
- proximity sensor-based smart bins are responsible for detecting the closeness of waste material within a certain level without any physical contact with the waste.

- thermal and temperature sensor-based smart sensors are basically used to monitor the inner temperature of a smart waste bin and help in generating a timely alert to the control room if an abrupt rise in temperature is discovered.

11.4.2 Benefits of Using Smart Bins

The main benefits of using smart bins are as follows:

- overflowing dustbins will stop as they are monitored in real-time;
- helps in optimizing the collection route and reduces the number of waste collection vehicles;
- an optimized path can be shared with collection vehicles in real-time thus minimizing the fuel consumption; and
- it saves both money and the environment.

11.5 Technologies for Disposal, Treatment, and Recycling of Solid Waste

The term "solid waste" can typically be described as the accumulation of unwanted materials usually obtained from trade, commercial deeds, household, sewage treatment solids, industrial, agricultural activities and other miscellaneous sources (Sasikumar 2009). With the rapid technological advancement around the world many countries are now facing the menace caused by the large accumulation of such waste on a daily basis. Amongst them the most dangerous waste types are those that don't decompose easily. It is therefore necessary for the concerned authorities of the area to develop smart green waste management systems that can help the whole city to deal with the fiasco caused by these waste. Some of the countries have taken a step forward to deal with this problem in a much more sustainable way. Countries like Greece, the UK and Austria are continuously recycling 10, 17 and 60% waste on a regular basis, respectively.

A similar concept has also been put forward by the European parliament where recycling of 70% of waste associated with construction sites and 50% of waste associated with households is mandatory for every person who has involvement with any of them.

In Germany, the Green Dot program is a testimonial of how citizens of the whole country have been able to dispose of 30 million tons of waste since the inception of the program a few years ago. Based on several factors like waste material types, use of land and available area, waste management methods can also widely vary on a regular basis. Their point of action depends on the range of disposal, treatment and recycling methods of the whole solid waste which is collected from a particular set of area or region.

11.5.1 Disposal Method

For several decades, waste disposal has been considered a vital matter of concern for the whole community that lives across a modern-day smart city. The main reason behind this are the ever-increasing population and pollution levels that hamper the life structure of the whole city in a much more critical way. However, to deal with it, various policies, strategies and schemes have been deployed by the concerned authorities around the world to dispose of waste in a much more effective manner. Amongst these waste disposal methods, two are well known: the landfill and incineration.

11.5.1.1 Landfill

This method of discharge is considered as one of the most common and accepted ways of disposing of solid waste in the world. It includes burial of solid wastes in vacant landfill sites that are usually located away from the city. In particular, the abandoned lands with unused mining pits or borrow voids are considered as the perfect sites for landfill solid waste generated by whole societies living in and around the cities. It has been proved to be the most economical and hygienic means of solid waste management when it is designed and maintained properly. On the other hand, it can also create havoc if it is not appropriately managed as it can become a source of pollution to the atmosphere. If poorly maintained, it can result in wind-blown litter which often invites various vermin such as flies, pests, rodents and other parasites. Apart from this, it also results in the production of leachate and other gaseous byproducts such as methane and carbon dioxide which further affects the environment in an adverse manner similar to the greenhouse effect.

The only way to save the environment from these consequences could be if the whole method is managed properly in a much more sustainable way. In such methods solid waste is usually spread in thin layers of about 10 feet followed by alternate layers of clean soil which is often compacted with the help of machines like bulldozers. These practices in turn help reduce the soil and ground water pollution to a great extent which is quite beneficial for the whole community.

11.5.1.2 Incineration

Another effective way of disposing of waste can be achieved through combustion of the waste in proper chambers. This whole process is generally termed as "incineration" and sometimes these solid waste disposal systems are also called "thermal treatment".

Though several dangerous forms of waste like biological and medical waste are disposed of using this technique, it is quite useful for all other solid, gaseous as well as unwanted liquid substances as well, mainly due to its ability to cut down the eye-opening size of waste that expands during the whole process. However, this technique is contentious concerning the environment due to the release of a various gas pollutants that are quite harmful for the people who work in and around it.

Despite its demerits, one of its peculiarities comes from the fact that the heat that is generated during the combustion is often employed as an effective source of energy to burn more such waste on a regular basis. This method is quite common where limited area is available for disposal.

11.5.2 Treatment Methods

There are some important methods for treating waste, each of which is described as follows:

11.5.2.1 Composting

When organic material present inside the waste is decomposed and recycled as a useful constituent into the soil, the natural process is termed as composting. The whole process of composting is a simple procedure which consist of three layers in general. Out of these three layers the first layer consists of waste, oxygen and heat, and the second layer involves the process of microbiological decomposition, whilst the third layer consists of the compost, water, carbon dioxide and heat (Golomeova et al. 2013). In this process the whole organic waste is usually reduced to much finer levels. Due to the release of carbon dioxide, oxygen, nitrogen, and water vapor in the atmosphere the organic matter is reduced to much further levels. The whole process of composting includes the formation of microorganisms, waste decomposition and generation of organic material which is further refined in the subsequent stages. In general, the whole process of composting is carried out in three main phases which is described as mesophilic phase, thermophilic phase and maturation phases.

- in the mesophilic phase under moderate temperature, the decomposition is carried out;
- in thermophilic phase under high heat, several thermophilic microorganisms perform the decomposition process; and
- in the maturation phase, the temperature starts decreasing due to the reduction in the number of high-energy compounds which results in the predominance of the mesophilic phase.

Depending upon the climatic situations, the decomposition under natural surroundings can go on from months to years.

Based on several studies and test cases, several aspects are found to play a significant role in the whole process of composting (Teira 2003), some of which are described below:

Organic Waste Particle Size

During the process when the whole structure of organic waste is cut down into smaller portions, the process of decomposition becomes quicker due to dedicated microorganism units on each of these surfaces.

Aerification

Aerification is accomplished by improving the compost stack with natural air. Due to the presence of an adequate measure of oxygen, decomposition happens very rapidly. Hence, at the beginning of the process, the soil heap ought to be routinely blended to fulfill the amount of outside air. In the main long stretches of treating the soil, the necessities of oxygen are most prominent.

Porosity

The space among the particles in compost is referred to as "Porosity". In the event that the organic matter isn't soaked with water, as air is being blown into these spaces that provide the microorganisms with oxygen. There could be consequences, however, as the soil heap with water could start decreasing the air spaces and then it approaches to backing off the way toward treating the soil.

Moistness Measure

Acceptable moisture content is 40–60%. If this level is less than 40%, movement of bacteria will be reduced and broken down and it then falls below 15–20%. Then again, if the content is over 60%, the air volume is decreased, making an unpleasant smell and slowing down the procedure of decomposition.

pH Value of the Material

The ideal pH count for microbiological movement is in the vicinity of 6.5 and 7.5. But at the beginning of the process, the pH estimation of the organic waste will be 7 which is a neutral value.

A Ratio of Carbon and Nitrogen

The ratio of carbon and nitrogen must be appropriate as both the constituents play a vital role: carbon used by microorganism as a source of energy and protein synthesis has been done by nitrogen. Hence, concerning weight, the ratio should be 30 shares of carbon and one share of nitrogen.

11.5.2.2 Gasification

In the process of gasification, a chain of chemical reactions takes place to convert the carbon holding feedstock through the use of limited oxygen into the syngas, that is, synthetic gas.

Though most of the time it seems like a combustion process, in reality it's entirely different as the gasification process makes use of sufficient oxygen which later merged with steam and is heated underneath through the extreme pressure. On the other hand, during the combustion process, a right amount of oxygen for burning to generate light and heat is always required. The gasification process in general triggers the sequence of reactions which is responsible for the formation of a hydrogen and carbon monoxide mixture.

This synthetic mixture of gases is often utilized as a preliminary stage to make manures, methane, pure hydrogen or fluid transportation fills. In particular, there are several types of methods for gasification which are usually described as fixed bed, fluidized bed and entrained flow.

Benefits:
- easy CO_2 sequestration;
- removal of most pollutants from the gas before the turbine (solids, sulfur, heavy metals);
- nitrogen dioxide, sulfur dioxide, and dust are emitted at a slower rate.

Drawbacks:
- quite similar to the combustion process;
- emission of hydrogen sulfide, benzene, i.e. hydrocarbon and dust;
- production of effluents.

Applications:
- Power generation in integrated gasification combined cycle (IGCC) power blocks;
- For chemical technology, synthesis gas formation;
- Artificial motor fuel creation (synthetic petrol and diesel);
- Synthetic gas generation (SNG).

11.5.2.3 Pyrolysis

The process of thermal degradation of material that contains carbon generally involves indirect heat from the external source, either with limited amount or in the exhaustive absence of oxygen in order to obtain pyrolysis oil, charcoal and ignitable gases (Puna and Santos 2010). In opposition to gasification and combustion methods, which include a complete or limited material oxidation process, a pyrolysis process is performed on heat without air, making the entire process typically an endothermic practice with a high-energy product at the end.

Apyrolysis products always produce solid, liquid and non-condensable gases where charcoal and biochar are usually categorized under the solid pyrolysis product whereas hydrogen, methane, carbon monoxide, carbon dioxide, nitrogen and CnHmare categorized as the gases that are non-condensable. In the entire course of pyrolysis, molecules of material are heated up at a defined temperature. However, until the process is completed, the pyrolysis unit conceals the equipment and it is transported at a defined speed by the screw conveyor. Composition and product yield (include pyrolysis oil, char, and syngas) are usually defined based on the selected temperature set during the process.

11.5.3 Recycling Methods

Another effective method of solid waste management is recycling. It can be defined as the practice of recovering the solid material from the waste that could be reused directly or turned into reusable forms after undergoing simple and economical processes. Some new methods are described as follows:

11.5.3.1 Physical Reprocessing

In general, physical reprocessing refers to the collection and reuse of daily life solid wastes like newspapers, magazines, cardboard, polyethylene, aluminum foils, food wraps and packaging materials, empty plastic and glass bottles of various types, pet bottles and jars, others household items on a regular basis. These materials are collected, and are often segregated from their composition to be recycled into new products. The separation can be done either at the point of collection by providing separate containers, or it could be done at the disposal sites alone. Recycling of the above solid wastes into a new product is comparatively easy as these are composed of the same kind of materials. On the other hand, recycling of e-waste like computers, mobile phones or other electronic items is very cumbersome as it requires additional processing of dismantling of such products for further segregation at the minute level.

11.5.3.2 Biological Reprocessing

Biological reprocessing includes the recycling of organic waste such as plant products, crop byproducts and food waste by using various decomposing methods. It results in the production of compost manure that could be utilized in growing crops and vegetation. Methane gas is another major byproduct of the decomposition process that could be used as green fuel for cooking, electricity generation and other industrial purposes. There are numerous technologies and methods available for composting and digestion of organic wastes that vary in complexity from simple home compost bins to large-scale industrial decomposers. The entire decomposition methods are usually categorized as aerobic, anaerobic or hybrid, depending upon the requirement.

11.5.3.3 Energy Recovery

The term energy recovery from the solid waste refers to produce energy either by directly burning the solid waste material as a fuel or indirectly by treating them to create another kind of flammable substance that can be used as fuel for turbines. In general, there are two methods required for solid waste to energy conversion: pyrolysis and gasification. Both of them are the thermal processes that use high temperature and low oxygen supply for the disposal of solid waste that occurs in closed containers. The municipal solid waste that undergoes the pyrolysis process is converted into solid, liquid and gas products where the liquid and gas products can be used as fuel for boilers and turbines to generate electricity and the solid product can be further treated to produce products such as activated carbon, required for a number of vital applications.

11.5.4 Challenges

Potential challenges that can be encountered while designing the smart waste management systems are listed below:

- Wi-Fi module dependency: the entire setup stops working completely if the Wi-Fi module gets damaged;
- connection problem: sometimes it becomes difficult to load the data onto the server due to some connection problem;
- slow network speeds: there must be fast and efficient Internet connection strength for the administrator to check the status of the bins in real-time on a regular basis;
- it requires a subscriber identification module (SIM) with enabled data plan: the Wi-Fi component requires SIM data to transmit live data to the server, for instance the status of a bin. As a result, Internet connectivity must be enabled on the smartphone and should continuously link to the central server.
- power supply: this is essential to ensure everything works well.
- knowledge of smartphones: people need to be up-to-date on the various apps and uses of smartphones.

Conclusion

New solutions for managing waste have emerged due to the rise of recent technologies like IoT. These solutions can also be applied in smart cities as they require the deployment of a cost-effective and low-power consumption sensor network which can be used to deploy the smart waste management system. Different wireless technologies can help to transfer the waste data to the concerned authorities. A good cloud-based solution can analyze the data and can send the different commands to manage that waste. This chapter describes all the important methods for a green smart waste management system.

References

Buchanan R. 1992. Wicked Problems in Design Thinking. *Design Issues* 8(2): 5–21.

Camhi J. 2016. Business Insider Intelligence Projects 34 Billion Devices Will Be Connected by 2020. Retrieved from: http://goo.gl/chQtVd.

Chourabi, H., Nam, T., Walker, S., Gil-Garcia, J. R., Mellouli, S., Nahon, K., Pardo, T. A. and Scholl, H. J. 2012. Understanding Smart Cities: An Integrative Framework. *IEEE, in the Proceedings of 45th Hawaii International Conference on System Science (HICSS)*, 2289–2297.

Constantino. D. 2014. Urban Smartness: Tools and Experiences. In *Smart Rules for Smart Cities*, edited by E. R. Sanseverino, V. Vaccaro and G. Zizzo (eds). Vol. 12. Springer, Cham.

Enevo – Waste Collection for Smart Cities, 2016. Retrieved from: http://enevo.com.

ESP8266. Retrieved from: https://en.wikipedia.org/wiki/ESP8266.

Ganesh E. N. 2017. Implementation of IoT Architecture for Smart Home using GSM Technology. *International Journal of Computer Techniques* 4(1): 42–48.

Golomeova S., Srebrenkoska, V., Krsteva, S. and Spasova, S. 2013. Solid Waste Treatment Technologies. Retrieved from: http://eprints.ugd.edu.mk/7733/1/SOLID%20WASTE%20TREATMENT%20TECHNOLOGIES.pdf.

Lundin, C. and Ozkil, A. G. 2017. Smart Cities: A Case Study in Waste Monitoring and Management, in *Proceedings of the 50th Hawaii International Conference on System Sciences* (conference report).

Puna J. F. and Santos T. M. 2010. *Thermal Conversion Technologies for Solid Wastes: A New Way to Produce Sustainable Energy*. High Institute of Engineering of Lisbon, Chemical Engineering Department: Lisbon.

Sasikumar, K. G. 2009. *Solid Waste Management*. PHI Learning Private Limited: New Delhi.

Shyamala, S. C., Sindhe, K., Muddy, V. and Chitra, C. N. 2016. Smart Waste Management System. *International Journal of Scientific Development and Research (IJSDR)*: 223–230.

Teira, R. M. 2003. *Class Notes of the Subject Environmental Engineering*. Department of Agricultural Engineering, University: Spain

Weber E. P. and Khademian A. M. 2008. Wicked Problems, Knowledge Challenges, and Collaborative Capacity Builders in Network Settings. *Public Administration Review*, 68(2): 334–349.

12
Green Smart Water and Sanitation System

Priya Singh, Garima Singh and Gurjit Kaur

CONTENTS

12.1 Introduction 239
12.2 Features of the Green Smart Water System and Green Smart
       Sanitation System..241
    12.2.1 Features of the Green Smart Water System.................................242
    12.2.2 Features of the Green Smart Sanitation System 243
12.3 Technologies to Design the Green Smart Water System and
       Smart Sanitation System 245
    12.3.1 Green Smart Water Management Technologies 245
    12.3.2 Green Smart Sanitation System Technologies 251
12.4 Biogas...256
12.5 Challenges to Design the Green Smart Water and
       Sanitation System..257
    12.5.1 Behavior Building and Modification..258
    12.5.2 Retrofitting 258
    12.5.3 Public Participation 258
    12.5.4 Funding 258
    12.5.5 Economical...258
Conclusion 259
References..259

12.1 Introduction

The water cycle (i.e. resource of water, water generation, water allocation, water dissipation, water accumulation and operation of waste water) is an important part of the city development system. It influences every part of urban and rural area of a city and its functionality, such as feeding people, production of energy, expanding tourism, maintaining the atmosphere and living beings' health and in productive advancement. But with time, water availability is decreasing due to plenty of use and lack of appropriate measures and precautions to conserve it for different applications. As

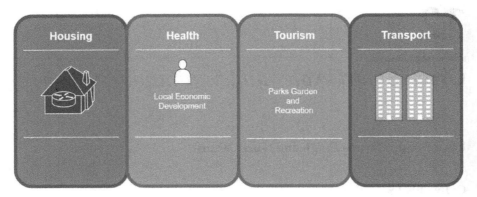

FIGURE 12.1
Connection of water system and city.

depicted in Figure 12.1, the presence and allocation of H_2O sources usually depends upon the city's functionalities like housing, health, tourism, economic advancement, transport, energy, recreation and waste management.

With the ever-increasing population and associated economic activities, high amounts of green gas pollution are simply giving rise to a number of hazards and unfavorable climatic variations that are detrimental to the residents of a smart city.

The recent urban growth has also led to a tradeoff of water between city and agricultural need. For improving these effects, GSMS is needed to be deployed properly with appropriate strategies.

As of now the whole world is heading towards these four main water level risks:

- shortage of water
- inadequate quality
- excess of poor water concentration
- freshwater resilience

All this demands an awareness for handling water-related challenges by considering all risks, as well as their potential impact on industrial, agricultural, economical and urban level (Tilley et al. 2014). Thus, proper management and work is needed to recycle, secure and allocate the water through proper channels.

To date, the following are the areas in which GSWM is working

- *service of raw water*: pure water is diverted and distributed at different parts of a smart city's area.
- *water supply service*: it provides purified and safe treated water to several areas related in the city, such as the resident area, market area and factory area.

- *drainage of water service*: proper drainage network with perfect pipe routing layout is mandatory to protect human health and to prevent floods.
- *purification service of waste water*: it provides purification of waste water where treated water can be further supplied to industrial/resident/market services, ensuring the safety of the public and the environment.
- *reused water service*: this is offered through the large and main sewage purification firms like an extra-perk investment to the factory users or customers for example electricity generators.
- *additional services of water supply*: engineering, procurement, and construction (EPC) equipment services can also help in removing salt from the ocean water and thus can make it purified for drinking.

In recent times, GSWM and GSSS in cities are now integrating with ICT and various sensors. These sensors ensure precise control over the data ingested from the water resources and thus allows an efficient and optimized management of water for various uses. Even the IoT smart water management has put a significant impact on the cost of the water distribution, forming a cost-efficient urban water distribution.

The agriculture sector is especially reaping the most of the IoT advantages in water management. These sensors are capable of performing various functions which include sensing and analysis of different maintenance issues in the urban sector of water supply. The data from different sensors are usually sent to the technician as well as the operator of the water network who provide the water service to the city with proper allocation.

In simple terms they analyze the data collected from these sensors and accordingly schedule the supply of water to appropriate channels to resolve various maintenance issues. Even in GSSS various sensors are equipped to sense the level of urinals in pipes and waste water. With the aim to overcome water-related challenges, the International Telecommunication Union (ITU) made Focus Groups on GSWM and GSSS for smart sustainable cities growth and safe environment. GSWM thus aims for a durable, properly synchronized development and management of water sources by combination of ICT products, tools and solutions for smart sanitation. Further sections of this chapter include features of GSWM and GSSS along with the techniques and technologies used to make it much more efficient for the modern-day smart cities.

12.2 Features of the Green Smart Water System and Green Smart Sanitation System

Smart technologies in today's smart cities are playing a crucial role in effectively and efficiently managing water resources, distribution and consumption. With the development of ICT infrastructure with green technologies, the work of measuring, monitoring, and control of water resources can be done at a lower cost and with greater precision. Some of the important features of GSWS and GSSS are as follows:

12.2.1 Features of the Green Smart Water System

A sustainable water plan is achieved by proper operational performance. GSWS offer great potential to improve the water system by improving the bill payments system and making it more convenient for customers, and the revenue collection system. It also reduces the administrative cost by proper utility. The whole system not only helps in reducing the potential theft of water (Di Nardo et al. 2013), but also helps in spotting possible leakage in pipes, thus decreasing pumping, maintenance and purification costs associated with the smart water system strategy. The three key features of the GSWS are as follows:

12.2.1.1 Smart Water System Concept

Technologies for implementing a smart water system include conventional treatment technologies for drinking water supply, sewage and industrial waste water which can be treated with proper membrane bioreactor systems. Apart from this, customer information management technology using smart meter systems also plays a vital role in preserving the overall water quality level.

12.2.1.2 Smart Water Loss Management System

Water loss management is becoming increasingly important as supplies are stretched by population growth and water scarcity. Many regions around the world are experiencing record droughts and faster depletion of aquifers, leaving no scope for their replenishment. Incorporating smart water technologies allows water providers to minimize non-revenue water (NRW) by finding leakages quickly and providing ways to channelize the flow to the destination with some other possible route.

12.2.1.3 Smart Meters

Smart meters can help in reducing the inefficiencies in water supply systems by detecting leakage and illegal connections remotely. It can also improve billing accuracy and can prevent corrupt practices related to illegal connections and meter hampering.

12.2.1.4 Mobile Banking

The concept of mobile banking provides a base for customer-friendly mobile payment/saving and billing solutions which could decrease the unnecessary

Green Smart Water and Sanitation System

transaction time for bill payment. This will eventually result in collection efficiency that will ultimately achieve customer satisfaction.

12.2.1.5 Standpipe Management Models

This model includes smart use of meter and mobile banking that leads to cashless and secure water points; thus any unconnected poorer residents could get benefit from social tariffs and the utility provider improves their revenue base. Smart technology also allows for standpipe performance checking, regulation and accurate data to generate cost-effective water point expansion.

12.2.2 Features of the Green Smart Sanitation System

Most of the times when we think sanitation network we are usually mistaken with thinking only about toilets. However, the toilet only served to be a single part in the complete sanitation system. Other elements such as collection, transport, purification and use of excreta also helps in building up this whole system for proper functionalities (Brdjanovic et al. 2015). All these elements are usually considered liable for organized and safe sanitation. This division of sanitation into the discussed five parts gives more flexible opportunity in designing a fully fledged sanitation system. One such system is shown in Figure 12.2.

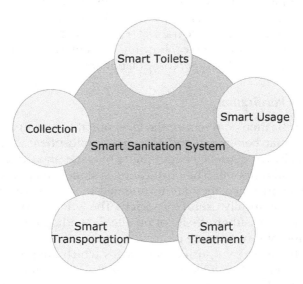

FIGURE 12.2
Parts of smart sanitation system.

12.2.2.1 Smart Toilets

In addition to the toilet facility such as hand washing, privacy, safety and comfort are always considered vital aspects in the proper working of the whole sanitation system. Due to this, hygienic safety is usually provided in smart toilet designs.

12.2.2.2 Smart Collection

A collection facility should prevent the contact and exposure of excreta with pathogens. The human excreta awaiting transportation also comes under collection. A few collection facilities include pre-purification of excreta. By including this necessary function, smart collection facilities make efficient use of large, open accumulated waste that are usually dumped or present in the open atmosphere.

12.2.2.3 Smart Transportation

The sustainable and proper working of the whole sanitation system greatly depends on proper management and organization responsible for transporting excreta and urine waste to associated spaces. For achieving an efficient transportation, the transport network is usually broadened into two parts, of which the first system comprises of the infrastructure, for example the sewerage networks, while the second system includes chain supply and logistics with the help of regular transportation facilities like trucks, vacuum tankers, carts and tricycles. A typical sewage network usually employs a large quantity of water which is often affected by conditions of soil, the water amount for flushing and capacity of sewage. Factors influencing the design and implementation of the waste transport system depend upon the amount of waste produced, haul distance, housing density, road conditions, street access, road gradient, traffic type, cost of labor and fuel.

12.2.2.4 Smart Purification

The aim of purification is to reduce the level of pathogens in excreta to stop infections to human beings and pollution in the atmosphere. A typical purification system always ensures the reusability of water and urine with proper extraction of vital nutrients. The purification system is therefore considered as green and smart when they are formulated in such a way that the end products are economically useful for society. The whole purification process is usually performed either off-site or on-site depending on the on the land location and reusability of excreta.

In general, there are two purification stages which are involved in every waste purification system. One is categorized as the primary purification stage while the other one is the second purification stage. In the whole purification system, the primary purification stage is mainly responsible in reducing pathogens, weight and volume for safe storage. On the other hand the second purification reduces pathogens to much lower limits using different chemicals to make it useful for a number of agricultural-based activities.

So, smart sanitation is not only beneficial for people and environment safety but it also plays a major role in the food industry market. Among the whole purification process the recovery system implies the extraction and usage of reusable material from feces or waste water. The feces that are collected as a part of the purification system usually contain a large volume of nutrition-heavy fertilization, which often results in reducing the demand for harmful artificial fertilizers which inhibit the proper growth of agricultural produce and also lead to widespread disease However on the other hand the purified waste helps in producing biogas and improves the condition of soil to much more appropriate levels. This biogas could prove to be a useful asset in the household work.

12.3 Technologies to Design the Green Smart Water System and Smart Sanitation System

Sanitation, along with clean water and food security, is a primary driver for improving public health. It reduces people's exposure to disease by providing a clean living environment. It is therefore a crucial element in breaking the cycle of infection-disease-recovery-infection, resulting from unsafe disposal of human waste containing pathogens. Amongst different aspects, behavioral and technical measures are the two most vital factors which are required to create a hygienic environment (Hashemi et al. 2015). Critical measures to achieve this goal often include hand washing before cooking, and boiling or chlorinating drinking water. To give a detailed glimpse about the section, some of the very important technologies to design GSWS and GSSS are discussed in the following sections.

12.3.1 Green Smart Water Management Technologies

Many new technologies are devised and developed in order to improve efficiency and performance and reduce the cost and hazardous environmental effects for green smart water management as depicted in Figure 12.3. Some of them are summarized in the following sections:

12.3.1.1 Smart Sensors and Pipe Networks

With the advent of new technologies in the water supply network, various parameters and essential key factors like strain, temperature, pressure

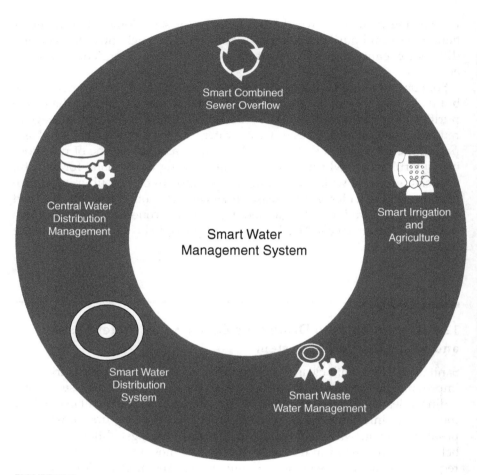

FIGURE 12.3
Green smart water management technologies and tools.

variation water flow and its quality are often measured by various sensors that are fitted across it. This facilitates continuous monitoring and detection of any infectious objects that are present inside the safe water supply network (Robles et al. 2015). By connecting these smart pipes to an antenna and wireless processor one could transfer the data to a common center for checking directly the key factors to analyze regarding the status of the water supply. This whole process helps water managers to detect and locate the water leakage easily and often helps them to call maintenance during any type of failure.

These wireless sensor systems in general help the cities to monitor the whole water supply network more accurately and efficiently. Some of these sensors are usually considered beneficial in detecting or checking the sound-related signature of the pipe or any potential pipe leakage, for example if the groundwater absorption is high or it is wet, it could be sign of a leakage from pipe; or if the smallest amount of noise created daily is high it also shows that a tiny leakage was recently formed. Nowadays these type of sensors are often employed in modern ICT architectures for a green smart water management system. Apart from this there are also some sensors available in the water management system which help in detecting the speed of running water 0 to 3 m^3/hr (5 liters/minute). These sensors are also considered helpful in detecting small leakages that are not easily observed in most other cases. The system which employs such sensors usually involves sending of real-time data related to flow inside the pipe to the associated cloud server where after processing the leakage is detected and a proper alert is conveyed to the appropriate authorities for maintenance.

Apart from measuring and monitoring water supply flow, such modern sensors also help in improving the quality of water used for irrigation by measuring parameters such as humidity, air temperature, soil temperature and moisture, wetness of leaf, atmospheric pressure, solar radiation, trunk/stem/fruit measurement, wind speed/direction and rainfall. Urban applications include park irrigation to commercial irrigation systems, providing better allocation of water between sectors. These type of sensors are also used for accessing the water content of surface water, and purified water sewerage inside a city. Presently lots of checking tasks are still conducted manually, which often result in a reduction in accuracy and increasing time management.

To cope with these kinds of issues, a more robust water monitoring system that can check the chemical substance in water, along with tiny solids, or organism tissue for accessing quality needs necessarily to be employed. Waste water management sensors are also utilized in these types of systems, where they are mainly responsible for the online checking and verifying of important chemicals present in the water. Apart from this it also helps in monitoring pH levels, conductivity, dissolved O_2 level, turbidity, NH_4, P, nitrate, chemical oxygen demand (COD) and metal ions and so on.

Various sensor techniques such as micro-electro-mechanical-system, electrochemical and spectrophotometric technologies are also employed to these systems to attain satisfactory measurements. Apart from this such sensors also prove useful due to their low power consumption and ability to cater to cost-effective needs. Some of the essential work involved in a smart sensing network for water quality monitoring includes the following:

- checking of quality of water by detecting small changes in parameters;
- collecting information to prevent pollution and sending vital information related to water supply in a timely manner to maintain and provide a solution instantly, for example pipes and sewage leaks; and
- checking whether the prime objective of ensuring proper water quality flow is met.

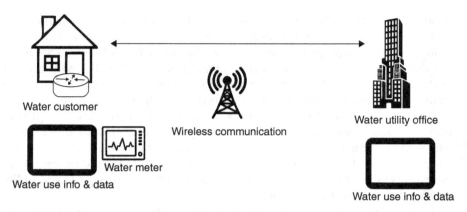

FIGURE 12.4
Smart meter technologies.

12.3.1.2 Smart Use of Meter

The concept of the advance use of meter infrastructure (AMI) generally involves involvement of an electronic device, which is responsible for supporting online checking and measurement of electricity, gas, heat and water usage through smart meters. These smart meters can often be used in GSWS to measure the data smartly (Clement, Ploennigs and Kabitzsch 2012). For water usage, smart meters contain a controller integrated with a sensor acting as a meter, along with a wireless transmitter and a high power battery. These meters usually involve a data logger which performs the continuous monitoring of water consumption of a location, or the area (Hauber-Davidson and Idris 2006) on a real-time basis. A typical smart meter always performs a two-way communication which plays an important role distribution between the central network system and the meter, as shown in Figure 12.4. In general the data to be transmitted is transferred through different channels.

12.3.1.3 Communication Modems

To make communication robust in the whole process, technologies like Bluetooth, global system for mobile communications/general packet radio service (GSM/GPRS), wireless M-Bus communication, and ethernet are often employed. These communication techniques in general play a vital role in providing data collected from the sensors and meters efficiently and effectively. The data which is collected through the different type of sensors is usually made available online to geographic information systems and customer information systems through cloud computing servers so that decision-making within the system could be improved to much better levels. These communication technologies help in creating effective alerts (e.g. leak alert, reverse flow, battery levels, fraud alarm, and water quality alarms) to the main care takers of the systems in a timely manner. Mainly these solutions of communication provide a large range of coverage with scope for effective and timely alert calls (Robles et al. 2014).

12.3.1.4 Geographic Information Systems (GIS)

It is a type of analysis system which is generally employed to capture, manage and analyze geographical information related to an area or a region. GIS usually seems to have a high area of application in many sectors like utilities, natural resources, public safety, transportation and defense.

12.3.1.5 Cloud Computing

The whole cloud computing architecture involved in such systems generally uses an outer analyzing and calculating system which is exterior to the limit of a user's self-designed area. In general, it enables the following key functions:

- manage computing and monitor broad network; and
- trace application and tenants for purpose of billing.

Cloud computing in a city gives better management of workload and calculation by the help of many applications. Besides these technologies, many other technologies such as supervisory control and data acquisition – SCADA systems, web-based expanded communication system – are also used for wide area data management and control of any condition unfavorable for water management.

12.3.1.6 Intelligent Sensing Using ICT

With the ever-increasing demand for water supply there are generally many challenges for a water management system. Under these conditions, new technologies are invented using modern ICT frameworks (information and communications technology) which often employ vast sensor networks for detecting and improving the water system-related issues faced by the management on a regular basis. Some of the important factors involved in this process are as follows (Stewart et al. 2010):

12.3.1.6.1 Visualization of Health Condition of Water Infrastructure and Hydraulic Conditions

For the proper functioning of the water system the quality of water content and its flow must always be noted and studied with utmost care. With technological advancement, modern-day ICT systems are giving birth to a new sensor technique which includes a sensing unit for the measurement of water pressure to much deeper levels. The main plus point of these kinds of systems is the collection of a comprehensive data report which includes measurement of all essential parameters and factors like sensitivity, resolution and frequency range at much better levels.

12.3.1.6.2 Capturing a Pre-failure Signs of Water Pipe Burst with Vibration Sensor

To monitor the whole process efficiently and effectively, experiments are usually performed on a regular basis by ITC to capture different signs of pre-failure of a water pipe before it burst. All this is generally done with the help of a vibration sensor which is always fitted across a pipe. The motivation behind this whole process evolved mainly from an experimental program which was carried out in the Hydrodynamics Lab of Imperial College London (Pipe Rig which simulates the operation of water supply networks). Based on this, the whole setup was develop for the water management systems. In a typical situation when water pressure is increased inside the pipe of a water supply network, signatures in the vibro-acoustic signals are generated which indicate discovery of a potential pipe burst in a particular network area. Advanced signal processing algorithms are then developed with the vibro-acoustic signals generated to check and know pre-failure sites, so that the pipe could be changed before a possible burst.

12.3.1.6.3 Understanding of Water Pressure and Pipe Vibration Changes Caused by Valve Operation

In general, different experiments are also performed using new kinds of sensors to check the water pressure and vibration of pipes when water valves are opened or closed. This whole process helps us depict that rapid alteration of water pressure ultimately results in a large mechanical load on pipes, frequently leading to their deterioration.

12.3.1.6.4 Smart Operation of Water Infrastructure with Virtual Modeling

Real-time simulation of the complex hydraulic conditions in large-scale networks is done by ITC where the essential data collected with sensors is processed with a number of big data techniques. This method in general is mainly used to describe the water infrastructure and hydraulic conditions in the real world with a mathematical model. For the analysis, the water supply network is often perceived as a communication network which is operated by remote and electronic control of pumps and valves. The whole experiment helps in giving insightful information regarding the efficient operation of water infrastructure and prevents water loss by optimizing the possible bursting of aged pipes. Apart from this, algorithms for pump and valve control are also developed by using virtual modeling of water infrastructure. These models play an important role in describing dynamic changes and transition of water pressure, which enable a higher level control of water networks which has not been possible so far. In the future, more of such technologies and techniques in different interrelated areas of civil engineering and ICT need to be merged so that challenges related to water management can be met effectively (Robles et al. 2015).

12.3.2 Green Smart Sanitation System Technologies

The concept of a smart city ensures availability of resources to all citizens while taking appropriate measures for good sanitation. Under sanitation, availability and maintenance of toilets, and the drainage system, new sustainable technologies demand major attention. Existing toilets are inefficient in their usage of water for flushing. In the context of scarce resources, it is vital to conserve them wisely so as to facilitate them to all citizens. As water is one of the major and must utilities of a city, its conservation is highly desirable in all possible aspects. Therefore, in this section, various technologies to design green smart sanitation are described.

12.3.2.1 Smart Dry Hygiene Toilets

The main goal of a hygienic dry sanitation system is to make a sanitation system solution which is not only cheap, modular, sustainable and manageable, but also usable according to rural conditions like water scarcity. In a typical dry toilet, dependence on water is not needed. Here the excreta is collected inside a deep pit below the seat. These dry toilets usually have the pedestal or squatting plate with a limited area to reduce soiling (Hassan et al. 2007). Such toilet structures are often made of fiber-enforced materials, ferro-cement or durable plastic which is hard enough to bear the pressure of a human. This whole system helps in creating the separation of solid waste, urine and washing water. It consists of a squat-type toilet pan which has three exclusive ports for solids, urine and washing water. It is therefore used mainly in rural areas with sufficient storage, purification and excreta systems that help in dealing with the waste in a smart and sustainable manner. These kinds of toilets are more suitable for areas of less water, flooding zones and hard soil areas.

Advantages:
- less water requirement;
- could be constructed by local elements;
- could be used indoors.

Disadvantages:
- cleaning with less water is a task;
- collected excreta have to be carefully handled because they contain pathogens;
- removal of waste must be done in a timely manner to stop any type of infection.

12.3.2.1.1 Pit Design

The pits for such solid waste are usually made up of a stepped depression in the ground covered with bricks/cement blocks. A bamboo mat is placed at an angle so that rolling of solid waste is not possible. To filter out washing water from solid waste, a bamboo sieve is used. The washing water flows through the mat and is collected in the shallow part of the pit. Eventually waste water would go to ground after getting filtered through sand, lime and gravel beds. The remaining solid waste inside the pit disintegrates over a period of time and converts itself in to manure which is useful in agricultural-related activities (Hashemi et al. 2015).

12.3.2.2 Waterless Urinal System

A typical sample of urine usually contains a high amount of excreted N (nitrogen content), phosphorus and potassium, whose processing are always considered easier than feces. Because of many other such vital factors, the separate collection of urine is always considered favorable. The handling of such excrement becomes easier by separation of urine from the solid excrement. Overall, urine is considered the perfect nutrient solution because the N (nitrogen content) in it is present in the form of urea, phosphorous as superphosphate and potassium in ionic form which all are quite useful for the growth of the plants. In addition, urine also contains micro nutrition. Using separate collection, the nutrition of urine could be directly recovered. If urine is not separated, runoffs and evaporation could lead to loss of nutritional value and the nutrition could end up in water bodies. Urine is therefore used either undiluted or diluted, depending on the particular type of application it is employed for. But, the diluted form is usually considered to be more advantageous because of the number of merits it offers.

To take off possible pathogens from urine it requires to be stored in closed containers before using. For using urine in agricultural land, it should be stored for at least six months so that the rotavirus infection and viral infection would be reduced and inactivation of parasitic worms can be done properly. While storing urine, air tight storage is considered crucial so that N (nitrogen content) cannot evaporate. Thus, storage plays an important role when we are using human urine. Urine is often considered a very high quality, cheap alternative to N (nitrogen content) -rich mineral fertilizer, which has a number of essential roles to play in agriculture-related activities.

These waterless urinals do not require water for flushing and can be used for saving water and energy. This quality of urine is based on the principles of environmental sanitation and helps in reducing environmental damage to the maximum possible extent.

12.3.2.2.1 Functioning of Waterless Urinals

The odor in urine highly affects the performance of urinals. The enzyme urea hydrolyzes the urea present in urine into ammonia and carbonate. That decomposes instantly to carbonic acid and a second molecule of ammonia. The overall reaction that is involved in this whole process is as follows:

$$NH_2(CO)NH_2 + 2H_2O \; NH_3 + NH_4^+ + HCO_3^-$$

However, these waterless urinals utilize odor control mechanisms for flushing. Waterless urinals need regular cleaning routines the same as conventional urinals, proportional to the number of users. However, waterless urinals could be cleaned more easily. Drains deposits such as hail and precipitation salts in urine could block the drain (Kueng et al. 2012). Even the use of hard water for flushing could aggregate the problem. These type of problems could prove to be quite useful in such waterless urinals. In addition, the installation and maintenance cost associated with them is usually low. Therefore, odor trap mechanisms are often fitted in waterless urinals to assist in the prevention of odor that develops inside the drainage lines to spread to the surrounding area.

12.3.2.3 Pour Flush Slabs

This U-trap serves as a seal which prevents problems of odor, mosquitoes and flies. After usage, excreta is flushed manually. It is used where flushing water requirement is high so it could be constructed in areas with plenty of water. Also, areas with a large population could use this due to the separation and prevention of infection techniques (Sadegh, Chaskel and Lebovits 1999).

Advantages:

- conveniently reusable;
- no need to handle fresh excreta;
- could be used indoors.

Disadvantages:

- U-trap could easily become blocked;
- pathogens are mixed in water so large volume of water gets infected while flushing.

12.3.2.4 Smart Collection System

This is a deep pit system containing a double pit, used for collection and composting of human waste from a dry toilet. Each pit is made adjacent to the other. The pit base has dry leaves as the first layer, and ash or soil are mixed after every defecation. If the volume decreases to three-fourths, the concrete slabs are covered over the second pit. It works not only for collection, but also for pre-treatment. The pit could be reused one after the other. After pre-purification, excreta is used like fertilizer. Pre-purification usually takes 12 months (Osathanunkul et al. 2016).

12.3.2.5 Smart Transportation System

MAPET and Vacutug are two of the many examples of mechanical emptying systems to empty pits and (septic) tanks.

There are two main systems that are often employed to empty pits and septic tanks mechanically. Both these systems are popularly known as manual pit emptying technology (MAPET) and the UN-Habitat Vacutug systems and generally involves the process of cleaning with mini tanks, hosepipes and pumps which are usually operated by small-scale private operators.

Out of the two, MAPET relies on a hand-operated pump which usually results in filling a 200 liter waste-collecting tank in 15–20 minutes. While on the other hand, Vacutug drives out 500 liters into a waste collecting tank using a gasoline-operated motor within half a minute. The sludge which is usually collected in nearby disposal units are carried further to the city treatment plants using vacuum tankers.

12.3.2.6 Smart Purification System

Co-composting is the purification process consisting of an aerobic degradation of organics, by using more than one feedstock, for example fecal sludge and organic solid waste. The feedstock containing fecal sludge has a large moisture and N (nitrogen content) content, whereas the biodegradable solid waste consists of high organic carbon. By merging these two key elements, the whole process of purification is optimized more efficiently. In general, there are two types of co-composting designs: open and in-vessel. In the open composting, the sludge and solid waste is piled into long heaps called windrows and generally left to decompose. These windrow piles are alternately turned to provide oxygen with each part having the same temperature. While in the in-vessel composting method, the requirements involve moisture, air contact and mechanical mixing. Due to the same reason, it is often not considered as a correct option for decentralized utilities. It requires careful planning for proper purification even though it is a simple process.

12.3.2.6.1 Co-Composting

Composting is a bacterial process in which bacteria and other aerobic creatures feed on waste material and decompose it fully to smoothen the process of purification. Composting represents one material while the process of co-composting represents involvements of two or more materials. Both of them are generally considered as the most accepted procedure to decompose excreta for purification. To begin the composting sequence, the blended compostable materials are placed in long or round piles known as windrowers. The process mixes high-carbon and high-N (nitrogen content) material that are present in the waste itself. To perfectly treat excreta and other organic material through windrows, the WHO recommended active windrow co-composting with other organic material for a duration of one month at a temperature of around 55–60°C, which usually takes around another 2 to 4 months' curing to maintain the compost. The whole process generally eliminates the requisite amount of pathogens to make it much safer for involvement with other procedures.

Using excreta, especially urine, for household organic makes compost have a higher nutrient value (N-P-K) compared to compost created only using kitchen or garden wastes. Co-composting merges excreta and solid waste, thus it helps in improving efficiency. It should be built close to the source of organic substances so that transportation cost can be minimized, and at the same time must located away from buildings and housing locations so that nuisance could not be caused to the people and environment. The buildup space should be covered to prevent the whole setup from rain and storm hazards. In this way it won't be able to hinder the process of purification at any cost. Apart from this, for dewatered sludge, a ratio of 1:2 to 1:3 of sludge to solid waste should be used. Liquid sludge should be used at a ratio of 1:5 to 1:10 of sludge to solid waste.

12.3.2.6.2 Dehydration

Dehydration is the process of adding long time dry organic material at a high ambient temperature to treat feces. This process ultimately enables them to transform into products that are safe for reuse or disposal. It could be done either on-site, in dehydration vaults or off-site, in dehydration beds or bags well isolated from humidity. Dehydration vaults generally collect, store and dry (dehydrate) feces on-site. These feces are considered to be dehydrated well when the vaults are properly ventilated. Such vaults should also remain water tight to prevent entering of moisture when urine and anal cleansing water are diverted away from the vaults. It usually requires several years of storage at ambient temperatures and conditions for pathogens to die off completely. These processes could significantly accelerate by increasing temperature, pH (alkaline purification) and minimization of the moisture content. This increment in temperature and decrement of moisture could be achieved by employing clever design and ventilation strategies which ultimately help in enhancing the process of drying. In this whole process sawdust can also be employed to absorb humidity completely. Feces purification from dehydration systems (such as urine diversion dehydration toilets, or UDDTs) generally need two levels of purification in order to convert them into compost, even if the distinction between these purifications is often diffuse.

The first level of purification takes place during the collection of the feces in the dehydration vaults. Its main work is to reduce the volume and maintain the appropriate hygiene. The factors that often affect the first level of purification are time, temperature, dehydration and pH values of the whole mixture which needs to be treated.

The second level of purification could take place in the toilet. The main aim of this purification is to render the feces hygienically safe and transform them into humanure.

12.3.2.6.3 Planted Soil Filter

Planted soil filter in general is a typical natural system which treats solid-free waste water. It is also often known as constructed wetlands / reed bed systems. This waste water could be pre-treated waste water coming from flushing toilets or fecal waste water coming from urine diverted toilets. Mostly, they could be either mixed with the kitchen waste water and bathroom waste water, or separated from it. A planted soil filter, in addition to a storage tank which is settled and watertight, contains a gravel and sand mixture layered at the bottom, planted with wetland plants like reeds (House et al. 1999). Solid less waste water is drained from the storage tank above the filter or via a below-ground inlet system and is allowed to pass through the filter. Though horizontal-pass soil filters are commonly used, and are generally considered easy to construct compared to vertical-pass filters, they are often considered less effective at debarring N (nitrogen content). In general, there are a number of key elements involved in a typical waste water treatment process, amongst which the bacteria tend to play a crucial role. After treating the whole waste water properly with the process, the continuous flow could be mixed into above-groundwater, which could be used for agricultural irrigation and groundwater restorage.

12.3.2.6.4 Anaerobic Digestion

In this digestion process, various organic matters from animal, human vegetable extracts are broken down by microorganisms through biological activities, without any air presence. The whole anaerobic process usually results in the production of a combustible gas, methane, which is considered a vital element of biogas. This digestion process usually takes a few weeks to a few months, after which it is left out so that the slurry can be removed, either all at once or in slots. In the whole process of anaerobic digestion there are a number of options which are present, including simple digestion technologies to technically combust designs on a house-like scale or municipal scale.

12.4 Biogas

A mixture of carbon dioxide (40%) with methane (40%), made by bacterial decomposition with the help of the aerobic process from organic wastes, like animal excreta, human excreta and leftover crops, are generally known as biogas.

Low level biogas manufacturers generate energy for low scale electricity, conditioning and cooking. High-level biogas decomposers produce sufficient gas to fuel engines to generate high amounts of lightning and power source for electricity. Nearly 6 kWh/m^3 amount of (thermal) energy was known to be available from biogas. This volume of electricity generally amounts from ½ liter of diesel oil and 5 kg of firewood together. In general, 1 kg of human excreta generates nearly 50 liters of biogas while 1 kg of cattle dung produces 40 liters of biogas, along with 1 kg of birds droppings that produces nearly 70 liters of biogas (Köttner 2002). Therefore, it is often seen that these animal and human excreta and further breakable organic material generates enough biogas that could cover future energy need without any degradation of environment and pollution effect. The important condition of biogas usage in large amounts is the presence of specially made biogas burners or developed consumer appliances. Purification of biogas should be done properly before it is used. Its purification usually involves removal of wH_2O, hydrogen sulfide or CO_2 by raw gas. Also, the risk of explosion or leakage is generally considered very high due to which various safety measures are often needed.

Advantages of Biogas Generation:

- clean energy supply;
- decreases respiratory diseases; and
- reduces non-renewable energy use, thus reducing pollution.

Disadvantages:

- efficiency is less as compared to present-day energy sources; and
- needs high security and storage.

12.5 Challenges to Design the Green Smart Water and Sanitation System

There are many challenges faced when designing the green smart water system and green smart sanitation system. Some are listed here:

Sustainability: the crisis is not about sanitation. It's all about ensuring sustainability criteria within its value chain model. Sustainable sanitation needs some basic criteria:

1. economic viability
2. social acceptability
3. technical competence

4. institutionally recognized
5. protects the environment and the natural resources.

12.5.1 Behavior Building and Modification

However, the biggest stumbling block is not the lack of enough toilets, but the difficulty in convincing people to start using them.

12.5.2 Retrofitting

Retrofitting the existing sanitation systems demands a holistic approach with an effort to develop technologies which could promote recycling and reuse of waste and other disposal.

12.5.3 Public Participation

Sustainable sanitation and hygiene is a challenge designed to inculcate healthy surroundings. Therefore, more involvement and inspiration from citizens could play a vital role, as they are the ones who end up with concrete benefits of having a decent quality of life.

12.5.4 Funding

Proper allocation of funds available with the Government of India is another serious issue that demands effective planning and supervision in this regard.

12.5.5 Economical

Catalyzing the creation of smart cities in various regions and parts of the country is yet another challenge which needs to be cost-efficient.

Smart cities in general are slow to achieve that realistic outlook. Ignorance of which often leads to the hampering of the conventional efforts by short-term planning and lack of sufficient funding. Lack of promotion of proper sanitation, education and training, coupled with a conservative mindset of most low developing countries, have made the crisis spring back with a much more negative impact. An example of one such negative aspect can be seen through open defecation practices that have been in process till now. The process of sustainable sanitation is an essential orientation towards behavioral change as it is a progressive step towards the smart social development of a city along with hygiene promotion. In a country like India, improved sanitation and hygiene practices are an investment rather than expenditure in the health of future generations. Consequently, the smart city is all about efficiency, optimization, inevitability, convenience and security. However, the most important aspect here is public participation in governance of the aforesaid measures rather than just a ceremonial participation. Only then will the national mission of being free of open defecation be possible in light of smart city projects.

Conclusion

With the increasing risk of water shortage as well as water need in every part of the urban and rural areas of a city, green smart water management and sanitation systems have emerged as a lifeline for survival in smart cities. This not only achieves surplus supply and reusability of water, but also helps the environment and growth of economic usage. This chapter explains these systems, utilizing emerging technology like IoT, cloud computing, GIS and ICT and merging with present scenarios for optimum results regarding risks involved in the water system. A smart sanitation system utilizes all the waste materials and extracts the nutrients for further use in agriculture, with proper precautions taken for preventing harmful bacterial contact with users. Nowadays, many countries have started implementing these green smart water and sanitation systems in which electronic systems are used to detect imminent clogging of drains and sewers, and automatically clean them by mechanical means. Thus, green smart water management and smart sanitation systems have evolved as strong pillars for building a smart city by providing adequate water supply and reuse options and utilizing excreta and urine as useful products.

References

Brdjanovic, D., Zakaria, F., Mawioo, P. M., Garcia, H. A, Hooijmans, C. M., Curko, J., Thye, Y. P. and Setiadi T. 2015. eSOS®–Emergency Sanitation Operation System. *Journal of Water, Sanitation and Hygiene for Development* 5(1): 156–164.

Clement, J., Ploennigs, J. and Kabitzsch K. 2012. Smart Meter: Detect and Individualize ADLs. *Ambient Assisted Living*, 107–122. Springer: Berlin, Heidelberg.

Di Nardo, A., Natale, M. Di, Santonastaso, G. F. and Venticinque S. 2013. An Automated Tool for Smart Water Network Partitioning. *Water Resources Management* 27(13): 4493–4508.

Hashemi, S., Han, M., Kim, T. and Kim, Y. 2015. Innovative Toilet Technologies for Smart and Green Cities. In *Proceedings of the 8th Conference International Forum on Urbanism (IFoU)*, Incheon, Korea, 22–24.

Hassan, S. 2007. Smart Toilet Seat. *US Patent* 7(216): 374.

Hauber-Davidson, G. and Idris E. 2006. Smart Water Metering. *Water* 33(3): 38–41.

House, C. H., Bergmann, B. A., Stomp, A.-M. and Frederick D. J. 1999. Combining Constructed Wetlands and Aquatic and Soil Filters for Reclamation and Reuse of Water. *Ecological Engineering* 12(1–2): 27–38.

Köttner, M. 2002. Dry fermentation–a new method for biological treatment in ecological sanitation systems (ECOSAN) for biogas and fertilizer production from stackable biomass suitable for semiarid climates. International Biogas and Bioenergy Centre of Competence (IBBC), German Biogas and Bioenergy society (GERBIO), Heimstraße 1, 74592 Kirchberg / Jagst, Germany Ecosan Projekt, R 1740, German Agency for Technical Cooperation (GTZ), Dag-Hammarskjöld-Weg 1-5 65726 Eschborn.

Kueng, G. 2012. Hybrid Waterless Urinal. *US Patent* 8(291): 522.

Osathanunkul, K., Hantarkul, K., Pramokchon, P., Khoenkaw, P. and Tantitharanukul N. 2016. Design and Implementation of an Automatic Smart Urinal Flusher. *International Computer Science and Engineering Conference (ICSEC)*, IEEE, 1–4.

Robles, T., Alcarria, R., Martín, D., Morales, A., Navarro, M., Calero, R., Iglesias, S. and López M. 2014. An Internet of Things-based Model for Smart Water Management. *28th International Conference on Advanced Information Networking and Applications Workshops*. IEEE, 821–826.

Robles, T., Alcarria, R., Andrés, D. M. de, Cruz, M. N. de la, Calero, R., Iglesias, S. and López M. 2015. An IoT-based Reference Architecture for Smart Water Management Processes. *Journal of Wireless Mobile Networks, Ubiquitous Computing, and Dependable Applications* 6(1): 4–23.

Sadegh, A. M., Chaskel, M. and Lebovits G. I. 1999. Apparatus for Automatic Washing, Sanitizing and Drying Toilet Seats. *US Patent* 6(003): 159.

Stewart, R. A., Willis, R., Giurco, D., Panuwatwanich, K. and Capati G. 2010. Web-based Knowledge Management System: Linking Smart Metering to the Future of Urban Water Planning. *Australian Planner* 47(2): 66–74.

Tilley, E. 2014. *Compendium of Sanitation Systems and Technologies*. Eawag: Dübendorf, CH.

13

Innovation Opportunities through Internet of Things (IoT) for Smart Cities

Rajalakshmi Krishnamurthi, Anand Nayyar and Arun Solanki

CONTENTS

13.1 Introduction 262
    13.1.1 Various Definitions of the Smart City 263
13.2 Applications of Smart City 265
    13.2.1 Traffic Monitoring System 265
    13.2.2 Air Quality Monitoring System 265
    13.2.3 Noise Monitoring Systems 265
    13.2.4 Energy Monitoring Systems 265
    13.2.5 Waste Management Systems 266
    13.2.6 Smart Parking System 266
    13.2.7 Smart Lighting System 266
13.3 Framework of the Smart City 267
13.4 Smart Services of Various Infrastructures of the Smart City 268
    13.4.1 Physical Infrastructure 269
    13.4.2 Social Infrastructure 269
    13.4.3 Institutional Infrastructure 270
    13.4.4 Economical Infrastructure 270
13.5 Standard Organizations for Smart Cities 270
    13.5.1 International Standard Organizations 270
    13.5.2 National Standards Organizations 271
13.6 Smart Cities – Deployment Stages 272
13.7 Challenges of the Smart City 274
13.8 Difference between ICT and IoT Technologies 276
13.9 Enabling Technologies of the Smart City 277
    13.9.1 Internet of Things 277
    13.9.2 Cloud Computing 278
    13.9.3 Big Data 278
13.10 IoT-based Architecture of the Smart City 280
    13.10.1 IoT Edge Systems 280
    13.10.2 Data Link Layer 280
    13.10.3 Network Layer 281

13.10.4 Transport Layer 281
13.10.5 Application Layer 282
13.11 Functional Opportunities for the IoT-based Smart City 282
13.12 Nonfunctional Opportunities for the IoT-based Smart City 284
13.13 IoT-based Smart City Systems 285
Conclusion 288
References 289

13.1 Introduction

According to United Nation Population Fund Forecast (UNPFF), in the past half century, there has been an exponential world population growth to 1.2% per year as the average rate. The world cities population shot up to 4.2 billion by 2018. As the world population tends to increase, the amount of people migrating to cities is also growing per day. It is to be noted that, in 2007, the livelihood of city people has surpassed the livelihood of rural areas. Further, the United Nation (UN) envisages that city populations will grow by 70% by 2050. Data shows that about 1.3 billion people migrated to small cities between 1950 and 2010. It is more than double the population, compared to medium cities with 632 million population and large cities with 570 million populations. Based on a socioeconomic perspective, urbanization offers huge employment opportunities for millions of inhabitants across the globe. Hence, cities are the major pillars of the growing economy of any country. However, currently, existing city resources and infrastructures are not sufficient enough to meet the growing demand of urban population.

In 2017, the International Data Corporation (IDC) predicted that for incorporation of a smart city more than $81 billion will be invested by 2018, and $158 billion will be invested by 2022. Eastern countries like Singapore, Shanghai and Tokyo are the top competitors. While in the West, cities like New York City and London are the top smart city competitors. As per IDCS's Smart Cities Spending Guide from their Worldwide Semiannual report, some of the high-end services offered by top class smart cities would be visual surveillance, smart public lighting systems, smart public transport systems and other integrated smart systems. Further, the report advocates prosperous employment opportunities and the technological enhancement of citizens under the smart city concept.

According to a BSI report, in 2014, the focus of urbanization was on enabling effective use of technologies for the benefit of individual citizens and business people. Rather, the focus of urbanization should be an amalgamation of technologies with government bodies, such that the people of the cities can be better served in sustainable, efficient and effective ways. That is, making these cities smarter through technologies and organizations would definitely promote an optimized way of utilizing urban resources and infrastructures. At present, people are bound by limited services through technology. Instead, there should exist a collaborative platform for citizens, business organizations and government to connect with each other to better understand the necessity of smart city concepts. Further, the focus of the smart city should basically be on leveraging intelligence through technologies by connecting aspects like physical, social, business and ICT infrastructures.

The key contributions of this chapter will be to explore:

1. the technologies and service requirements for smart city implementation;
2. the challenges and limitations that exist in the present ICT-based smart city concept;
3. awareness regarding principle components of IoT like heterogeneity, interoperability, resilience, cognition, proximity, security and relevance;
4. discussion on various IoT-based frameworks that exist for the smart city and the challenges and limitations imposed by IoT in the smart city;
5. in-depth convergence of various technologies that leverage the growth of smart cities like mobile computing, cloud computing and big data; and
6. discussion on technologies and services offered by real-life existing smart city models based on case studies.

Accordingly, in this chapter, various aspects of the smart city are explored in order to guide researchers, software developers, industrialists and government organizations. The chapter has been broadly classified into two parts. The first part focuses on the ICT-based concepts, namely framework applications, smart services, standards, deployment stages and the smartness enhancement process of the smart city. In addition to this, the various challenges surrounding the present smart cities and the necessity of IoT-based smart cities is discussed. The second part focuses on smart cities based on IoT. The strategic concepts like enabling technologies for a smart city and the architecture of the IoT-based smart city are discussed. Further, the functional and nonfunctional opportunities are presented. Finally, various existing IoT-based smart cities across the globe are discussed.

13.1.1 Various Definitions of the Smart City

The concept of the smart city is still at the growing stage, hence there is no concrete formulation of concepts and standardization as yet. The taxonomy, perspective and significance of the concept of a smart city are different across the geographical distribution. There exist various definitions for smart cities as proposed by various authors (Dohler 2011; Schaffers 2011).

According to BSI, a smart city is defined as the "Effective integration of physical, digital and human systems in the built environment to deliver a sustainable, prosperous and inclusive future for its citizens."

According to Gartner (2015), "A smart city is an urbanized area where multiple public and private sectors cooperate to achieve sustainable outcomes through the analysis of contextual real-time information shared among sector-specific information and operational technology systems."

According to the Indian government (2014), a "Smart City offers sustainability in terms of economic activities and employment opportunities to a wide section of its residents, regardless of their level of education, skills or income levels."

According to the Institute of Electrical and Electronics Engineers (IEEE), a smart city can be defined as "A city that brings together technology, government and society to enable the following characteristics: smart cities, a smart economy, smart mobility, a smart environment, smart people, smart living, smart governance" (Bellavista 2013).

The concept of a smart city can also be visualized via definitions proposed with regard to stakeholder, technology, performance and intelligence perspective as discussed in the following sections:

13.1.1.1 Stakeholder Perspective

A smart city is the platform that targets integration of all the stakeholders for the benefit of the people and communities. The people, governance, management, organizations, natural environment, infrastructures, technology, economy and policies are the key stakeholders of the smart city concept.

13.1.1.2 Technology Perspective

The smart city is a technology platform that provides and integrates different heterogeneous solution systems that will support a variety of application domains. Examples of technologies are sensor network, wireless network, mobile network, security and privacy, embedded systems and so on.

13.1.1.3 Performance Perspective

A smart city focuses on performance parameters like awareness, synergy, self-decisiveness, self-regulation, strategic activities and the transformability of different systems involved in enhancing the quality of life of smart city citizens.

13.1.1.4 Intelligent Perspective

The new intelligence of the smart city primarily dwells on effort, increasing the efficient conglomerate of digital communication networks, ubiquitous embedded systems intelligence, capillary network of sensors and tags and the cognitive and awareness capability of software.

13.2 Applications of Smart City

According to Uusitalo (2006), Zygiaris (2013) and Silva (2013) there is huge scope and opportunity for application of a smart city. The following are various applications of smart cities:

13.2.1 Traffic Monitoring System

Zhang (2018), Liu (2017) and Djahel and collegues (2015) discussed the traffic monitoring system for an IoT-based smart city environment. In this, several cameras are installed across the main cities in order to perform the operation of capturing videos of the vehicle and human traffic at various locations. Later, the data traffic captured by camera are gathered, transmitted and processed for further analysis and decision-making by traffic management.

13.2.2 Air Quality Monitoring System

Dutta (2017) presented an IoT-based air quality monitoring system that monitors the intensity of several chemical air particles and stringent pollutants coagulated in the atmosphere of the smart city. The measured intensity is digitally gathered and analyzed at a centralized control center under the smart city concept. On the basis of the analyzed data, authoritative decisions are made by the smart city authorities in real-time.

13.2.3 Noise Monitoring Systems

Zappatore (2016) and Rana (2010) discussed the smart city solution for aural pollution monitoring via noise monitoring systems. The smart noise examining system gathers and calibrates the noise produced at particular location and time in the smart city. Over a period of time and using data analytics, this noise calibration is mapped to examine unusual activities like hostile conditions and accidents in real-time across the smart city.

13.2.4 Energy Monitoring Systems

The smart energy monitoring system provides details about consumption of energy by individual homes in a smart city (Zhou 2018; Al-Hader 2009).The energy consumption monitoring includes commodities like electricity, water and cooking gas resources distributed to every citizen via pipelines. These commodity pipelines are monitored via smart sensors installed in the pipeline and on the basis of data gathered, the energy consumption of every home can be measured. In addition to the services, public lighting systems, including traffic lights, heating and air conditioning systems and surveillance cameras of buildings are also monitored. Based on data caught by these monitoring systems the smart city administration can plan electricity generation and circulation across the resource generating plants not only to conserve energy, but to integrate green energy sustainable solutions for a better future in the smart cities.

13.2.5 Waste Management Systems

The waste management system provides smart monitoring of garbage bins across the streets of the smart city (Anagnostopoulos 2017). This kind of smart waste management system is highly effective in keeping the city clean, economical and environmentally friendly. The sensing devices installed on the garbage bin monitors the load of each bin, alerts the garbage collector trucks and also optimizes the route of collecting such garbage bins. The overall cost of waste gathering is significantly reduced and also optimizes the waste collection system.

13.2.6 Smart Parking System

The present manual parking system depends solely on driver experience and luck in finding a free slot to park (Pham 2015). This system of parking is highly ineffective and time-consuming considering the scenario of heavily populated cities. In recent research, various wireless technologies such as radio frequency identifier (RFID), near frequency communication (NFC), ZigBee and the Internet are integrated to propose a smart parking solution (Lee 2008). A smart parking system involves the requirement of data like vehicle GPS locations, distance of vehicle from free parking slot and cost of parking. Whenever the driver needs to park, the information can be fetched by request, from the smart parking service providers using smartphones or tablets. This way the parking slot can be reserved in advance using smartphone applications. The advantages of the smart parking system are the reduction of time searching for a vacant parking slot, reduction in CO_2 levels, less traffic congestion and satisfaction of people. The elderly or disabled persons driving their vehicles also benefit from this type of smart parking system via pre-booking of parking spaces in advance of time.

13.2.7 Smart Lighting System

According to Green House Gas (GHG) report, about 19% of global energy consumption is through the lighting system, which also entails 6% of environmental pollution via emission of greenhouse gases. Smart lighting systems being equipped with wireless sensor network (WSN) and LED-based luminaries promises potential benefits with regard to energy conservation, controlled sensing, 24/7 monitoring of lighting systems and dynamic demand management of power consumption (Shelby 2010). The smart city provides a facility to the smart lighting system, which is dimming public lights, based on different weather conditions and the number of people around. In recent research, a WSN-based ZigBee network is the suitable choice for smart energy-efficient lighting systems due to using low power and being less complex and more reliable. The public smart lighting systems are monitored, controlled and managed by smart city administratives remotely, using interactive software interfaces and luminaries networked through a ZigBee mesh network and the Internet (Budhiputra 2016).

Several other applications of the smart city include building health monitoring systems and the introduction of public green buildings, for example.

13.3 Framework of the Smart City

ISO TC 268 (sustainable development of communities), in 2012, proposed a three-layer functionality model for the smart city. The framework is comprised of layers: life service layer, facility layer and urban infrastructure layer as shown in Figure 13.1 below.

Community services layer / life service layer: the life service layer is also known as community service layer and it provides the necessary primary services for

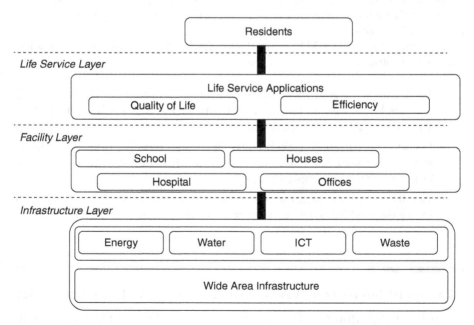

FIGURE 13.1
Framework for smart city.

leading a quality urban lifestyle. These services consist of education system services, healthcare system services and police and security forces. These services are majorly provided by various city administrative departments. As these community services are within the scope and control of city administrative services, they require further improvements, additional functionalities and policy decisions to be taken at local legislative level administrative bodies.

Community facilities layer: according to the ISO framework, the community facility layer is also known as facility layer. The community facility layer is further classified into two types. First, facilities in which services like transportation junctions, rail stations, fuel bunks and so on are provided to the community. Second, the facilities in which various services are used by the community, such as homes, corporate buildings, and commercial complexes.

Community infrastructure layer: this layer is also known as an urban infrastructure layer. This layer provides basic daily requirements for citizens such as electricity, gas supply, clean drinking water, transportation, information and communication infrastructure (ICI), waste and sewage management systems and so on to lead a quality life within smart city. The urban infrastructure layer is monitored and managed by public or private infrastructure administrative operators.

Wide area infrastructure layer: the lowest layer is known as wide area infrastructure. This layer provides the foundation for the above three community-based layers. The layer primarily connects and provides inter-urban cities with wide area facilities and services like a power grid, transportation across cities, bridges connecting cities and so on.

According to an ISO 37120 smart city framework report, the following three major observations are observed:

- First, under the concept of smart city implementation, in the real world, there is large-scale requirement for renovation and sophisticated facilities at every level of functional entities.
- Second, all different stakeholders and their defined roles are unique across infrastructure layers in terms of functional entities and the corresponding services provided.
- Finally, the role of administrative bodies or governing bodies varies at every defined infrastructure layer in terms of functions and services.

13.4 Smart Services of Various Infrastructures of the Smart City

According to the Ministry of Urban Development (MoUD) of India, the infrastructure of the smart city can be categorized into four areas: physical, social, institutional and economical.

13.4.1 Physical Infrastructure

The physical infrastructure includes transportation in and around the smart city, architectural construction and historical buildings, energy conservation, waste and sewage management, housing systems, disaster and emergency management, water management, public work developments like subways, bridges, public lighting systems, dams, flyovers and so on.

For transportation, the various services offered by a smart city include smart ticketing, smart vehicle parking systems, user interactive bus stands, traveling route finders and optimization, smart inter-transport facilities like local commute, taxis, buses, airport, highways, bridges, ports and smart sign posts. From the construction and buildings point of view, the services that can be offered are smart building maneuver systems like elevators and escalators, smart video surveillance and security systems and smart building map guides. From the energy conservative perspective, services are smart power management, smart power grid infrastructure, smart electricity metering, smart gas consumption monitoring, gas pipeline management and smart electrical appliances and lighting systems. From the perspective of a waste and sewage management mechanism, there is the smart waste garbage bin tracking system, smart sewage maintenance systems, clean surrounding services, public gardening services and smart public convenience services.

From the disaster and emergency management system, the services are mostly integrated with parallel services like police departments, security forces, fire department, hospitals and administrative departments. Water management services in the smart city include water metering, water storage and leaking system maintenance, smart water pipes, sensor-based water management systems, smart agriculture and irrigation system management. Also, the public work development services include a smart public lighting system and technology-integrated construction of dams, subways, canals and so on.

13.4.2 Social Infrastructure

The social infrastructure constitutes the safety and security of people and their community, the healthcare system, education system, environment monitoring, entertainment system and home security system. From the safety and security aspect, services include a smart monitory system for personal safety and security, surveillance services and remote monitoring services. Further, smart healthcare, remotely reachable education and public open digital entertainments are included under social infrastructure services. From the environmental point, services like smart lighting and smart irrigation are of utmost significance. From a home and building automation social infrastructure perspective, services like remote controlling appliances, a security system and weather monitoring are included.

13.4.3 Institutional Infrastructure

The institutional infrastructure involves confidential monetary transactions, a judicial system, advisory committee and decision-making systems. The services offered by the monetary transaction infrastructure involve smart digital monetary transactions, optimization and modeling of digital accountability. The infrastructure makes every citizen participate in the election through smart voting, smart governance, social networking, grievances and petition handling mechanisms and smart alert services during crises or epidemics. The smart legal systems involve handling of crime and civic disciplinarians as well as implementation of fast and efficient judicial services to the citizens to facilitate the smooth enforcement of judicial law.

13.4.4 Economical Infrastructure

Economic infrastructure includes the creation of job vacancies for people, enabling livelihoods through social activities, tax collection and transparent tax management system services. The creation of job vacancies involves software platforms to post job vacancies by different organizations, generating a profile database for suitable candidates, managing digital accounts for job seekers and providing assistance for different careers counseling through suitable counselors. The livelihood of the citizens is enabled via residential social welfares, NGOs and social activities across the smart city. The smart city enables the growth of citizen welfare through taxes paid by every individual. The monitoring and transparent management of such taxes collected through citizens are essential aspects for the proper functioning of smart city.

13.5 Standard Organizations for Smart Cities

13.5.1 International Standard Organizations

In recent years, the smart city concept has pioneered several new technical resolutions in the area of smart utilization of technologies. However, the innovation of technology alone cannot drive the diverse necessities of smart city. Rather, there is need for a suitable standardized framework that enables the large-scale operation of the smart city concept. Hence, there is an essential need for standards that can interleave between entities, technologies and functioning systems of the smart city.

According to BSI, three major standards are required for a smart city, namely management standards, data standards and technical standards. The management standards provide benchmarking, quality checking, collaboration and assessment of different service modules and smart city service providers. The data standards ascertain standard data formats which are used for different requirements along with the necessary security aspects of the smart city concept. The technology standards ensure connectivity between different components, deployment in the market and generating opportunities within the smart city.

Various national and international standards surrounding smart cities are as follows:

IEC–SEG: the International Electrotechnical Commission (IEC) has initiated a setup called the "System Evaluation Group (SEG1)". The primary objective of this group is to conduct reviews and evaluations regarding the standardization status of the smart city. The system proposed strategies for enabling new standardizations that will meet future smart city development requirements.

ISO TC 268/SC: the Technical Committee for Smart City (TC 268/SC 1) was established by ISO in 2012. This committee has reviewed the various activities carried out so far with respect to metrics. Particularly, the document ISO 37120:2014 published by TC 268/SC 1 defines various indicators for measuring and steering the performance of smart city services and quality of living (QoL).

ITU: the International Telecommunication Union (ITU) initiated the technical group focusing on sustainable smart cities called SG-5. The objective of SG-5 group is to promote awareness and knowledge exchange towards the integration of ICT for a sustainable smart city. Further, the group identifies and takes the initiative for standardizing various smart city frameworks.

ISO/IEC: the Joint Technical Committee (JTC) is formed by two major players, namely ISO and IEC, and has commenced a study group focusing on smart cities (SG 1). The objective of JTC is to analyze and synthesize technical documents with respect to the societal and economical requirements for incorporating ICT standards into smart cities.

CEN-CENELEC-ETSI: the European Committee for Electrotechnical Standardization (CENELEC) focuses on developing European standards for various electrical engineering systems. The specialized groups, namely European Telecommunications Standards Institute (ETSI) and Smart and Sustainable Cities and Communities Coordination Group (SSCG-CG), are analyzing the needs for European citizenship. Further, the standardization of sustainability of European smart cities is a focus of these standard groups.

13.5.2 National Standards Organizations

UK – BSI: in 2014, with the joint forum by the UK-based British Standards Institution (BSI), and Department of Business, Innovation and Skills (DBIS), the documentation defining a framework for smart cities was released. The document elaborates regulations and frameworks for designing smart cities.

German – VDE: in 2014, the German-based standard coordinator, Deutsche Kommission Elektrotechnik Elektronik Informationstechnik (DKE), in collaboration with Deutsches Institut für Normung (DIN) and Verband der Elektrotechnik, Elektronik und Informationstechnik (VDE), was formed. The group published the roadmap for German-based standardization for smart cities in the near future.

US – ANSI: the United States-based ANSI proposed required standardization and road maps for sustainable smart cities as the standard solution for designing smart cities.

Spain – AENOR: the Spanish Association for Standardization and Certification is the standardizing organization to figure out the needs for smart city incorporation in Spain.

Poland – SSCC: the organization group, Smart and Sustainable Cities and Communities was established in 2014.

13.6 Smart Cities – Deployment Stages

Considering the standard deployment of smart cities, there are three stages of smart city deployment, namely the bootstrapping phase, growth phase and wide adoption phase (Chen 2016).

Bootstrap phase: there are two major objectives of this phase. The first objective is to offer services such that revenue is generated by these services and utilities. The technologies deployed and their related services for a smart city should not be limited to the utility.. It provides advanced smart services to the community and people of a smart city. Further, the services and technologies of the smart city should provide a good profit return proportional to the heavy investment involved in their deployment. The second objective of this phase is to lay down the foundation for further growth of revenue and development.. The focus of the smart city should not be limited, only to set up successful application and services. Rather, the deployment of the smart city services should guarantee revenue generation.

Growth phase: the objective of this phase is to investigate the barriers and new requirements of the community and people based on the previous deployed smart city services. Further, the cash flow from a previous deployed smart city has to be monitored. It is necessary to observed that the growing phase of the smart city technologies and services lead to three important points: first, large investment will be required for further growth because of new ideas and services for improving quality of life; second, there may not be instant short-term returns, however, there can be long-term returns on specified investments; third, there may not be many direct benefits or gains associated with the deployed infrastructure of smart city, however, there can be possible secondary profits and revenues.

On the other hand, to anticipate the prospective growth of investment during the growth phase there are two ways: first, the previously invested infrastructure during the bootstrap phase generates initial short financial gains; and second, with the already existing infrastructure and initial short financial gains, the larger investments are attracted from private capitalization and thus the smart city concept is promoted on a larger scale. Such a process will ease out the limitations and barriers at the entry level deployment of the smart city-based infrastructures, technologies and services.

Wide adoption phase: in this phase, the smart city services are widely popular and easily offered by stakeholders. This ease of services is possible based on the stability and resilience of the previously deployed methodology, achieved financial gains and advantages of existing foundational infrastructures. These factors are irrespective of the volume of revenues and scalability of utilities provided by earlier phases of deployment. A wider adoption phase will depend on the data generated in the previous phases. The value chain process involving standards and APIs provides a platform for the further penetration of the smart city across boundaries of the globe. During this phase, the different API standardizations are made concrete and stabilized. The standardization during the wider adoption phase provides third-party investors and developers with several harmonized and heterogeneous services easily for the community and people of the smart city. During this phase, the objectives like sustainability and interoperability of developing services are of utmost priority.

13.6.1 Smartness Enhancement in Smart Cities

The major factors for laying the foundation for smartness are a data centric approach and technological integration. According to a BSI report, the data centric approach of a smart city requires a bottom-up approach of systems integrated with a top-down approach of services. Similarly, the technological integration involves the vertical collaboration of sensors, communication networks, data analysis and control systems and horizontal collaboration of isolated technological systems with people-based services and functionalities as depicted in Figure 13.2.

According to a BSI report, the primary challenge in the smart city concept is data collection, handling and knowledge generation based on sensor data gathered from various smart city infrastructures. Vertical integration through four step-by-step technological evolutions, namely pervasive computing based on sensors, low-cost communication, real-time management systems and advanced data analytics, will provide smartness to the city. Pervasive computing includes various sensors, actuators and devices such

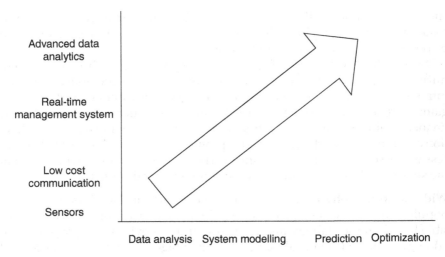

FIGURE 13.2
Growth of smartness in smart cities.

that it is possible to measure or sense the smart city environment. The data gathered through sensors provides information regarding energy usage, water consumption, transportation facilities and infrastructure systems. The networking technologies are based on wireless protocols like Bluetooth and ZigBee provide low-cost, less complex and more reliable communications. Further, a real-time management system based on optimization modules and control systems is required. Advanced data analytics involve the application of intelligence on the huge volume of data collected from various sensors. Integration of all these factors in city infrastructures highly leverages the smartness through public services and functionalities to uplift citizens' quality of living to a greater extent.

The horizontal integration involves collaboration of isolated technological systems like data analysis and system modeling, simulation-based prediction, problem-solving thorough optimization methods. For example, the collaboration of geographical information systems (GIS) and customer information systems can help customers check the weather forecast.

13.7 Challenges of the Smart City

The concept of the smart city has huge potential for development and economic growth (Zanella et al. 2014; Nam, 2011). However, apart from being highly advantageous, smart cities are surrounded by technical challenges, for example political, technical and financial.

13.7.1 Political Challenges

The political challenges of the smart city involve how to assign various responsibilities such as decision-making power, controlling authority, advisory committees among the different types of stakeholders. The various smart city stakeholders are government, public administration, private administration, citizens, builders, technological proprietors and service providers. The major objective of the political challenge is to dedicate and dictate the role of each of the stakeholders and then obtain integrated resolution among the various decisions and development of individual stakeholders. Achieving such an objective involves huge human effort and understanding of the objectives thoroughly and also the dedicated involvement of stakeholders for a common cause. The effective way is to set up standardized institutions responsible for a decision-making process. In this way, the execution of each process will be monitored and analyzed by these standardized institutions. In addition to this, the strategic planning and management of every smart city component will be encapsulated under a single dedicated institutional organization.

13.7.2 Technical Challenges

The technical challenges of the smart city include factors like interoperability of the technologies, heterogeneity of the sensors, computing devices, software and hardware that are all involved in smart city implementation. Under a smart city, different smart concepts are integrated. Each of these smart concept services has different specifications and requirements of technology. Most of the time, the interoperability of these different sets of smart services are very tedious and incur huge costs. For example, the smart power grid implementation cannot be interoperable with most of the other smart concept services like smart waste management, smart resource management systems and so on. On the other hand, there are huge possibilities of interoperability between technologies like smart air quality monitoring and smart noise monitoring systems for the overall enhancement of citizen life in smart cities. Hence, exploring and utilizing such possibilities of these technical challenges is still under development in the smart city concept.

13.7.3 Financial Challenges

The financial challenge of the smart city involves the lack of a proper business model. The financial aspect of a smart city is totally dependent on the global economy. As the global economy has shown unsteadiness in recent times, it has a direct impact on the business development and growth in a smart city. Public investment in the smart city is still not particularly favorable considering the present weak global economies. It is observed that the smart city financial applications show up with only steady short-term returns to the investors. For this purpose, strategic planning and deployment of the smart city is essential. This type of of setup will surely require long-term investments from both public and private sectors for concrete development of the smart city concept.

13.8 Difference between ICT and IoT Technologies

13.8.1 Generic ICT Technologies

The Information and Communication Technologies (ICT) are restricted to electronic integrated circuits of nanotechnology, which is a part of disruptive technology. ICT-based devices are totally different from conventional transistor-based technology. ICT incorporates wireless communication and networks for connecting various devices. Wireless communication uses radio frequency-based waves for transmission. It is comprised of fixed, portable device applications and mobile communications. However, wireless communication suffers from high latency, low data rate and less security as millions of devices are connected in the smart city in real-time. Next, ICT does not really address the low power and restricted resource capability of devices. The computing devices are assumed to be resource-rich in terms of computing capability and energy availability. But, the devices that are used in the smart city are limited in resources like power, memory, computation, storage and networking. Further, ICT depends on the middleware adaptive mechanism and cognitive computing systems. These mechanisms are highly complex and involve heavy computation, and hence are not suitable for the constrained computation of a smart city. Finally, ICT-based technological solutions lack support for applications that are required to analyze dynamic changes in the context of physical factors.

According to Chen (2012), ICT focuses on seven major factors for establishing a smart city:

- to improve the quality of life for people and enhance their lifestyles;
- to offer infrastructures for development and various services to people;
- to focus on the needs of the people and community of the smart city;
- to provide sustainable resources and a healthy environment;
- to incorporate governance of policies, management of institutions and administration of public and private services offered to the citizens of the smart city;

- to improve the economy and the financial status of the people; and
- to provide mobility features to the citizens.

13.8.2 IoT Technology

IoT technology exhibits the amalgamation of ICT technology with enabling technologies like cloud computing, big data and cyber-physical systems. The objectives of Leadership in Enabling & Industrial Technologies (LEIT) is anticipated through the implementation of IoT along with enabling technology. The IoT incorporates multiple stakeholders within its framework. Hence, the IoT-based smart city demonstrates multidimensional instances of applications and services in improving the quality of life of the smart city.

13.9 Enabling Technologies of the Smart City

It is significant to observe that the initial focus of the smart city is the digitization of various functioning systems (Nam and Pardo 2011). But now there are several essential needs and changes required in the smart city concept (Santana 2017). Hence, the current focus of the smart city is to provide sustainable living conditions for improving the quality of every citizen's life.

Under this section various components are addressed like: 1. What are the different enabling technologies? 2. How can these different enabling technologies be incorporated to achieve the goal of sustainable software platforms for the smart city? With this aspect, this section discusses four main enabling technologies for the sustainable smart city, namely Internet of things, bid data, cloud computing, and cyber-physical systems

13.9.1 Internet of Things

According to Bellavista (2013), the Internet of things interconnects billions of devices with different technical characteristics by means of the Internet. The IoT architecture comprises three major components, namely edge hardware, processing middleware and application services. The edge hardware consists of a variety of sensors, actuators and embedded devices. The processing middleware performs data gathering, storing data and processing data for knowledge generation according to the smart city user service requirements. IoT enhances the concept of the smart city, and in this way, the edge hardware devices of IoT collect data via sensors deployed across the city. The middleware enables the functionalities that process the data gathered from various devices. After this, the specialized service components of IoT middleware serve the specific service requirement of the smart city.

13.9.2 Cloud Computing

Cloud computing offers three major cloud services, namely Infrastructure as a Service (IaaS), Software as a Service (SaaS) and Platform as a service (PaaS) (Bonetto et al. 2012). Cloud computing overcomes the imitation of storage and computation capabilities of resource constraint IoT systems. Further, cloud computing enhances the IoT-based smart city by providing infrastructure, software and platform as services. In fact, Distefano (2012) discussed "Sensing as a Service" to facilitate the IoT systems. Fortes (2014) described the combination of cloud computing and IoT as the "Cloud of Things". According to Galache (2014), several purposes of smart city applications can be resolved through the integration of IoT and cloud computing such as City Platform as Service (CPaaS) and Smart City Application Software as a Service (CSaaS).

13.9.3 Big Data

According to Cheng (2015), conventional techniques such as relational databases are meant for a limited size of data and are not efficient for processing voluminous IoT data. In an IoT network scenario, a lot of data has been collected through the billions of IoT sensors that these conventional mechanisms cannot handle. Thus, big data systems possess four major features, namely volume, variety, velocity and veracity which are capable of handling this IoT data as depicted in Figure 13.3 below.

Volume: the volume refers to the enormous amount of data collected from different sensors involved in IoT systems. Specifically, in the smart city concept, massive amounts of data are gathered from different sensory sources associated with the smart city people and infrastructures.

Variety: the variety in data refers to the different types of sensor data that are gathered from different sensors of the IoT system. In the smart city environment, the examples include the temperature of the surroundings as recorded by temperature sensors, pressure sensed by pressure sensors and proximity calibrated by proximity sensors.

Velocity: the velocity represents the rate at which the gathered data is processed. In the smart city, the application services require real-time processing of data and offline data processing. For example, in situations like traffic management, disaster management and resource management systems, the data has to be monitored, gathered and processed instantaneously in real-time. On the other hand, under the smart city concept, the

Innovation Opportunities through IoT

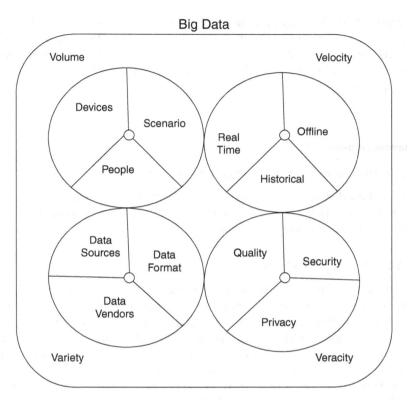

FIGURE 13.3
Four Vs characteristics of big data.

authorities, governments and policymakers may monitor the infrastructural requirements, population growth and penetration, educational systems and any other development requirements when offline for a particular history of the period.

Veracity: the veracity refers to the quality of data, security and privacy of the information gathered. The data quality indicates error and free data collected from the smart city IoT systems. Data security involves a procedure to avoid data misuse or tampering by malicious attackers and intruders. The data privacy provides originality, trust and authenticity of information gathered from the smart city environment.

13.9.3.1 Cyber-physical Systems

The primary objective of the cyber-physical system (CPS) is to provide a common platform for IoT smart city physical objects and the functional components of enabling technology (Cassandras 2016; IBM Brazil 2011;

Gurgen, 2013). The major advantages of CPS in the smart city model include efficient resource management and modernization of the smart city system through the smart grid systems and further to enhance conventional value-added smart city systems through public infrastructures like hospitals, schools and community centers.

13.10 IoT-based Architecture of the Smart City

The IoT-based smart city consists of five major layers, namely edge systems layer, data link layer, network protocols layer, transport protocols layer and application protocols layer (Al-Hader 2009).

13.10.1 IoT Edge Systems

The objective of the IoT edge systems devices include smart city realization, in terms of communication with other IoT edge devices. Hence, the IoT end systems are further classified according to their purpose and the location of the edge devices within the IoT protocol. The IoT edge nodes are capable of monitoring and sensing the environment and thus generate and forward data to centralized control systems. The property of edge devices includes low cost with computational resource constraints. IoT edge nodes are further classified according to the characteristics of the computing device, energy input method, networking transmission capability, type of sensors/ actuators involved and the standard of data link layer mechanism for these edge devices. The IoT gateways interconnect with the IoT edge nodes. Further, gateways facilitate nodes to communicate with the remote computing and controlling infrastructure of the IoT system. The IoT backend servers perform sensory data collection, storing, pre-processing and analyzing. Further, these servers are intended to perform value-added services as per the user requirements

13.10.2 Data Link Layer

The data link layer objective is to provide network communication support for information flow among a wide range of edge devices (Zanella 2014). In the case of the smart city, the data link layer protocols are classified as unconstrained transmission and constrained transmission protocols. Unconstrained transmission protocols comprise WAN, MAN and LAN network types that involve communication protocols such as Wi-Fi, WiMAX, Power Line Communication (PLC), Wired Ethernet, and mobile LTE/4G technologies (Kinney 2015). The basic architecture of IoT-based smart cities is shown in Figure 13.4.

Innovation Opportunities through IoT

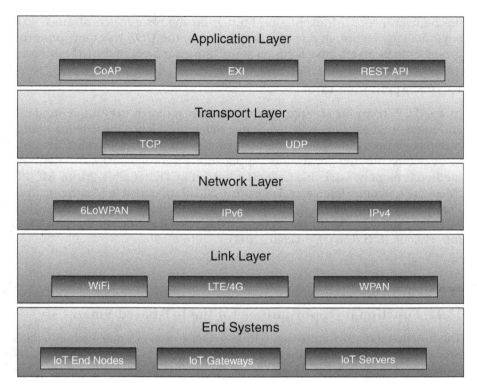

FIGURE 13.4
Basic architecture for IoT-based smart city.

13.10.3 Network Layer

The IPv6 protocol is capable of addressing uniquely millions of IoT devices based on a 128 bit addressing scheme (Mehmood 2017). As envisaged, the IoT network comprises of billions of edge devices and each of these devices needs an IP address. However, these IoT edge devices are resource constrained. Hence, the processing of a 128 bit IPv6 network address would cause huge computing overheads for IoT edge devices. Thus, there is need for a 6LoWPAN efficient low power networking protocol. In 6LoWPAN, the network border router assigns each edge device with a 6LoWPAN address with respect to its corresponding IPv6 address.

13.10.4 Transport Layer

To access the web contents, an HTTP application protocol is used to request and fetch response from remote servers (Santucci 2010). However, the HTTP protocol involves a huge overhead of three-way TCP handshaking mechanisms for reliable communication at the transport layer. This mechanism incurs a transmission burden on the resource-constrained IoT edge devices. Hence, in most IoT applications, the UDP transport protocol promises to be an effective solution at the transport layer of the IoT network.

13.10.5 Application Layer

In the TCP/IP protocol suite, the application protocol exhibits redundant and huge overheads in order to achieve reliability. Hence the existing application layer protocols are not favorable for resource-constrained IoT devices. To overcome this, the Constrained Application Protocol (CoAP), and RestAPI methods, namely GET, PUT, POST, and DELETE (Shelby, 2010), MQTT (Message Queuing Telemetry Transport), XMPP (Extensible Messaging and Presence Protocol) are used. Similarly, computationally intensive Extensible Markup Language (XML) are replaced using open data format Efficient XML Interchange (EXI.(Castellani et al., 2011).

13.11 Functional Opportunities for the IoT-based Smart City

The functional opportunities are wireless sensor network management, data processing and analysis, data access mechanism, software engineering and on-demand management.

Wireless Sensor Network Management: wireless sensor network forms the foundation layer in the IoT-based smart city architecture. WSN gathers the necessary data by monitoring the natural environment. Hence, the management of wireless sensor network and devices that are deployed across the smart city is pivotal. The software platform for the management of such sensors, actuators and devices should be able to organize, manage and monitor them. The activity life cycle of devices such as inserting new nodes, removing any nodes and monitoring need to be optimally performed by the software platform. The scalability of the sensor network requires to be enhanced to meet the future growth of devices when added onto the IoT smart city network. Similarly, the localization of these sensor devices deployed at various places in the smart city is also a major factor to be handled by the software platform.

Data Processing and Analysis: a huge amount of data is gathered from the IoT devices such wireless sensors, actuators, and devices. The software component of the IoT-based smart city is responsible for data processing need to exhibit efficient mechanism to handle the huge data sets. The software component performs workflow-based data processing, handles appropriate engines for providing inferences as per the smart city service requirements and provides tools for processing big data. The objective of the data analysis component includes verification, validation, aggregation, filter and analyzing of the data gathered from the various devices in the smart city.

Similarly, the real-time processing and analyses of data needs development of efficient and optimal software systems. Hence, there are lots of opportunities under data processing and analysis depending on requirements and specification of various services of the smart city.

Data Access Mechanism: it is noted that the end devices of the smart city exhibit resource constrain aspects such as computation capability, storage space, memory capability and network transmission. Hence, the huge data generated by these end devices are stored in various external sources like cloud computing, open data systems and external standalone data platforms. The smart city systems have to provide efficient mechanisms to access data from various external systems that store the huge data. The application of such systems allows different APIs like ReST, CoAP and EXI to access data. Similarly, the mechanism for accessing data from the cloud, big data management systems like Hadoop, Spark requires a lot of enhancement in terms of optimization and performance.

Software Engineering: the IoT-based smart city needs software platforms and tools for developing and maintaining several services and user interfaces specific to the requirement of user applications. The major objectives of the software engineering for the IoT.based smart city concepts are to provide connectivity, interoperability and security. As per the literature, the smart city ecosystem continues to expand beyond billions of connecting devices. However, the existing software models for the networking of these devices are centralized and within a limited scale. Hence, the first objective of software engineering is to develop efficient software components and service applications that can provide uninterrupted connectivity among a large scale of end devices and computing nodes. Next, within the IoT-based smart city, the connecting devices and computing nodes are highly heterogeneous, distributed, autonomous and involve different technologies and specifications and are owned by individual proprietors. Hence, software applications must be developed with the ability to interconnect and interact with a wide range of connecting components, while leveraging the dynamic advantages of individual components. To address connectivity and interoperability issues among different IoT systems, the multimodal radio propagation is required at the physical layer of each device. In this view, different radio technologies are integrated through software engineering solutions like system-on-chip. The third objective of software engineering for the IoT-based smart city is to provide a security platform, as the IoT-based smart city is rampant with known and unknown attacks and vulnerabilities. The software development process for security must incorporate dependable and robust functionalities within the applications. Further, visual-based tools are one aspect that eases the developers to define and implement various user applications. Similarly, software tools need to address the designing of workflow, reporting generation mechanism and data analyzing techniques for various applications.

On-demand Service Management: the IoT-based smart city promotes service-oriented architecture (SOA) for developing software services and its applications. Further, these services are of two categories, namely offline services and online services. The offline services include accessing the sensor data, analyzing the collected data and influencing the result of workflow and further data processing to meet the additional requirements of applications. For example, in order to provide the smart transport system, it is essential to know the demography of the city, number of people traveling every day, type of vehicles preferred for traveling, time of traveling and so on. For this purpose, the data are gathered offline, and then the analysis is carried out to meet the requirement of the smart transport system.

On the other hand, on-demand requirements and specifications of the service need special techniques and tools, compatibility and interoperability with other enabling services. For example, the smart energy monitoring system involves real-time demand of a particular smart service customer. Here, the complexity lies in on-demand requirements and consumption rate of users, which are highly dynamic in nature. Hence, real-time decisions have to be performed according to the on-demand needs of the customers. In addition to this, the optimization of energy allocation to every other customer under a smart energy monitoring system has to be carried out.

13.12 Nonfunctional Opportunities for the IoT-based Smart City

The nonfunctional opportunities focus on the performance of the smart city components and are identified as interoperability, scalability, resilience, cognition, security and privacy.

Interoperability: the smart city consists of several types of end devices, access level communication technologies and service level software components. Both type and diversity are required to meet the smart city service specifications and requirements. However, there is a need to enhance the interoperability of these components, in order to deliver the best services for improving the quality of life of people. The smart city concept is complex as it involves various vendors, systems, programming platform different methods of data sharing exist in the smart city. Also, there is a huge role of different legacy software and systems involved in the IoT-based smart city. Hence, the current focus of the IoT-based smart city includes developing interoperable components, standardizing software interfaces and integrating different development platforms.

Scalability: it is envisaged that there will be a huge increase in the number of end devices, quantity of data collected, number of users, number of services and applications over the next few years. Hence, this nonfunctional component, namely scalability, leads to several innovation opportunities in terms of device network management, data management and services management.

Resilience: the IoT-based smart city has to provide a countermeasure mechanism for threats in the system. The threats can harm user identity, data protection, service representation and real-world objects. Hence, there is the requirement for handling the dynamism and real-time processing using intelligent systems, which provides mechanisms in such a way that the services and functionalities are always available. Similarly, the system must provide a survivability mechanism in case of faulty or failure of service components.

Cognitive: a smart city consists of several service systems and must provide the intelligence upon minimizing the intervention of human resource. Hence, the system must provide the ability to automatically choose the relevant potential software interfaces, optimizing mechanisms and processing techniques. Also, the discovery of suitable devices, selection of software platforms and networking techniques must also be part of this cognition-enabled smart city. Further, AI, along with pervasive computing provide solution to effective cognition and intelligent processing models for smart city concept.

Security and privacy: the security aspect of the IoT-based smart city involves the identification and prevention of illegitimate users to use the sensitive data offered by the services. Hence, there is huge opportunity to develop systems to handle security issues and avoid imposing security attacks both at real-time as well as on stored data. At present, techniques like light cryptography offer such a security facility, however it needs further enhancement. Similarly, the privacy policies regarding the sensitive information of people, such as health-related data and localization issues, identify how to handle information appropriately. Hence, a smart city system must provide the platform to collect and process the user data in very effective way without manipulating or malicious intent.

13.13 IoT-based Smart City Systems

The major focuses of IoT-based smart city systems are to develop platforms that are capable of automatic reaction and response to the dynamic changes happening within the environment of smart city, and establishing communications and make collective decisions among different smart systems. According to Galache (2014), the IoT-based smart city must include systems with self-properties like self-organizing, self-descriptive, self-configurable, self-discoverable, self-optimizing, self-healing, self-protecting.

From this view, in this section, several such IoT-based smart city systems that have been implemented across the world in recent years are discussed and a summary is depicted in Table 13.1.

Padova smart city: the system proposed by Bui (2011) focuses on two domains, namely the healthcare system and smart lighting system. Under healthcare, the major services offered are patient health monitoring, transmission of vital health records to physicians and emergency call services during any critical situations. Under the smart lighting system, the major services offered are gathering sensor data from different public street light posts installed across the smart city, and managing the various functionalities of the smart street lights depending on the weather and the number of people around.

GAMBAS: the GAMBAS software platform for the smart city focuses on smart transportation across the streets of Spain (Foell 2014). The software platform was proposed by Apolinarski (2014). The major services offered are the context-aware roadways find the best travel route through bus transport, passenger count and seat availability-monitoring in the bus. Further, an efficient bus travel visualization and navigation tool is provided through this software platform.

ClouT: the ClouT software platform was proposed by Galache (2014) and provides a smart disaster management system across Japan (Tei 2014). The system offers services like ICT-based monitoring and alert messaging for earthquake and storm indications. Further, it includes a notification dissemination system, which informs about various events or health activities to be held across the city.

Scallop4SC: the Scallop4SC software platform for smart city across Japan was proposed by Takahashi (2012). The main focus of Scallop4SC is energy management. The services offered are a visualization tool for consumers about electricity and gas resources in smart homes.

Smart Santander: the Smart Santander software platform was proposed by Vakali (2014) and offers two domains of services, namely smart traffic management and air pollution monitoring across Spain city. Under smart traffic management, the services offered are monitoring of available parking slots across the city, like hospital buildings, entertainment halls, shopping malls and schools. Under the air pollution monitoring system, services like psychological analysis of citizens at various commonplaces of smart city, monitoring the level of air pollution and the visualization of air quality index through interactive software tools are included.

WindyGrid: the WindyGrid software platform provides service to analyze environment and condition of building infrastructures across the Chicago city. The software platform for the IoT-based smart city was proposed by Thornton (2013).The services offered include a situational alert system and monitoring of critical incidents like accidents and natural calamities. The

TABLE 13.1

Summary of Smart City Software Platforms and Services

Software Platform	Author & Year	City	Domain	Services
Padova Smart City	Bui [2011]	Padua	Healthcare	• patient health monitoring • personalized health record transmission to physicians • emergency call services
			Smart lighting system	• gather data from the public street light posts • manage the functioning of street lights
GAMBAS	Handte [2014]	Spain	Smart transport	• context-aware roadway navigation • find the best route for bus travel • passenger count monitoring in buses
ClouT	Galache [2014]	Japan	Disaster management	• alert system for earthquakes • storm alert system • city event notification for citizen • health activity across city notification
Scallop4SC	Yamamota [2014]	Kobe City, Japan	Energy management	• visualization of electricity, gas consumption at house • analyse data for monitoring city and nearby places • detection of energy wastages at smart homes
Smart Santander	Vakali [2014]	Spain	Smart traffic management Air pollution monitoring	• monitoring of smart parking • emotional monitoring of citizens, air quality monitoring and visualization
WindyGrid	Rutkin [2014]	Chicago	Environment awareness Salubrity of building	• situational alert system, monitoring of incidents • historical data analysis, traffic monitoring, CCTV monitoring
OpenIoT	Anagnostopoulos [2015]	Galway, Ireland	Waste management City sensing	• prioritized garbage bin monitoring • CO_2 level monitoring services

historical data analysis is performed over vehicle traffic across the city and also monitoring and managing of cameras installed in public places.

OpenIoT: the European Union proposed the OpenIoT software platform to offer IoT-based smart city services (Anagnostopoulos 2017; Galache 2014). The services include waste management and city sensing services. Under the waste management system is included the monitoring, prioritizing and cleaning of garbage bins and garbage trucks across important places like hospitals, schools, tourist spots and other public places. Under the city sensing system, the services offered are monitoring the CO_2 level and other micropollutant level across the smart city. The city sense data is publically available through various web services.

The following Table 13.1 highlights the summary of Smart City Software platforms and services

Conclusion

The smart city concept has significantly gained attention among researchers, government and business sectors. In this chapter, we analyzed the various solutions associated with the current available concept of smart city. The smart city-based ICT technologies alone are insufficient to meet the growing demand of resources and infrastructures. Hence, in this chapter, optimized solutions and enhancements that are required within the existing smart city concept are discussed.

For this purpose, the various associated problems and gaps with an ICT-based smart city are identified. Rather than using the available technologies for citizen requirement, the focus of the smart city needs to be on providing software platform-based smart city services. In this way, more sustainable solutions and services can be offered to the smart city citizens. Further, the discussion of enabling technologies like IoT, big data, cloud computing and cyber-physical System are contributing towards the growth of smart city. The amalgamation of IoT with other enabling technologies improves the performance of integrated services within the smart city concept.

Various examples of smart city concepts that are implemented and deployed across the world are also presented in this chapter. These smart city initiatives are mainly in a nascent stage and possess multiple opportunities and challenges that are yet required to be focused. The most solid solution is to develop a common platform to conglomerate the research groups, industries, government and NGOs, such that the multitude problem of political, technical, services and financial problems related to a smart city can be established. Thus, the ultimate goal of the smart city such as enhancing the quality of life of the citizens is achieved.

In this chapter, the major functional and nonfunctional opportunities of an IoT-based smart city have been described. These opportunities would definitely pave the way to industries, researchers and software developers of the smart city platform to identify the critical components of the smart city. Further, they can focus on the implementation of these components to enhance the quality of life of smart city citizens.

References

Al-Hader, M., Rodzi, A., Sharif, A. R. and Ahmad N. 2009. Smart City Components Architecture. In *CSSim 2009 – 1st International Conference on Computational Intelligence, Modelling, and Simulation*. Retrieved from: https://doi.org/10.1109/CSSim.2009.34.

Anagnostopoulos, T., Kolomvatsos, K., Anagnostopoulos, C., Zaslavsky, A. and Hadjiefthymiades S. 2015. Assessing Dynamic Models for High Priority Waste Collection in Smart Cities. *Journal of Systems and Software* 50(6): 78. Retrieved from: https://doi.org/10.1016/j.jss.2015.08.049.

Anagnostopoulos, T., Zaslavsky, A., Kolomvatsos, K., Medvedev, A., Amirian, P., Morley, J. and Hadjieftymiades S. 2017. Challenges and Opportunities of Waste Management in IoT-Enabled Smart Cities: A Survey. *IEEE Transactions on Sustainable Computing*. Retrieved from: https://doi.org/10.1109/tsusc.2017.2691049.

Apolinarski, W., Iqbal, U. and Parreira J. X. 2014. The GAMBAS Middleware and SDK for Smart City Applications. In *IEEE International Conference on Pervasive Computing and Communication Workshops, PERCOM WORKSHOPS 2014*. Retrieved from: https://doi.org/10.1109/PerComW.2014.6815176.

Bellavista, P., Cardone, G., Corradi, A., and Foschini L. 2013. Convergence of MANET and WSN in IoT Urban Scenarios. *IEEE Sensors Journal*. Retrieved from: https://doi.org/10.1109/JSEN.2013.2272099.

Bonetto, R., Bui, N., Lakkundi, V., Olivereau, A., Serbanati, A. and Rossi, M. 2012. Secure Communication for Smart IoT Objects: Protocol Stacks, Use Cases and Practical Examples. In *IEEE International Symposium on a World of Wireless, Mobile and Multimedia Networks, WoWMoM 2012 – Digital Proceedings*, 1–7. Retrieved from: https://doi.org/10.1109/WoWMoM.2012.6263790.

Budhiputra, P. M. and Putra K. P. 2016. Smart City Framework Based on Business Process Re-Engineering Approach. In *International Conference on ICT for Smart Society, ICISS 2016*. Retrieved from: https://doi.org/10.1109/ICTSS.2016.7792851.

Bui, N. 2011. Internet of Things Architecture Project Deliverable D1. 1 – SOTA Report on Existing Integration Frameworks / Architectures for WSN, RFID and Other Emerging IoT Related Technologies. *Architecture*, IoT-A, 32–34.

Cassandras, C. G. 2016. Smart Cities as Cyber-Physical Social Systems. *Engineering* 2(2): 156–158. Retrieved from: https://doi.org/10.1016/J.ENG.2016.02.012.

Castellani, A. P., Gheda, M., Bui, N., Rossi, M. and Zorzi M. 2011. Web Services for the Internet of Things through CoAP and EXI. In *IEEE International Conference on Communications*. Retrieved from: https://doi.org/10.1109/iccw.2011.5963563.

Chen, N. and Du W. 2016. Spatial-Temporal Based Integrated Management for Smart City: Framework, Key Techniques and Implementation. In *International Conference on Geoinformatics*, 1–4. Retrieved from: https://doi.org/10.1109/GEOINFORMATICS.2015.7378628.

Chen, N. and Hu C. 2012. A Sharable and Interoperable Meta-Model for Atmospheric Satellite Sensors and Observations. *IEEE Journal of Selected Topics in Applied Earth Observations and Remote Sensing* 5(5): 1519–1530. Retrieved from: https://doi.org/10.1109/JSTARS.2012.2198616.

Cheng, B., Longo, S., Cirillo, F., Bauer, M. and Kovacs E. 2015. Building a Big Data Platform for Smart Cities: Experience and Lessons from Santander. In *Proceedings IEEE International Congress on Big Data, BigData Congress 2015*. Retrieved from: https://doi.org/10.1109/BigDataCongress.2015.91.

Distefano, S., Merlino, G. and Puliafito A. 2012. Enabling the Cloud of Things. In *Proceedings – 6th International Conference on Innovative Mobile and Internet Services in Ubiquitous Computing, IMIS 2012*. Retrieved from: https://doi.org/10.1109/IMIS.2012.61.

Djahel, S., Doolan, R., Muntean, G. M. and Murphy J. 2015. A Communications-oriented Perspective on Traffic Management Systems for Smart Cities: Challenges and Innovative Approaches. *IEEE Communications Surveys and Tutorials*. Retrieved from: https://doi.org/10.1109/COMST.2014.2339817.

Dohler, M., Vilajosana, I., Vilajosana, X., and Llosa J. 2011. Smart Cities: An Action Plan. In *Barcelona Smart Cities Congress*, Barcelona, 1–6.

Dutta, J., Gazi, F., Roy, S., and Chowdhury C. 2017. AirSense: Opportunistic Crowd-sensing Based Air Quality Monitoring System for Smart City. In *Proceedings of IEEE Sensors*. Retrieved from: https://doi.org/10.1109/ICSENS.2016.7808730.

Foell, S., Kortuem, G., Rawassizadeh, R., Handte, M., Iqbal, U. and Marrón P. 2014. Micro-Navigation for Urban Bus Passengers: Using the Internet of Things to Improve the Public Transport Experience. Retrieved from: https://doi.org/10.4108/icst.urb-iot.2014.257373.

Fortes, M. Z., Ferreira, V. H., Sotelo, G. G., Cabral, A. S., Correia, W. F. and Pacheco O. L.C. 2014. Deployment of Smart Metering in the Búzios City. In *IEEE PES Transmission and Distribution Conference and Exposition, PES T and D-LA 2014 – Conference Proceedings*. Retrieved from: https://doi.org/10.1109/TDC-LA.2014.6955278.

Galache, J. A., Yonezawa, T., Gurgen, L., Pavia, D., Grella, M., and Maeomichi, H. 2014. ClouT: Leveraging Cloud Computing Techniques for Improving Management of Massive IoT Data. In *Proceedings – IEEE 7th International Conference on Service-Oriented Computing and Applications, SOCA 2014*. Retrieved from: https://doi.org/10.1109/SOCA.2014.47.

Gurgen, L., Gunalp, O., Benazzouz, Y. and Galissot M. 2013. Self-Aware Cyber-Physical Systems and Applications in Smart Buildings and Cities. Retrieved from: https://doi.org/10.7873/date.2013.240.

IBM Brazil. 2011. *Rio Operations Centre – Rio de Janeiro – Brazil. Talks at Google: Ethan Zuckerman, Digital Cosmopolitans.*

IBM, and Solutions Government. 2011. *IBM Smarter City. IBM Industry Solutions*.

IEEE Computer Society. 2012. *IEEE Standard for Local and metropolitan area networks – Part 15.6: Wireless Body Area Networks IEEE Std 802.15.6*.

ISO/TC 268 – *Sustainable Cities and Communities*. n.d. Accessed April 3, 2019: www.iso.org/committee/656906.html.

Kinney, P., Jamieson, P. and Gutiérrez J. 2015. IEEE 802.15 WPAN™ Task Group 4 (TG4). *Task Group 4 (TG4)*.

Lee, S., Yoon, D. and Ghosh A. 2008. Intelligent Parking Lot Application Using Wireless Sensor Networks. In *2008 International Symposium on Collaborative Technologies and Systems, CTS'08*. Retrieved from: https://doi.org/10.1109/CTS.2008.4543911.

Liu, Z., Jiang, S., Zhou, P. and Li M. 2017. A Participatory Urban Traffic Monitoring System: The Power of Bus Riders. *IEEE Transactions on Intelligent Transportation Systems*. Retrieved from: https://doi.org/10.1109/TITS.2017.2650215.

Mehmood, Y., Ahmad, F., Yaqoob, I., Adnane, A., Imran, M. and Guizani, S. 2017. Internet-of-Things-based Smart Cities: Recent Advances and Challenges. *IEEE Communications Magazine*. Retrieved from: https://doi.org/10.1109/MCOM.2017.1600514.

Nam, T. and Pardo T. A. 2011. Conceptualizing Smart City with Dimensions of Technology, People, and Institutions. Retrieved from: https://doi.org/10.1145/2037556.2037602.

PAS 181:2014 *Smart City Framework. Guide to Establishing Strategies for Smart Cities and Communities*. n.d. Accessed April 3, 2019: https://shop.bsigroup.com/ProductDetail/?pid=000000000030277667.

Pham, T. N., Tsai, M. F., Nguyen, D. B., Dow, C. R. and Deng D. J. 2015. A Cloud-based Smart-parking System Based on Internet-of-Things Technologies. *IEEE Access*. Retrieved from: https://doi.org/10.1109/ACCESS.2015.2477299.

Rana, R. K., Chou, C. T., Kanhere, S. S., Bulusu, N. and Hu, W. 2010. Ear-Phone: An End-to-End Participatory Urban Noise Mapping System. In *Information Processing In Sensor Networks*. ACM, 105–116.

Santana, E. F. Z., Chaves, A. P., Gerosa, M. A., Kon, F. and Milojicic D. S. 2017. Software Platforms for Smart Cities. *ACM Computing Surveys*. Retrieved from: https://doi.org/10.1145/3124391.

Santucci, G. 2010. The Internet of Things: Between the Revolution of the Internet and the Metamorphosis of Objects Gérald. *Vision and Challenges for Realizing the Internet of Things*. Retrieved from: https://doi.org/10.2759/26127.

Schaffers, H., Komninos, N., Pallot, M., Trousse, B., Nilsson, M. and Oliveira A. 2011. Smart Cities and the Future Internet: Towards Cooperation Frameworks for Open Innovation. *Lecture Notes in Computer Science (Including Subseries Lecture Notes in Artificial Intelligence and Lecture Notes in Bioinformatics)*. Retrieved from: https://doi.org/10.1007/978-3-642-20898-0_31.

Shelby, Z., Hartke, K. and Bormann C. 2010. *The Constrained Application Protocol (CoAP)*. No. RFC 7252.

Silva, W. M. da, Alvaro, A., Tomas, G. H. R. P., Afonso, R. A., Dias, K. L. and Garcia V. C. 2013. *Smart Cities Software Architectures*. Retrieved from: https://doi.org/10.1145/2480362.2480688.

Smart Cities Look to the Future – Smarter With Gartner. n.d. Accessed April 3, 2019. Retrieved from:www.gartner.com/smarterwithgartner/smart-cities-look-to-the-future/.

Smart City Standards and Publications | BSI Group. n.d. Accessed April 3, 2019: www.bsigroup.com/en-GB/smart-cities/Smart-Cities-Standards-and-Publication/.

Takahashi, K., Yamamoto, S., Okushi, A., Matsumoto, S. and Nakamura, M. 2012. Design and Implementation of Service API for Large-Scale House Log in Smart City Cloud. *CloudCom 2012 – Proceedings 4th IEEE International Conference on*

Cloud Computing Technology and Science. Retrieved from: https://doi.org/10.1109/CloudCom.2012.6427590.

Tei, K. and Gurgen L. 2014. ClouT: Cloud of Things for Empowering the Citizen Clout in Smart Cities. *IEEE World Forum on Internet of Things, WF-IoT 2014*. Retrieved from: https://doi.org/10.1109/WF-IoT.2014.6803191.

Thornton, S. 2013. Chicago's Windy Grid: Taking Situational Awareness to a New Level. *Data Smart City Solutions*. Retrieved from: https://datasmart.ash.harvard.edu/news/article/chicagos-windygrid-taking-situational- awareness-to-a-new-level-259.

Unidas, Organización de Naciones. 2017. The World Population Prospects: The 2017 Revision. *World Population Prospec ts T he 2017 Revision*. Retrieved from: https://doi.org/10.1017/CBO9781107415324.004.

Uusitalo, M. A. 2006. Global Visions for the Future Wireless World from the WWRF. *IEEE Vehicular Technology Magazine*. Retrieved from: https://doi.org/10.1109/MVT.2006.283570.

Vakali, A., Anthopoulos, L. and Krco S. 2014. Smart Cities Data Streams Integration: Experimenting with Internet of Things and Social Data Flows. *ACM*. Retrieved from: https://doi.org/10.1145/2611040.2611094.

Worldwide Semi-annual Smart Cities Spending Guide. n.d. Accessed April 3, 2019: www.idc.com/getdoc.jsp?containerId=IDC_P37477.

Zanella, A., Bui, N., Castellani, A., Vangelista, L. and Zorzi, M. 2014. Internet of Things for Smart Cities. *IEEE Internet of Things Journal*. Retrieved from: https://doi.org/10.1109/JIOT.2014.2306328.

Zappatore, M., Longo, A. and Bochicchio, M. A. 2016. Using Mobile Crowd Sensing for Noise Monitoring in Smart Cities. *2016 International Multidisciplinary Conference on Computer and Energy Science, SpliTech 2016*. Retrieved from: https://doi.org/10.1109/SpliTech.2016.7555950.

Zhang, R., Newman, S., Ortolani, M. and Silvestri S. 2018. A Network Tomography Approach for Traffic Monitoring in Smart Cities. *IEEE Transactions on Intelligent Transportation Systems*. Retrieved from: https://doi.org/10.1109/TITS.2018.2829086.

Zhou, Y., Wu, Q. and Yan T. 2018. A Hierarchical System for Energy Consumption Monitoring and Information Management. *Proceedings of the 2017 12th IEEE Conference on Industrial Electronics and Applications, ICIEA 2017*. Retrieved from: https://doi.org/10.1109/ICIEA.2017.8282887.

Zygiaris, S. 2013. Smart City Reference Model: Assisting Planners to Conceptualize the Building of Smart City Innovation Ecosystems. *Journal of the Knowledge Economy*. Retrieved from: https://doi.org/10.1007/s13132-012-0089-4.

14

Application of Smart City Concept to the Leading Destination: Evidence from Istanbul – A Case Study

Aysegul Acar, Eda Kocabas, M. Fevzi Esen and Fatih Canitez

CONTENTS

14.1 Introduction 293
14.2 Smart City and its Characteristics 296
14.3 Smart Tourism and Smart Tourism Destinations 299
    14.3.1 Smart Tourism Applications Based on ICTs.............................301
    14.3.2 Smart City and Smart Tourism Applications in Istanbul.........304
14.4 Case Study Analysis 309
Conclusion and Future Directions..314
References...315

14.1 Introduction

The concept of smartness found its roots back in the 1990s, but it has grown significantly since 2008. First, the term was invented as a complex technological infrastructure placed in urban areas to promote social, environmental as well as economic well-being. More specifically, in order to allude to information and communication technologies (ICTs) application to the cities' infrastructure system, the concept of "smart" was firstly utilized in the 1990s by the California Institute of Smart Communities, one of the primary agencies that investigate how a community could be considered smart. A few years later, the Centre on Governance at the University of Ottawa expanded the application of the term to include a governance-focused approach rather than merely emphasizing ICTs. Since the "smart" concept is linked to the academic literature, it has swiftly become a phenomenon for the private sector and governments.

In the twenty-first century, the accrual of ICTs is the most extensive change to a society. Furthermore, due to the ascent of ICTs, local governments have progressively sought programs and strategies to unify technologies between systems and city services – a pursuit of setting up different classifications, including: "wired", "cyber", "digital", "intelligent", "virtual" and "universal". The integration of ICTs enhances interconnected subsystems and processes to eventually cope with the social, economic and ecological difficulties caused by urbanism. This connotation of the latest technology has triggered concepts like the smart city, the smart planet and, in the near future, the smart tourism destination.

The notion of smart has been attached to cities to define the attempts to use innovative technologies in order to achieve optimization of resources, fair and effectual governance and to sustain a high standard of living. Briefly, the idea of the smart city represents an environment in which technology is placed in the cities. More specifically, the concept of smart cities is often associated with the embedded ecosystem of technology that aims to establish synergies with their social elements to improve the life quality of citizens and increase the effectiveness and efficiency of the city products or services. Cities can be called smart when they provide a higher quality of life, sustainable economic growth, human capital investment, satisfactory government contribution and higher standard in infrastructure systems that support the accurate dispersion of information across the city. Hence, smart cities must rely on their smartness in three areas: knowledge, infrastructure and human capital.

ICTs enable cities to be more enjoyable and available for both locals and visitors via interactive services that connect all of the local organizations to ensure real-time activities and services as well as to utilize the data centrally for a better city system. This innovation will be synchronized with the city's social parts to enrich residents' quality of life while enhancing the productivity of city services, like optimization of energy usage and enhancement on traffic monitoring. In fact, ICTs promote cities that address social challenges. Smart city development also provides access to value-added services, such as access to real-time information on the public transport network, both for the city locals and visitors. In addition, the smart city has enabled the cities to connect their stakeholders through the Internet of things (IoT), allowing the city to interact progressively with their stakeholders. Today, tourism destinations are affected by emerging technologies and face some new difficulties originating from both environmental changes and clients' expectations. To manage these difficulties, destinations must first detect the types of difficulties that they encounter.

From a tourism point of view, while enhancing efficiency and the computerization of support process for relevant institutions, ICTs can contribute to the creation of value-added experiences for tourists. So, smart city development can also promote the formation of smart tourism as well as smart tourism destinations which are smart cities' particular cases: they implement smart city infrastructure systems and principles in rural and urban areas via not only regulating big data from locals, but also from tourists in their efforts to promote availability, resource mobility, sustainability, allocation of services and life quality. The ultimate goal of smart tourism is to use the systems to improve the tourism practice and to increase the efficiency of resource management aimed at maximizing competitiveness of destination, consumer satisfaction of the target and to follow a sustainable path for a longer period of time.

With the latest technology placed in the destination, it can enhance touristic experiences and increase the competitiveness of the destination. The smart city concept can also be applied to smart tourism destinations. This concept provides tourists with better interaction, communication and building closer relationships with local government, enterprises, city attractions and also the residents. Nonetheless, smart tourism attributes to a new smart tourism economy with new players, new exchange models and new resources. Ultimately, it supports the city services and developments in various ways. Continuously, the meaning of the innovation in the application of software, hardware and development of networks is that the smart tourism city can respond efficiently, quickly and effectively to the needs of locals and tourists and can perform better than its competitors and can maintain long-term well-being. Smart tourism application makes a lot of sense especially when considering the cities requiring more infrastructural investments and attracting millions of visitors who benefit from the city's resources and services.

It can be concluded that smart tourism destinations are focusing on the needs of tourists by closing ICTs with the tourism innovation industry and traditional culture to improve the quality of service in tourism, to enhance tourism administration and to broaden the scale of industry to a wider extent. The basic smart tourism destination priorities are: 1. enhancing the tourist's travel experience, 2. providing a smarter platform for distributing and collecting data within a destination, 3. facilitating tourism resources by effective and efficient allocation and 4. integrating tourism suppliers at both the macro and micro level to procure excellent distribution of tourism sector to the local residents. Generally, tourists have less awareness and limited knowledge on the destinations they visit. They have distinct behaviors, features and needs. Creating crowd-sourced applications via utilizing tourists' input can provide invaluable information to the destinations that capture the requests and complaints of tourists in a timely manner. This chapter will throw some light on the literature on smart tourism destinations. Case studies can be more appropriate to further examine the best practices of smart tourism destinations and to create a deeper understanding within this framework.

This study specifically aims to show how Istanbul, as a leading tourism destination, creates high-tech systems to transform itself into a smart city tourism destination and contend in the worldwide knowledge-based economy while addressing the research questions as follows:

> Question 1: How can Istanbul and its governmental authorities direct this smart city transformation?

Question 2: What can be fundamental drivers and restrictions for this change?
Question 3: What could be the primary handicap that Istanbul faces? (any suggestion could be said)
Question 4: What can be the expected terms or conditions for smart city transformation?
Question 5: Which infrastructures are needed to become a smart city or a smart tourism destination?
Question 6: What kind of collaboration strategies should be developed with the city stakeholders (NGOs, governmental authorities and private bodies) during the process of becoming a smart city?

To achieve the research objectives of this chapter, a case study approach is applied. This approach consists of various analyses and interviews of sites by acquiring views from observations and interviews. Various data are collected and secondary data are used at the same time through in-depth interviews with high-level managers of the city authorities such as the manager of Istanbul Electric Tramway and Tunnel Establishments (IETT), leaders from different NGOs being active specifically for smart city applications and also secondary data are utilized. To achieve the stated goals, the chapter has the following structure: first, the existing literature on smart cities, smart tourism and smart tourism destinations is reviewed. With a detailed definition of the case study analysis, the main elements of Istanbul's smart tourism destination strategy are examined. Afterwards, the advantages and disadvantages of being a smart tourism destination and its future objectives are stated. Last, the chapter concludes with different views produced from the case study.

14.2 Smart City and its Characteristics

Studies have indicated that urban areas are home to around half of the population of the world (Dirks, Gurdgiev and Keeling 2010 and Dirks, Keeling and Dencik 2009). The rural-urban migration is predicted to continue for the next couple of decades. With the growing urban population, myriads of problems are witnessed in various sectors, including physical, technical and material problems such as aging, inadequate and deteriorating infrastructures, traffic congestion, human health concerns, air pollution, scarcity of resources and challenges in waste management (Borja 2007; Marceau 2008; Toppeta 2010 and Washburn et al. 2010). Moreover, there are other challenges within the urban setting, which are rather organizational and social in nature than physical, material or technical. These challenges are linked to the various factors such as social and political complexities, competing objectives and values, high levels of interdependence as well as multiple and diverse stakeholders.

Consequently, the problems associated with the urban living are usually tangled and dehumanizing (Dawes, Cresswell and Pardo 2009; Rittel and Webber 1973; Weber and Khademian 2008).

Therefore, the concept of a smart city is an idea for contemporary times. It is an idea that the city planners must deeply understand because it will perhaps be a panacea to the problems of the urban areas through ensuring livable conditions even with the population explosion witnessed in these urban spaces. The challenges that the cities around the globe pose have jolted all the stakeholders into thinking of smarter ways to overcome such challenges. All these ways have been inculcated into the concept of a smart city. As a result, a smart city is defined as a city in which ICTs have been integrated into the traditional infrastructure through the utilization of technological advancements. A smart city has the main objective of developing an effective system in addressing the urban challenges in a feasible manner through coordinated technological frameworks, techniques, models for using urban data across temporal and partial scales as well as conscious development of new technologies to achieve dissemination and communication ends. Moreover, the smart city targets enhancing urban governance, organizational structures and identifying common problems relating to the cities' infrastructure, energy, risks, transport, hazards and other problems (Batty et al. 2012). The concept of a smart city is a term in high circulation within the scholarly world but a thorough understanding of what it really entails has not been clearly understood among the academia and the professionals. This is evidenced by a few existing studies that systematically explore this area. However, it must be noted that the concept of a smart city is still new and its definition and conceptualization is still developing (Boulton, Brunn and Devriendt 2011; Hollands 2008).Various definitions nonetheless have been put forth both in academic and practical spheres in an attempt to effectively define this term (see Table 14.1 below).

A smart city has a unique set of features that drive the attainment of objectives. Such features have been explored by various literatures and identified to include interactive and creative platforms that allow information exchange, effective utilization of data and effective deployment of ICTs to engage effectively with all the stakeholders. In the identification and provision of smart solutions, the utilization of open data is an emerging idea (Berrone et al. 2016). In a smart concept, an emphasis on infrastructure and human capital is a key priority (Bakıcı, Almirall and Wareham 2013). Through effective use of ICTs and efficient governance systems, the smart cities achieve their goals and thus facilitate transparency and customer satisfaction (Storper and Scott 2009). In these smart cities, the inhabitants will have high levels of interaction due to effective communication networks. The smart cities' pillars include people, processes and technologies that enable people to access essential services such as healthcare, education, tourism, sustainable environment and safe buildings. Even control systems have web-enabled mechanisms, which provide an integrated approach to managing outcomes. In essence,

TABLE 14.1

Definitions of a Smart City

Definitions	Sources
A smart city is a city that coordinates, integrates and monitors the entire critical infrastructure such as water and power, seaports, communications, airports, subways, rails, tunnels and road bridges within its perimeters in order to optimize efficiency and performance.	Hall (2000)
A smart city is a city that shares a culture, inspirations, life and knowledge. This is a city that enables its inhabitants to develop and utilize their creativity to progress and live comfortably.	Porter et al. (2002)
A smart city is a city that utilizes ICTs to enhance its inhabitants' freedoms (such as of speech) and liberties as well as facilitates access to services and information.	Partridge (2004)
A smart city is a city responsive to the people's needs and promising a better future for its inhabitants in economy, governance and people, mobility, better living and environment because it is premised on a combination of activities and strengths such as building self-decisive and informed citizenry.	Giffinger, Kramar and Haindl (2010)
A smart city is a city that employs smart commuting technologies to develop more sustainable and livable conditions for its inhabitants by integrating the critical infrastructural components and services including utilities, transportation, real estate, public safety, healthcare, education, and city administration.	Washburn et al. (2010)
A smart city is a city that integrates Web 2.0 technology and ICTs with other design, organizational and planning efforts to speed up the systems as well as demystify bureaucratic processes so that new and innovative solutions to the challenges posed by urban complexities are facilitated in order to better the living standards of the inhabitants.	Toppeta(2010)
A smart city is an interconnected, instrumentalized, and intelligent city.	Harrison et al. (2010)
A smart city is a city in which there is a conscious and dedicated commitment to invest in social and human capital as well as in both modern communication (ICTs) and traditional infrastructures to enhance a high quality of life and fuel economic growth through prudent management of natural resources and involvement of all stakeholders.	Thite (2011)
A smart city is a city with geographically defined area with advanced technologies such as energy production, ICTs and logistic production among others, which work collaboratively to better the lives of citizens in areas such as inclusion, well-being, participation, intelligent development and quality of environment. All these elements are coordinated through a well-established system in terms of rules and policy.	Dameri(2013)
A smart city is a city in which technological advancements make a bold statement through elaborate ICT systems, which have the capacity to provide high quality services to its citizens with a view to improving their life.	Piro et al. (2014)
A smart city is characterized by a number of factors such as a focus on economic and creative activities to promote urban development, an emphasis on environmental sustainability, a focus on social inclusivity and social capital and effective and networked infrastructure that allows for social, political and economic expediency.	Albino et al. (2015)
A smart city has the ability to attract human capital and to move this human capital based on needs basis in conjunction with other factors such as individual and organized actors with heavy reliance of ICTs.	Meijer and Bolivar (2016)

the features of a smart city can be narrowed down to contain six elements or aspects: smart economy, smart environment, smart mobility, smart living, smart people and smart government (Dassani, Nirwan and Hariharan 2015; Buhalis and Amaranggana 2014). The six aspects are linked to customary regional and eighteenth-century urban growth and developmental theories. Specifically, the elements are centered separately on principles such as competitiveness among the regions, money, ICTs and transport economies, physical resources, social and human capital, standard of life and the contribution of members of the society toward the cities' progress.

14.3 Smart Tourism and Smart Tourism Destinations

Smart places aim at increasing attractiveness and promote the standard of life of all its stakeholders, comprising of inhabitants and visitors (Caragliu, Del Bo and Nijkamp 2011; Buhalis and Amaranggana 2014). To realize the outcomes, a wide array of elements ought to be incorporated. Until today, most of the studies on smart tourism and smart tourism destinations are theoretical and mainly concentrated on the importance of tourism industry-led growth and co-establishment of activities to promote the experience of tourists (Wang et al. 2013; Buhalis and Amaranggana 2014; Gretzel et al. 2015a). Exercising smartness in tourism destinations has been vital because the better-educated and involved tourists are vigorously interrelating with the targeted places, co-establishing tourist attractions for value creation (Neuhofer, Buhalis and Ladkin 2012). Smart tourism means the incorporation of technological infrastructure like communication using mobile phones, Internet and augmented reality, that incorporates software, hardware and network skills to give real-time information that allows better intelligent judgment amongst every stakeholder (Gretzel et al. 2015a; Hunter et al. 2015; Tu and Liu 2014). Molz (2012) is one of those who tried to list the features of smart tourism. He pointed out that smart tourism is connected with (1) web-abilities, using applications based on web; (2) the visitors as co-producers of targeted products; (3) promoting knowledge and experience using new skills (augmented reality); (4) linking and sympathizing with local inhabitants and other tourists who visited the destinations; and (5) enhancing sustainability of ecology and community (Molz 2012). In addition to these, various theoretical contexts concerning smart tourism situations have been suggested. According to Koo et al. (2017), smart tourism is composed of six strata: 1. IT infrastructure for the incorporation of systems, instruments and IoT; 2. aggregator of data for the incorporation of diverse data; 3. platforms such as cloud, free API and apps; 4. structure signifying policies, process and procedure, and authority; 5. citizen, who is the driver of innovation as well as users of cultural and entertainment activities as a member in the society and economy; and 6. the tourist, being the main consumer of smart tourism. Generally, structure is an expansion of the current frameworks relating to smart cities. The five layers appearing at the initial position have similar characteristics as smart cities do (Koo et al. 2017). Similarly, the major constituents of smart tourism consist of three layers comprising smart knowledge, smart industry environment and smart tourist destination (Gretzel et al. 2015b) apart from three components; data was added as the fourth, which entails processes, interchange and accumulation.

The idea of the smart tourism destinations came from the smart city and, specifically, smart tourism notions. Baggio and Del Chiappa (2014), while relating the notion of smartness (Nachira 2002) to the tourism industry, explained a smart tourism destination as an integrated organization of participants providing goods and services to travelers, perfected by a technological substructure intended to establish a digital and modern setting that enhances teamwork, sharing of knowledge and open innovation. Incorporating smartness into tourism destinations means vigorously communicating with all stakeholders by means of a technological arena (Buhalis et al. 2014).

Smart tourism targets the restructuring and creation of the tourist experience with the tourism industry and its practices (Wang et al. 2013). According to Gretzel et al. (2015a), a tourism system benefits from smart technology which handles intelligent tourist services and is considered the co-creation of values in a tourism system. Similarly, the main objective is to make use of the system to promote tourists' experience and enhance resource management efficiency relating to maximization both of touristic destination competitiveness and satisfaction of tourists, whilst also proving sustainability for a specified timeframe (Buhalis and Amaranggana 2015). Smart tourism destinations are a benefit to: (1) technology entrenched settings; (2) responsive procedures at both macro and micro levels; (3) devices for the end-user in several touchpoints; and (4) involve participants who utilize the platform vigorously as a central system. The main concern of smart tourism destinations is to promote an experience to travelers; to avail additional smart platforms so as to collect and disseminate data to the destinations; to enable the effective distribution of tourism facilities; and to incorporate tourism providers both at macro and micro level targeting (Rong 2012). In most cases, travelers usually have inadequate experience and limited information on destinations they travel to for leisure. They always have varied requirements and personality traits. The creation of crowd-originated applications by means of tourist participation may give invaluable understanding to meet tourists' requests and grievances in an appropriate way (Haubensak 2011). Smart tourism destinations are the center of tourism facilities and data resources, sustained by IoT and cloud computing, aimed at promoting tourists' knowledge by a process of intelligent recognition and supervision. The true logic concerning smart tourism destinations is to aim at tourists' requirement by merging the ICTs with spontaneous beliefs so as to enhance the standard of tourism services quality, promote tourism governance and expand industry further (Huang et al. 2012).

14.3.1 Smart Tourism Applications Based on ICTs

The new era of ICTs which permit large amounts of data to be converted into value propositions has unlocked a number of unusual tools concerning the tourism industry. Since the tourism industry is among the most appropriate sectors where ICTs are applied comprehensively from functioning and industry perceptions, it is not unexpected that the concept of smart tourism destination has evolved quickly (Koo et al. 2016). In addition, ICTs can enable cities to be more accessible and appealing, for both visitors and residents, since they take part in the creation of communicating services which intersect with local establishments, thus users can promptly obtain data and services. Contextually, a new type of destination has surfaced: "smart tourism destination". This notion is applicable to a destination where technology affects the experience of tourism, promotes the destination's competitiveness and enhances the growth of tourism projects (Boes et al. 2015; Presenza et al. 2014). Particularly, with the commencement of ICTs, smart tourism is centered on cloud computing, AI, big data and mobile interaction. IoT is an area in which every object or instrument could be converted into smart and distinguishable tools. These instruments can interact with several smart instruments by a network. The interaction produces a vast quantity of data that could be kept and computed using the cloud services available. Implementations of analyzing big data could be extremely important in examining the patterns and tendencies in the data. IoT develops platforms which are capable of transmitting various sorts of data through a participatory sensing system (Gutiérrez et al. 2013). In a tourism situation, the tourist can just utilize his/her phone to. find their destination and events through in-situ data gathering and commentaries. These events create enormous proportion of digital hints causing multidimensional groups of data that are referred to as big data. Through big data management, tourism establishments can obtain valuable insights and important understanding to improve tourists' experiences and they can promote how they associate with consumers (SOCAP International 2013). Owners of these types of technological advances have adequate competitive benefit in comparison to opponents.

Buhalis and Amaranggana (2014) point to three major aspects which make ICTs essential in promoting smartness in a tourism destination: IoT, cloud computing and end-user Internet service system. The cloud computing services are created to give an effective means to reach firms' web platforms and the data stored by the network. The application of cloud computing will result in a reduction in fixed charges and transform them as a variable cost centered on the requirements (Etro 2009). It also motivates the sharing of data which is essential to carry out projects on smart tourism destinations.

For instance, an advanced system for a tour guide can work for a large population of tourists (Zhang et al. 2012 as cited in Wang et al. 2013). Second, the IoT can sustain smart destinations through supplying data and examining it and also via automation and management (Chui et al. 2010). For instance, chips implanted into entry tickets permit suppliers of tourism products and services to trace tourists' whereabouts and their behavior of consumption which enables promotions based on location to be assessed (Lin 2011). Particularly for automation and management, the system can regulate the number of visitors in a given area through the use of several detectors that specify destinations' carrying capacity (Mingjun et al. 2012). The Internet service system for end-users is the third element of a smart destination. Equipment and Internet applications which provide support to cloud services as well as the IoT at different meanings for end-users (Huang et al. 2012). To make it clear, the individual payment systems are grounded on individual communication instruments like tablets and smartphones.

Wireless networking and screens which are touch-sensitive are installed in the tourism destination spots to help visitors. Providers of tourism services and government institutions are resourced with websites and links to the cloud amenities. Conversely, convenient connectivity is the key element to actuate these three important aspects of ICTs in smart tourism destinations. Bearing this in mind, it is of paramount importance for the government assisted by different interested parties to provide strong network coverage in the large towns so as to bridge the gap between countryside and commercially dense areas. The inventions motivated by the IoT indicate crucial returns for the growth of the tourism industry. For instance, wearable technologies which can be put on smart watches play a pivotal task in the tourism industry. Furthermore, they assist in data collection by use of their sensors and video recorders, but also by linking with the network and possibly the IoT and having a high ability to connect without any skills required. At the administrative level, this kind of system could be appropriate in controlling the number of visitors in a given tourism destination through the use of several types of sensors, each having the destination's carrying capacity as a point of reference. The social aspect equally has to be acknowledged as one of the smart objects fixed in these sites could instinctively stimulate the propagation of messages to relatives as well as friends to assist them to understand what they did in the past or what they are doing now, like relocating from one place to the other or interacting with friends. Likewise, the key issue relating to smart tourism sites is the consolidation of IoT and ICTs in a physical substructure. For example, Barcelona provides tourists with interactive bus shelters which, apart from providing relevant information to tourists and departure times, also have USB ports meant for mobile device charging. Additionally, it provides bicycles within the city and visitors can confirm their locations using a smartphone application, thus providing environmentally friendly transportation within the city (http://smartcity.bcn.cat/en/bicing.html).

The city of Brisbane has more than 100 beacons fixed at important points so as to relay information to travelers through a mobile application when they are still in a certain radius of the destination (Koo, Park and Lee 2016). Amsterdam applies beacons to enable tourist signs converted from a given language into other languages. Amsterdam Arena is analyzing sensors to enable control crowd more easily. Seoul has spent a considerable amount offering free Wi-Fi and smart mobile devices to visitors in South Korea's Island Jeju. It has just confirmed that the destination is a smart tourism center which applies advanced technology for providing services to visitors (Koo, Park and Lee 2016) and in Taiwan's Sunmoon Lake, location-based information is offered on the buses to tourists.

Smart tourism destinations should espouse the idea of a smart city through adopting relevant tourism technologies that match with the expectations of a smart city. Some of the smart services include augmented reality (AR), which allows visitors to access and experience time travel and digital recreation of tourism destinations (Chillon 2012). In addition, the vehicle tracking system offers real-time information of a transportation network that is traceable to the end-user's device (Arup 2010). The hotels should be sustainable in terms of energy demands and energy audits should be carried out by referring to environmental management decisions (Metric Stream 2013).

Moreover, there should be an application dispensing services in many languages electronically in various packages such as an electronic travel guide (Jordan 2011). The services should also be accessible through mobile applications such as QR codes and NFC tags that allow tourists to access information relating to areas of interest (GSMA 2012). A very efficient 24/7 customer service should be put in place in order to facilitate customer complaints, especially through a complaints management system supported by ICTs such as mobile applications and SMS, which could effectively direct them to the appropriate person(s) (Metric Stream 2013). In addition to these mobile applications, equipping the tourist destinations with technologies like cameras, sensors and other smart devices such as touch screens and so on is very critical in a smart city so that data on the tourists can be collected. This data can be stored in a wireless network through cloud computing. This data can also be examined by the diverse stakeholders in the tourism industry. For ease of services, tourists should be issued with travel cards, which they can use in all transport services as well as when they go shopping and so on.

Tourist activities in a smart city will be facilitated through developing mobile applications and touch screens installed in different destinations (Lee et al. 2011). Often, the mobile applications need Internet connection. Websites also play an important role in easing tourists' bookings. These applications will give invaluable information to the tourists, such as the shortest route, or alternative routes in case of traffic jams (Gretzel et al. 2015b). Mobile tours could enable visitors to find tourist attraction points, regional restaurants and so on. However, in smart tourism, there are a number of concerns that need to be addressed such as the privacy and security of the data of the tourists. Ethically speaking, the data collected during the tourists' stay should not be used for other intents except those it was intended to. The other concern is the overreliance on network services and technologies such that rendering services in a smart city may not be possible without the help of high-end infrastructure and use of smartphones. Moreover, serious systematic changes must be conducted on the industry to benefit from such technological advancements. Furthermore, the technologized world in a smart city is overreliant on knowledgeable as well as trained personnel. Therefore, an industry must collaborate with other industries such as telecoms in order to produce customer-centered services (Gretzel et al. 2015a).

14.3.2 Smart City and Smart Tourism Applications in Istanbul

Having a rapidly growing population (which is currently around 15 million people), Istanbul grapples with many urban challenges such as urban transport, energy, infrastructure and environmental pollution. The smart city concept is considered by Istanbul Metropolitan Municipality (IMM) a useful approach to tackle these issues. IMM undertook the smart city project at the beginning of 2017. A variety of stakeholders including public agencies, private companies, local communities, NGOs, and academia are included in the project. A subsidiary municipal company which is responsible for the provision of ICT solutions for Istanbul, ISBAK, is assigned as the leading coordinating agency for the project. One global and one local private technology company support ISBAK in delivering the project. Setting out the vision, strategies and roadmaps are the starting phase of the smart city project. Aiming to make Istanbul a smart city tourism destination, the needs, demands and expectations of the inhabitants as well as the visitors are sought to be incorporated. Around 11 million international tourists visited Istanbul in 2017 and the tourism sector is critical in Istanbul to boost the local and national economy.

As part of the smart city project, several sub-projects are planned. A big data management center is planned to collect and make use of the data generated from different sources of urban activities such as mobility, energy and waste management. An ecosystem for the start-up firms which can utilize the open data provided from this center is planned to foster economic development and innovation capacity of Istanbul. Additionally, IoT projects, smart communication networks and operation centers, energy automation systems, smart lighting and sensors, artificial intelligence solutions for energy, mobility, health and security are the other sub-projects.

Istanbul Convention and Visitors Bureau, as the official destination marketing office, launched a mobile application called OneIstanbul in 2016. The application provides all visitors with a variety of alternative routes within the city. It has four different language options including English, German, Arabic and Turkish. This mobile advantage enables visitors to discover the city in a self-organized way and creates smart individuals or visitors who want to experience the destination by themselves.

The local governments in Istanbul, both metropolitan and districts, are primarily responsible for implementing, coordinating and monitoring the smart city initiatives. They are located at the center of the stakeholder map, expected to cooperate with other stakeholders such as businesses, communities, NGOs and universities. They also finance the projects through various public-private partnerships. Learning from the best practices from the cities around the world, Istanbul has set out its vision, strategies and roadmap.

Besides, Fatih Municipality, one of the district municipalities of Istanbul, the local government is responsible for the historical peninsula which is the most visited area of Istanbul covering the key historical and cultural attraction points such as Topkapı Palace, Blue Mosque, Basilica Cistern and Hagia Sophia Museum. The municipality developed an in-house web-based GIS platform. An augmented reality platform, called FatihAR, is developed and used for visualizing the historical, cultural, public places and land parcels in four different languages in picture, audio and video formats. In the area of virtual reality, research and development activities are undertaken to make greater use of hologram, wearable technologies and beacons to enhance the tourist experience.

CEVKO (Environment Protection Foundation of Turkey) recycle mobile application is used for real-time monitoring of the waste collection and disposal processes. Fatih mobile application provides information about the municipal services. Fatih Municipality also aims at enhancing its governance and management processes through its updated system architecture, data mining and business intelligence products. Process analysis and automation systems help establish a digital platform to carry out its municipal functions.

Cekmekoy Municipality, on the other hand, seeks to improve citizen participation channels by developing a digital archive, called the Smart Pen project. The citizens not having adequate technological capability are specifically addressed and training programs tailored for them are developed. Facilitating property sales through user-friendly digital platforms is aimed at both visitors as well as local people. Provision of efficient and fast municipal services are the objectives of the set of projects undertaken by Cekmekoy Municipality. Increasing the use of locally developed technologies and decreasing the dependence on foreign technologies is another aim of the smart city concept adopted by this district municipality.

Data analytics is the first priority of the smart city project of IMM. Establishing a data management center to manage all municipal services through one center using cloud computing aims at realizing a smart city management system. Incorporating environment, mobility, energy, people, governance, economy and information aspects of urban management, this system is planned to deliver urban services in a more efficient way. A fleet monitoring center connects with the IoT devices installed on the vehicles; an environment control center monitors the waste management processes; a traffic control center monitors the traffic flow of Istanbul with 610 cameras; and the mobile application IBB NAVI which provides an online traffic and mobility navigation tool for the inhabitants and visitors are some of the smart city applications currently in use in Istanbul.

ISBAK defines the smart city objective as "enhancing the quality of life, using resources more productively and efficiently, utilizing the technology, engaging with all the stakeholders in the city to achieve a sustainable city". The smart city is considered by ISBAK as an all-encompassing urban framework to increase the participation of the citizens, grow the local economy, provide data-based urban management and establish an agile governance structure. ISBAK lists the major themes of the smart city initiative as the following: sustainable productivity, urban entrepreneurship, qualified labor force and an attractive city image. In this respect, a neoliberal logic is said to arguably permeate the driving force of the smart city concept in Istanbul. The other targets listed by ISBAK reiterate the underlying neoliberal mindset (Fortune 2018):

- integrating with global business networks and tracking the development with global performance indicators;
- supporting the start-ups and investors;
- facilitating the investment potential of the city to attract direct foreign capital by mitigating the perceived risks and threats;
- attracting skilled labor and enhancing the labor market through ensuring the participation of women into the labor force;
- helping establish brands in financial services;
- fostering innovation and competition by developing production-based sectors;
- setting up crowd funding platforms to facilitate financing;
- enhancing the tourism sector to increase the GDP, which is the most important indicator for a smart city.

The role of digital technologies should be clearly defined in achieving the goals related with becoming a smart city. What a smart city entails is not a city saturated with all kinds of new technologies, but a city effectively benefiting from the new technologies to increase the quality of life for its citizens and visitors. Criticized as the techno-centric view in the literature (Mora, Bolici and Deakin 2017), the smart city project in Istanbul seems to have a techno-centric orientation, which downplays the role of social, governance, organizational and environmental dimensions of the smart city concept. The smart technologies increase the efficiency of the services provided; in this sense, it is not an aim in itself but a means to achieve an end. The local government should put the needs, demands and interests of its citizens first rather than the interests of the technology companies seeking to market and sell their technology products and solutions.

Various smart city projects targeting tourists and visitors can be suggested to enhance their experience of the city. Kiosks providing information in foreign languages about the locations, tourist hotspots and attraction points, mobile applications giving information about the city, events and activities, traffic information, journey planners, mobility assistance, wireless technology to provide uninterrupted Internet connection throughout the public places such as public squares, bus stops and stations, parks and so on, smart health assistance tools via mobile applications, interactive mobile maps and guides are all examples that can potentially enhance the tourist experience in Istanbul.

The urban transport problems in Istanbul not only cause inconvenience for the inhabitants, but also for the visitors. Navigating through Istanbul's roads, streets and public transport is quite challenging, especially for the newcomers. Some of the challenges include finding information about the public transport network, schedule and fares, getting stuck in the traffic jam, finding information about how to get to tourist hotspots and finding locations for buying and tapping the smart cards used in Istanbul's public transport system. Excessive focus on using smart technologies in enhancing the mobility experience of tourists can lead to overlooking the other urban mobility problems, such as poor pedestrian environment, inadequate cycling facilities and reckless driver behavior, especially in minibuses and taxis.

Developing a more cycling-friendly urban mobility environment has the potential to enhance the tourist experience. Smart technologies and projects can play an important role here. Bike-sharing facilities using mobile application, electronic card payment and GPS monitoring devices can increase the use of bikes in experiencing the sights the city offers. IMM, through ISPARK, a municipal subsidiary company which is responsible for managing parking lots and establishing bicycle-sharing facilities and bicycle paths, is seeking to increase the role of cycling in urban mobility. ISBIKE is a bicycle-sharing scheme providing 1,500 bicycles and 140 stations throughout the city, especially along the coastlines of Bosphorus and Marmara Sea which offer a pleasant and unique view of Istanbul.

IETT, the municipal public bus operator of Istanbul, offers two mobile applications in 12 foreign languages ranging from English, French and German to Japanese, Greek and Chinese. The first one is MOBIETT which provides information about bus routes, estimated arrival times and schedules, and the second one is "HOW TO GO THERE", which provides journey planning assistance for the travelers. Currently, there are separate mobile applications designed by different companies and agencies for separate mobility functions. However, an integrated mobility application which will include all of the transport modes, for example bus, metro, ferry, minibuses, shared taxis and also the location of parking lots and capacity information, traffic information, alternative routes, can facilitate the visitor navigating through the complex and labyrinthian streets of Istanbul.

Taxis are one of the most frequently used transport modes in Istanbul. The problems related with taxi use by tourists have a negative influence on the urban image. Rude taxi drivers, charging more for tourists, selecting passengers based on their destinations are some of the problems causing a poor taxi travel experience for the tourists. The "iTaksi" project, undertaken by IMM, is part of the smart city projects which aim to improve the taxi travel experience of Istanbul. The taxis are to be monitored from the Transport Management Center of IMM. The project aims at using taxis more efficiently and reducing the waste of time, energy and fuel consumption. Through a mobile application, travelers can find the closest available taxis around them. The communication between taxi drivers and travelers are recorded to improve service quality and traveler satisfaction. Traffic congestion caused by idle circulation of the taxis on the streets to find passengers is expected to be reduced. The taxis are equipped with in-car surveillance cameras to improve the safety of the travelers and drivers. "iTaksi" provides alternative payment options such as credit cards or IstanbulKarts, the smart card for the public transport, besides cash for the travelers.

"IBB Navi", the mobile application developed by IMM for journey planning and finding the way to places also supports the system. Giving scores for the travel experience through the mobile application provides a customer-centered service opportunity. The project was launched initially with 4,000 taxis and is planned to gradually include all of 17,395 taxis in Istanbul. One of the most noteworthy points is the participatory approach adopted for the implementation of the project. The taxi associations, chamber of taxi operators and IMM cooperated throughout the entire project planning, implementation and control phases. The improvement in taxi travel experience by using ICTs is expected to contribute to an enhanced urban transport experience, which is one the weakest points in the tourist experience in Istanbul.

Whereas generating data through various channels is relatively advanced in Istanbul, especially for mobility, processing, using and sharing this data to address the need of residents and tourists has some limitations. The legislation for open data is still lacking and the technical capacity to set up such platforms is not yet ready. Although local government agencies are aware of the potential opportunities and benefits offered by open data, a national approach should be developed to address the legislative needs in this area. Open data is likely to foster the innovativeness of start-ups as they often have limited opportunities to access urban data collected by government agencies.

Last, but not least, major infrastructure projects have transformed the landscape for tourism, which are expected to continue with ongoing projects. These projects aim at ameliorating the urban mobility predicament. The third airport of Istanbul, the Third Bridge, or officially the Yavuz Sultan Selim Bridge over the Bosphorus, the underground rail tunnel across the Bosphorus called Marmaray, the road tunnel under the Bosphorus called the

Eurasia Tunnel all aim at improving the deteriorating transport conditions and facilitate the Bosphorus Crossing, which is also important for the tourist experience. The planned Channel Istanbul, the artificial sea-level waterway, will bring about another waterway besides the Bosphorus and will have implications for the tourist experience. Due to the extensive impact of those infrastructure projects on the urban landscape of Istanbul, the smart city transformation should incorporate the effects of these projects. The ICTs and IoT systems should be integrated to ensure an all-encompassing smart city transformation. Airports are especially important in transforming the tourist experience as these locations are the first contact points which can influence the tourist experience.

14.4 Case Study Analysis

Within this research, five different in-depth interviews were conducted with the municipality representatives and international non-governmental organizations such as the International Road Union (IRU) and the International Centre for Integrated Urban Planning and Transport (ICIUPT).

As mentioned in the introduction, the following questions were asked to the city professionals:

Question 1: How can Istanbul and its governmental authorities direct this smart city transformation?
Question 2: What can be fundamental drivers and restrictions for this change?
Question 3: What can be the primary handicap that Istanbul faces? (any suggestion could be said)
Question 4: What can be the expected terms or conditions for smart city transformation?
Question 5: Which infrastructures are needed to become a smart city or a smart tourism destination?
Question 6: What kind of collaboration strategies should be developed with the city stakeholders (NGOs, governmental authorities and private bodies) during the process of becoming a smart city?

Various answers to question 1 were collected and gathered in one paragraph. Representatives from Istanbul Metropolitan Municipality and non-governmental organizations suggest that strategic partnership and collaboration with different stakeholders should be developed. IMM launched a smart city transformation project at the beginning of 2017. Various stakeholders are involved in this process. A subsidiary municipal company, ISBAK, which provides ICT solutions for Istanbul, is assigned as the coordinator for this project. It cooperates with one global and one local technology company to deliver the projects planned for smart city transformation. After the studies analyzing the current situation of Istanbul, smart city vision, strategies and the roadmap for Istanbul are still in their development phase. A big data management center is planned to manage the data generated from various urban activities such as mobility, energy consumption, waste management and so on so that this data can be utilized to provide higher quality urban services. The development of start-up firms which can use the open data provided from this center is also planned to bring about urban development and innovation. IoT projects, establishment of smart communication networks and operation centers, provision of energy automation systems, smart lighting and sensors, artificial intelligence solutions for energy, mobility, health and security are some of the examples for the planned projects.

The following point is to identify the fundamental drivers and restrictions of Istanbul to be a smarter destination.

The key drivers for smart city initiatives for Istanbul can be listed as the following:

- a growing urban local economy with many links with national and international scales;
- fairly favorable financial and economic conditions on the part of IMM, the leading actor of the smart city initiative;
- growing urban sustainability concerns such as air pollution, resource depletion, chronic mobility and traffic problems, environmental degradation, urban sprawl which calls for smarter intervention mechanisms and frameworks;
- Istanbul's unique place as a cultural and historical center, which is important from a national and global perspective, and thus makes it imperative to preserve its unique identity through smart policies and measures;
- a dynamic and young population of Istanbul who are open to change and novelties;
- Istanbul has two crucial advantages: one is Turkish Airlines as the national flag carrier, and the other is being a cultural destination with a wealth of attraction points.

The restrictions or barriers, on the other hand, are the following:

- the main obstacle is obviously being an overpopulated destination; there are not so many cities in the world that have more than 14 million people;
- resistance to change on the part of local stakeholders. For example, taxi operators might be opposed to the technological novelties such as mobile applications due to fear of losing their power and interests;

- a lack of adequate participation from other government agencies, communities, private actors or NGOs;
- excessive focus on technology projects and overlooking the social, environmental and governance aspects of smart city concept;
- a lack of political support. The risk of remaining as a bureaucratic/technocratic project dealing only with the project implementation rather than a wider urban transformation;
- without a consultancy board including industry professionals, tourist guides, local population engagement and so on, Istanbul cannot gain success in smart city movements. It is also difficult to reconcile the interest of local residents;
- an international press image is important as well. The city is still considered to offer traditional or old-style services and facilities in the eyes of the visitors.

To the question of the primary handicap that Istanbul faces, all the interviewees agree that there are two main issues. The first one is the population. Providing services to almost 15 million Istanbulites while developing new projects is not an easy task for the municipality. The other obstacle may be that "working together or collaboration with other stakeholders" is not on the agenda of the city representatives. A lack of collaboration generates most of the issues that the city faces. It causes failure to ensure the participation and engagement of people, and lack of interest from the local community. Besides, the city council prefers a technocratic approach to a smart city project rather than a people-centered democratic approach. That also leads to miscommunication.

To thoroughly apply the concept of the smart city, there are some terms and conditions the city needs to follow. One representative from the International Road Union stated that Istanbul seems traditional in the eye of a visitor: according to him, the first step should be to change the image of the city and a traditional and modern smart technological city should go hand in hand. An ISBAK high-level manager said there should be a serious development in the working-together approach and in the end this should generate applicable smart projects created in a multidisciplinary environment. Supporting entrepreneurs is another point that should be considered if the city needs an innovative atmosphere. Local government is a key player to form a sustainable relationship between public-private sector and academia. The municipality has established a "Zemin Istanbul" platform to pioneer this cooperation. In this platform, there are three different areas of service: incubation, experiment and training centers.

The following points should also take into account achieving a wider awareness and consensus of the smart city concept amongst the various stakeholders:

- identification of the problems to be addressed and outcomes to be expected is key to achieve a real transformation toward being a smart city;
- setting out the vision and strategies clearly, and being transparent about the project steps, aims and outcomes;
- adopting a bottom-up approach by incorporating the expectations and contributions of the citizens rather than a top-down approach imposing a predetermined and preconceived project plan;
- refraining from aiming for quick gains by technical fixes and technological projects but aiming at wider transformation of urban living through smart city concepts;
- setting up open platforms for participation, engagement and negotiation with all stakeholders.

From an infrastructural point of view, to achieve the status of a smart tourism destination, there are a few different views from the interviewees. Identifying the areas such as urban mobility, energy, waste management, governance and so on can help facilitate the implementation. Some examples would be:

- setting up a data management center to manage, coordinate and direct all the urban transport from one center;
- using big data to come up with innovative solutions for urban problems such as traffic congestion and journey planning;
- identifying the tourist hotspots, events and locations and sharing information with tourists through one authoritative mobile application (example "Visit London" app.);
- identifying the problems where tourists have the most difficulty (e.g. finding Istanbulkart, learning about the urban transport, finding their way and so on) and developing smart solutions through mobile applications;
- demonstrating the sensitivity for sustainability through images, marketing campaigns and projects. Developing an image of "greener Istanbul" to target the sustainability objective of a smart city concept.

Providing visitors as well as residents with easy and safe access to transportation, services and goods in the city is the main factor to boost the image of Istanbul on international channels. In this manner, both enabling smart applications showing touristic areas and free Wi-Fi spots in the most visited locations will increase visitor satisfaction.

The last questions that the interviewees answered is about what kind of collaboration strategies should be developed with the city stakeholders

(NGOs, governmental authorities and private bodies) during the process of becoming a smart city.

The first contribution has come from the representative of the International Road Union. He suggested that Istanbul should create an advisory board consisting of the private tourism sector (travel agencies, hotels, restaurants and so on), governmental bodies (chamber of commerce, municipality, ministry and so on), consultancy and technology creators. And this advisory board should counsel for the creation of marketing content in several languages. For example, developing a OneIstanbul mobile application with more interactive content and direction.

Another approach has been stated by the Istanbul Municipality representative. He suggested that collaboration should be bottom-up, not top-down. Having a top-down strategy alienates the other stakeholders. They can feel that they are only contacted as part of a procedural project step. Their expectations should be effectively incorporated in all steps of the project, not just at the beginning. Specific segments of citizens such as the disabled, young people, women, seniors, tourists, refugees and so on should be included not only through NGOs, but also through actual groups of people from these segments. An effective transformative leadership should be put forward by the local government or its responsible agency, ISBAK.

Also, the role of private technology companies must be clearly defined. Their ideas and suggestions should be consulted but their guiding and leading the project should not be allowed. Otherwise, smart city projects can turn into a technology implementation project guided only by technology companies according to their interests rather than an urban transformation project guided by the interests of the citizens and people living in the city.

Building Technoparks, Living Labs, Fab-Lab, Zemin Istanbul is a good example for the creation of a smart city environment and also the number of projects created through these centers are expected to increase within the next years. That assists Istanbul to be an innovative, smart and tourist-friendly destination, on its way to become a smart tourism destination.

In order to look at the broader perspective, the smart city project in Istanbul, to a large extent, can be identified with a techno-centric view, which overlooks the social, cultural, organizational and environmental dimensions of smart city application. Assigning the responsibility to one agency, ISBAK, can increase the efficiency of implementation. However, it can exclude the other stakeholders' contribution even if they are sought to be included through workshops and meetings.

The overall results of the interviews suggest that stakeholder collaboration is an important factor to successfully manage smart city transformation. However, on the other hand, the municipality gives all responsibility to only one subsidiary company and it excludes other city partners from the "being a smarter city" process.

Conclusion and Future Directions

To conclude, there are many concerns and factors that the city authorities need to consider on the way to create a real smart city. As a first step, Istanbul needs to collect all the smart city projects and works that have been done so far to create an integrated and innovative roadmap to move further towards being a smart tourism destination.

Another significant point that needs to be taken into consideration is that strategic partnerships with different stakeholders should be developed. Very few destinations have a good strategic long-term plan on how to become a smart tourism destination. Istanbul, as one of the leading tourism destinations, should improve its organizational structure and it should create dedicated policies on how to achieve this by building an advisory board consisting of smart technological professionals, city officials and so on.

Another concern with the smart city vision is the potential loss of Istanbul's unique identity. Istanbul has a rich culture and history and an identity based on being "the capital of civilizations". Changing this identity with a futuristic smart city vision might lead to damaging the unique identity of Istanbul. This has also implications with regard to being a smart city tourism destination. Defining Istanbul solely through such descriptions is not compatible with its uniqueness which also increases its touristic value. However, this does not mean that the smart digital technologies cannot be beneficial. Rather, it means that the unique identity of Istanbul should not be compromised by adopting the smart city concept as a vision encompassing all aspects of the identity of a city. Therefore, the local government should be selective in shaping its smart city implementation pathway and related projects. The projects should support IMM's vision to be a local government "where sustainable and innovative solutions facilitate urban living and global value in terms of urbanism and civilization is generated" (Istanbul Metropolitan Municipality 2015.).

The progress in smart city implementation should be measured with objective performance indicators as well as subjective perception or satisfaction surveys. The outcomes of the sub-projects should be measured together with the performance scores so as to understand whether the projects achieved the targets. In terms of project management, the components of the smart city project should be prioritized and the relationships amongst them should be clearly defined. For example, smart urban mobility can be prioritized over the other dimensions such as energy, environment or waste management, depending on the opinions of the stakeholders. To achieve this, several multi-criteria decision-making methods such as analytical hierarchy process (AHP), TOPSIS or ELECTRE can be used to determine the ranking among the dimensions (Canitez and Deveci 2018). Employing such a method provides a participatory framework which is in alignment with the core values of the smart city concept. Moreover, it facilitates the decision-making process for the policy makers in choosing the most feasible course of action and also for the project practitioners in implementing the planned project steps. It is critically important to take into account the appropriate application framework besides the content of the smart city project. Unless such a framework is in place, the targets may fail to be attained even if they are well designed and well planned.

Setting up inter-organizational collaborative platforms for a more effective use of skills and resources is another critical point in smart city applications. However, the fragmented governance structure in Istanbul can hinder the realization of this objective. The boundaries, roles and responsibilities of the local government agencies can sometimes get blurred which leads to conflicting and duplicated projects. Ensuring the active participation of the people and citizen groups is essential for incorporating their demands and needs in setting out smart city implementation. Transparent, open, people-centered and efficient governance mechanisms are key to engagement at a higher level. However, the current smart city project in Istanbul has a risk of turning into a technocratic/bureaucratic project excluding the participation of the wider community. The awareness of the public about the smart city project is quite limited, confined to one-off workshops where experts from various government agencies attend. Limited information published about the progress of the project is another indicator of low openness to public participation and scrutiny.

Lastly, to briefly mention the future direction of the smart city implementations of Istanbul, this includes autonomous and connected vehicles, IoT, car sharing, open data platforms, artificial intelligence, cyber security and smart buildings.

References

Albino, V., Umberto, B. and Rosa, M. D. 2015. Smart Cities: Definitions, Dimensions, Performance, and Initiatives. *Journal of Urban Technology* 22(1): 1723–1738.

Arup Consulting. *Smart Cities: Transforming the 21st Century City via the Creative Use of Technology*. 2010. Accessed July 15, 2018: http://ww.cisco.com.

Baggio, R. and Del Chiappa, G. 2014. Real and Virtual Relationships in Tourism Digital Ecosystems. *Information Technology and Tourism* 14(1): 3–19.

Bakıcı, T., Almirall, E. and Wareham, J. 2013. Smart City Initiative: The Case of Barcelona. *Journal of the Knowledge Economy* 4: 135–148.

Berrone, P., Ricart, J. and Carrasco, C. 2016. The Open Kimono: Toward a General Framework for Open Data Initiatives in Cities. *California Management Review* 59: 39–70.

Boes, K., Buhalis, D. and Inversini, A. 2015. Smart Tourism: Ecosystems for Tourism Destination Competitiveness. *International Journal of Tourism Cities* 2: 391–403.

Borja, J. 2007. Counterpoint: Intelligent Cities and Innovative Cities. UniversitatOberta de Catalunya (UOC). *Knowledge Society5*. Accessed September 3, 2018: www.uoc.edu/uocpapers/5/dt/eng/mitchell.pdf.

Boulton, A., Brunn, S. D. and Devriendt, L. 2011. Cyberinfrastructures and Smart World Cities: Physical, Human, and Soft Infrastructures. In *International Handbook of Globalization and World Cities*, edited by P. Taylor, B. Derudder, M. Hoyler and F. Witlox. Edward Elgar: Cheltenham, UK. Accessed August 23, 2018: www.neogeographies.com/documents/cyberinfra structure_smart_world_cities.pdf.

Buhalis, D.s and Amaranggana, A. 2015. Smart Tourism Destinations Enhancing Tourism Experience through Personalisation of Services. In *Information and Communication Technologies in Tourism*, edited by Z. Xiang and L. Tussyadiah. Springer: Dublin, 377–389.

Buhalis, D.S. and Amaranggana, A. 2014. Smart Tourism Destinations, in *Information and Communication Technologies in Tourism*, edited by Z. Xiang and L. Tussyadiah. Springer: Dublin, 553–564.

Canitez, F. and Deveci, M. 2018. A Smart City Assessment Framework: The Case of Istanbul's Smart City Project. In *Economic and Social Development (Book of Proceedings), 32nd International Scientific Conference on Economic and Social Development*, 369–374.

Caragliu, A., Del Bo, C. and Nijkamp, P. 2011. Smart Cities in Europe. *Journal of Urban Technology* 18(2): 65–82.

Chillon, P. S. 2012. *Urban 360*. Accessed July 8, 2018: http://urban360.me/2012/02/08/this-place-worths-a-visit-intelligent-destinations-smart-city.

Chui, M., Löffler, M. and Roberts, R. 2010. The Internet of Things. *McKinsey Report on IoT* 2: 1–9.

Dameri, R. 2013. Searching for Smart City Definition: A Comprehensive Proposal. *International Journal of Computers & Technology* 11(5): 2544–2551.

Dassani, N., Nirwan, D. and Hariharan, G. 2015. A New Paradigm for Smart Cities. Dubai. *KPMG International*. Accessed August 10, 2018: https://assets.kpmg.com/content/dam/kpmg/pdf/2016/04/Dubaia-new-paradigm-for-smart-cities-uae.pdf.

Dawes, S., Cresswell, A. M. and Pardo, T. A. 2009. From Need to Know to Need to Share: Tangled Problems, Information Boundaries, and the Building of Public Sector Knowledge Networks. *Public Administration Review* 69(3): 392–402.

Dirks, S., Gurdgiev, C. and Keeling, M. 2010. *Smarter Cities for Smarter Growth: How Cities Can Optimize Their Systems for the Talent-Based Economy*. IBM Global Business Services. 2010. Accessed July 27, 2018: www.ibm.com/se/smartercities/pdf/GBE03348USEN.PDF.

Dirks, S., Keeling, M. and Dencik, J. 2009. How Smart is Your City?: Helping Cities Measure Progress. Somers, NY: IBM Global Business Services. Accessed June 19, 2018: www.ibm.com/downloads/cas/KLEYQE6Z.

Etro, F. 2009. The Economic Impact of Cloud Computing on Business Creation, Employment and Output in Europe. *Review Business and Economics* 54(2): 179–208.

Fortune, 2018. *Smart Cities*. Accessed August 29, 2018: www.fortuneturkey.com/akilli-sehirler-icin-akilli-gelecek-tasarimi-44346#popup.

Giffinger, R., Kramar, H. and Haindl, G. 2010. The Role of Rankings in Growing City Competition. *Urban Research & Practice* 3: 299–312.

Gretzel, U., Hannes, W., Chulmo, K. and Carlos, L. 2015a. Conceptual Foundations for Understanding Smart Tourism Ecosystems. *Computers in Human Behavior* 50(5): 58–63.

Gretzel, U., Marianna, S., Xiang, Z. and Chulmo, K. 2015b. Smart Tourism: Foundations and Developments. *Electronic Markets* 25(3): 179–188.

GSMA. 2012. *Finland: Forum Virium Helsinki*. Accessed July 27, 2018: www.gsma.com/iot/wp-content/uploads/2012/12/cl_forum_virium_12_12.pdf.

Gutiérrez, V., Galache, J. A., Sánchez, L., Muñoz, L., Hernández-Muñoz, J. M. and Fernandes, J. 2013. Smart Santander: Internet of Things Research and Innovation through Citizen Participation. In *The Future Internet*, edited by Alex Galis and Anastasius Gavras. Springer: Berlin Heidelberg, 173–186.

Hall, R. E. 2000. The Vision of a Smart City. *2nd International Life Extension Technology Workshop*. Paris, France. Accessed August 21, 2018: www.osti.gov/servlets/purl/773961/.

Harrison, C., Eckman, B., Hamilton, R., Hartswick, P., Kalagnanam, J., Paraszczak, J. and Williams, T. 2010. Foundations for Smarter Cities. *IBM Journal of Research and Development*, 54(4): 1–16.

Haubensak, O. 2011. *Smart Cities and Internet of Things. Business Aspects of the Internet of Things, Seminar of Advanced Topics*, ETH: Zurich.

Hollands, R. G. 2008. Will the Real Smart City Please Stand Up? *City* 12(3): 303–320.

Huang, X., Yuan, J. and Shi, M. 2012. Condition and Key Issues Analysis on the Smarter Tourism Construction in China. In *Multimedia and Signal Processing*, edited by F. L. Wang, J. Lei, R. Lau and J. Zhang. Springer: Berlin Heidelberg, 444–450.

Hunter, W. C., Chung, N., Gretzel, U. and Koo, C. 2015. Constructivist Research in Smart Tourism. *Asia Pacific Journal of Information Systems* 25(1): 105–120.

Istanbul Metropolitan Municipality. 2015. *Strategic Plan of 2015–2019*. Accessed August 28, 2018: www.ibb.gov.tr/TR/.

Jordan, B. 2011. *Corbin Ball Associates*. Accessed July 27, 2018: www.corbinball.com/articles_technology/index.cfm?fuseaction=cor_av&artID=8591.

Kanter, R. M. and Litow, S. S. 2009. *Informed and Interconnected: A Manifesto for Smarter Cities*. Harvard Business School General Management Unit Working Paper. Accessed July 7, 2018: http://papers.ssrn.com/sol3/papers.cfm?abstract_id=14 20236.

Koo, C., Park, J., Lee, J. N. 2017. Smart Tourism: Traveler, Business, and Organizational Perspectives. *Information & Management* 54: 683–686.

Koo, C., Shin, S., Gretzel, U., Cannon, W. Chung, N. 2016. Conceptualization of smart Tourism Destination Competitiveness. *Asia Pacific Journal of Information Systems* 26(4): 561–576.

Lee, C. K., Lee, J., Lo, P. W., Tang, H. L., Hsiao, W. H., Liu, J. Y. and Lin, T. L. 2011. Taiwan Perspective: Developing Smart Living Technology. *International Journal of Automation and Smart Technology* 1(1): 93–106.

Lin, Y. 2011. The Application of the Internet of things in Hainan Tourism Scenic Spot. *Hainan: Seventh International Conference on Computational Intelligence and Security*, 1549–1553.

Marceau, J. 2008. Introduction: Innovation in the city and innovative cities. Innovation: Management. *Policy & Practice* 10(2/3): 136–145.

Meijer, A. and Bolivar M. P. R. 2016. Governing the Smart City: A Review of the Literature on Smart Urban Governance. *International Review of Administrative Sciences* 82(2): 392–408.

Metric Stream. *Smart Cities Solutions*. 2013. Accessed July 17, 2018: www.metricstream.com/solutions/smart_cities.htm.

Mingjun, W., Zhen, Y., Wei, Z., Xishang, D., Xiaofei, Y., Chenggang, S., Xuhong, L., Fang, W. and Jinghai, H. 2012. A Research on Experimental System for Internet of Things Major and Application Project. *3rd International Conference on System Science, Engineering Design and Manufacturing Informatization (ICSEM)*, Chengdu, 261–263.

Molz, J. G. 2012. *Travel Connections: Tourism, Technology, and Togetherness in a Mobile World*. Routledge: Abingdon.

Mora, L., Bolici, R. and Deakin, M. 2017. The First Two Decades of Smart-City *Research: A Bibliometric Analysis. Journal of Urban Technology* 24(1): 1–25.

Nachira, F. 2002. *Toward a Network of Digital Business Ecosystems Fostering the Local Development*. Bruxelles: Directorate General Information Society and Media of the European Commission. 2002. Accessed June 11, 2018: www.digital-ecosystems.org (working paper).

Neuhofer, B., Buhalis, D. and Ladkin, A. 2012. Conceptualising Technology Enhanced Destination Experiences. *Journal of Destination Marketing & Management* 1(1/2): 36–46.

Partridge, H. L. 2004. *Developing a Human Perspective to the Digital Divide in the Smart City*. Paper presented by Partridge, Helen Australian Library and Information Association Biennial Conference, Gold Coast. Accessed July 12, 2018: https://core.ac.uk/download/pdf/10873550.pdf.

Piro, G., Cianci, I., Grieco, L. A., Boggia, G. and Camarda, Pietro. 2014. Information Centric Services in Smart Cities, *Journal of Systems and Software* 88(1): 169–188.

Porter, D. R., Dunphy, R. T. and Salvesen, D. 2002. *Making Smart Growth Work*. Urban Land Institute: Washington, DC.

Presenza, A., Micera, R., Splendiani, S. and Del Chiappa, G. 2014. Stakeholder e-Involvement and Participatory Tourism Planning: Analysis of an Italian Case Study. *International Journal of Knowledge Based Development* 5(3): 311–328.

Rittel, H. W. J. and Webber, M. 1973. Dilemmas in a General Theory of Planning. *Policy Sciences* 4: 155–169.

Rong, J. A. 2012. *China Economic Net*. Accessed July 14, 2018: http://en.ce.cn/Insight/201204/12/t20120412_23235803.shtml.

SOCAP International. 2013. *Now Arriving: Big Data in the Hospitality, Travel and Tourism Sector*. Accessed June 3, 2018: http://invattur.aimplas.es/ficheros/noticias.

Storper, M. and Scott, A. J. 2009. Rethinking Human Capital, Creativity and Urban Growth. *Journal of Economic Geography* 9(2): 147–167.

Thite, M. 2011. Smart Cities: Implications of Urban Planning for Human Resource Development. *Human Resource Development International* 14(5): 623–631.

Toppeta, D. 2010. *The Smart City Vision, How Innovation and ICT Can Build Smart, Livable, Sustainable Cities*. The Innovation Knowledge Foundation. Accessed June 12, 2018: www.thinkinnovation.org/file/research/23/en/Top peta_Report_005_2010.pdf.

Tu, Q. and Liu, A. 2014. A Framework of Smart Tourism Research and Related Progress in China. *International Conference on Management and Engineering*, October 24–26. Destech Publications, 140–146.

Wang, D., Li, X. and Li, Y. 2013. China's Smart Tourism Destination Initiative: A Taste of the Service-dominant Logic. *Journal of Destination Marketing and Management* 2(2): 59–61.

Washburn, D., Sindhu, U., Balaouras, S., Dines, R. A., Hayes, N. M. and Nelson, L. E. 2010. *Helping CIOs Understand Smart City Initiatives: Defining the Smart City, Its Drivers, and the Role of the CIO.* Forrester Research: Cambridge, MA. Accessed June 9, 2018: http://public.dhe.ibm.com/partnerworld/pub/smb/smarterplanet.

Weber, E. P. and Khademian, A. 2008. Wicked Problems, Knowledge Challenges, and Collaborative Capacity Builders in Network Settings. *Public Administration Review* 68(2): 334–349.

Zhang, L., Nao, L. and Min, L. 2012. Basic Concept and Theoretical System of Smart Tourism. *Tourism Tribune* 27(5): 66–72.

15

Green Roof Garden Concept for Smart Cities – A Case Study

Carlos Alberto Ochoa and Aida Yarira Reyes Escalante

CONTENTS

15.1 Introduction 321
15.2 Roof Gardens 322
15.3 Cultural Algorithms (CA) 323
15.4 Relationship of Variables in the Construction of Green Gardens 325
15.5 Methodology 326
    15.5.1 Mathematical Analysis by the Loads Exerted by the Construction Elements on the Roof 327
    15.5.2 Experiments 328
15.6 Analysis of Results 330
15.7 Experiment 332
Conclusion 338
Future Investigation 339
Acknowledgments 341
References 341

15.1 Introduction

The concept of sustainable development is established in a balanced relationship between economy, society and the environment. In April 1987, the report of the World Commission on Environment and Development (WCED), titled "Our Common Future", in which the concept of sustainable development says that it should be committed to meet the needs of the present generation without compromising the ability of future generations to meet their own needs (Brundtland 1987). In order to present advances towards sustainability there should be applied more efficient operational forms, where science and technology (S&T) become the key elements in order to have better results. The changes generate improving the results towards sustainability, modifications, knowledge, in almost every discipline. To do this, science and technology must be allowed to approach more complex problems generating innovating solutions.

In this regard in 2002, the Economic Commission for Latin America (ECLA) raised new challenges for S&T, which indicated a generation of new analytical approaches; this is due to the multidisciplinary methodology and the prioritization of activities on issues related to S&T in all world (Gallopín 2003).

Sustainability is a topic that directly influences the design, construction, technology and operational processes of any entity, for example buildings, urban blocks, neighborhoods, networks, centers, cities, regions of the countries (Hernandez 2008). The influence comes from the types of materials, spaces in which they are built, the designs and distribution of the spaces, materials, equipment, attachments, and the application of S&T to generate better comfort.

Adhya, Plowright and Stevens in 2010 mentioned that in the urban areas growth is not uniform in many situations. Growth is occurring in certain areas of the big cities. In addition, these authors state that although there is enormous growth of cities, there is still disarticulation, deterioration, and devaluation of the inner core cities. As such, these big cities present a scenario that provoke opportunities for improvement. The problems of the growth of the cities were addressed by the different organizations, for example faculties of architecture, urban planning and engineering, seeking to find new positions related to the topic of population growth, pollution and the environment. In order to solve these problems new opportunities should be explored.

15.2 Roof Gardens

The new proposals in the areas of architecture and engineering through the various projects of construction and redesign is around the theme of re-use of roofs as a way to use new alternative energy to optimize spaces. In this theme, Germain and colleagues in 2008 defined a new guide to help others and designed the roof garden. The document specifies that in increasingly dense and extensive cities, it is necessary to look for new ways to take advantage of the spaces, as well as the way to recover those that are not used or underutilized. The latter are considered spaces that are not useful, such as roofs, patios and balconies. Roof gardening means taking an inspiring, ecological and productive activity, and developing new links with the food chain, the seasons, the environment and the community. The vision of their ideal locations is to turn the city into a garden and its inhabitants in gardeners.

New designs in urban and urban development face many issues related to the sociocultural, economic, environmental and technological elements and all of these must be combined to determine the best option to solve the issues related to design and the implementation of various garden models on roofs. Sustainable urban development is a concept that is applied in the visions of the smart city (economic, social, physical and environmental, including a smart roof garden), framed in S&T strategies that guarantee improvements to population welfare. The main aim in the smart cities is to improve long-term welfare levels, without compromising the development possibilities of the surrounding areas and helping to reduce the harmful effects of the environment.

The constant growth of cities has caused a social-spatial segregation, to which designers, urban engineers and architects have contributed in the resolution of this phenomenon. The population has increased in both urban and rural areas, but urban growth lacks in sustainable strategies to be more environmentally friendly spaces; however strategies have not been able to guarantee these attributes.

Due to inefficiencies in urban development strategies, there is a need to seek the integration of sustainability principles in the development of new growth zones such as neighborhoods, economic zones, commercial plazas and public areas. Integrating the environment in new design proposals is the fundamental principle to reduce inefficiency. The constructions, whether housing, areas, departments, business, institutional buildings, among others, must consider the climatic, cultural, social and economic variables of the environment where they are located. Furthermore, these constructions must be compliant with local, national and international regulations.

Also, according to national and international policies it should be taken into consideration the provision of materials, equipment, labor, technological innovations and use of alternative energies in order to create the welfare and satisfaction of clients. Based on the problems in the relationship between city, urbanism and sustainable development, this research is established in order to articulate sustainability elements through new technological proposals that allow better decisions in the intelligence system of the roof garden.

15.3 Cultural Algorithms (CA)

Striphas (2015) indicated that algorithm emanates from various idioms. The development of CA can be attributed to Reynolds (1994) when he first developed an evolutionary algorithm, which is established on the basis that knowledge is not a technique integrated a priori, but that it is extracted during the same search process, which are known to this day as cultural algorithms.

Decision-making depends on the inherited culture, as well as the beliefs of the current population. Its origins are established in the applications

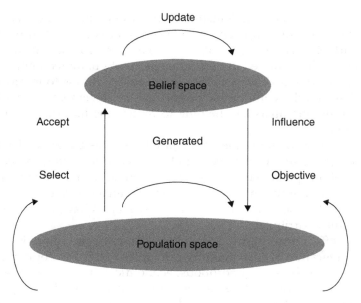

FIGURE 15.1
Spaces employed by the cultural algorithm (based on Reynolds 1994).

of sociologists and archaeologists when it is necessary to try to model the cultural evolution. There are several ways to apply the CA: Seaver (2017) researched cultural algorithms and defined tactics for the ethnography of algorithmic systems. Likewise, Dourish (2016) established the algorithms in context to algorithmic culture. Yang and colleagues (2012) created an efficient function optimization algorithm based on culture evolution. All offer different perspective algorithms that allow access to the different ways of learning and improvement in decision-making.

Reynolds (1994) mentioned the phenomenon of inheritance in cultural algorithms, with the aim of increasing the rate of learning or convergence, and in this way, the system is capable of responding to a large number of problems, through the components established in Figure 15.1.

Figure 15.1 demonstrates the interaction between the space of belief and the population space where the relationship is bidirectional. The population space works in a similar way to that of an evolutionary algorithm, and consists of a set of individual data where each has an independent feature used to determine their suitability (fitness).

A code description of the cultural algorithms is described by Yan and colleagues (2012), as follows:

a. BEGIN
b. t = 0

c. Initialize population P(t)
d. Initialize belief space B(t)
e. Repeat
f. Evaluate P(t)
g. Update (B(t), accept P(t))
h. Generate (P(t), influence B(t))
i. Select P(t) from P(t-1)
j. t+ = 1;
k. Until (termination condition)
l. END.

15.4 Relationship of Variables in the Construction of Green Gardens

Urbanization developed in a city applying the roof garden concept generates the construction of a structural equation based on the relationship of the variables and their incidences. To achieve this goal, the next structural equation (15.1) is considered, based on the data generated in the Table 15.1, designate the value related to the viability of a specific roof garden.

$$V = \left\{ \frac{CRG}{IRG} - \left(\frac{\#S * SRG}{PLA} RA \right) \right\} \pm SCA \quad (15.1)$$

TABLE 15.1

Multivariable Analysis with the Information Related with Each Roof Garden and its Components

Model of Roof Garden	Roof Garden Increase Index	Commercialization	Services	Recreation and Smart Farming Areas	Social Cost-Benefit in a Smart City	Equipment in the Roof Garden
Barbados	8	0.814	0.765	0.863	0.799	0.678
Crimea	14	0.795	0.811	0.835	0.847	0.715
Djibouti	5	0.747	0.838	0.842	0.817	0.818
Macao	8	0.816	0.794	0.803	0.858	0.902
Pitcairn	9	0.947	0.836	0.828	0.807	0.794
Tahití	12	0.877	0.819	0.842	0.805	0.816
Tuvalu	7	0.954	0.797	0.783	0.912	0.828

where

>   CRG = Cost of roof garden
>   IRG = Increase roof garden in the time
>   #S = Number of species in the roof garden
>   SRG = Size of roof garden
>   PLA = People accessing roof garden
>   RA = Recreational areas in roof garden
>   SCA = Social cultural aspects related to the location of the roof garden (Space to park, access to smart farming, security and others).

15.5 Methodology

The study was defined through exploratory and descriptive research using quantitative data analysis. The space was in the city of Querétaro, Mexico. For the study, a multiple matching method is used and a series of evaluations with different combinations of roof gardens (seven in total) was carried out. The study required a total of 50 runs of data using different scenarios and adding a variety of factors for its final result. The analysis of the roof garden needed to comply with the following indications:

a. priority will be given to similarity zone;
b. classification and ranking of the results is generated to group the similarities; and
c. all proposals are required to undergo the evaluations.

The specifications for the economic evaluation phase are specified by the following requirements. The preferences will be given to those results with greater similarity, later these aspects will be selected to complete the final results. The most important specification is that each roof garden must participate in the seven tests. All the roof garden projects must be ranked using the result evaluation. According to the results after the tests, a final multiple matching evaluation needs to be completed.

The hybrid CA is established by the following indications:

a. the rights are established for customers to evaluate according to the needs of the organization;
b. a comparison list of the results of the test will be generated before a new cycle begins;

Green Roof Garden Concept

c. each evaluation will have all the roof gardens scheduled in a group of 17 tests;
d. the hybrid algorithm will be programmed to set the time for comparison using a complete match analysis to roof garden;
e. the roof gardens are qualified for selection based on a model which established the base to gives priority;
f. the organization for each roof garden needs to match for each round in the algorithm;
g. roof gardens must indicate their participation for their evaluation in each of the series. In case one of the roof gardens refuse to participate in the series, the algorithm can nominate a roof garden to be established as a replacement and this roof garden should be ranked among the best roof gardens in the concentrated list;
h. on the basis of an average calculation of two decimals, the qualification list in the series of comparisons, before starting a new cycle, three qualifiers will be selected (excluding the seven roof gardens that will be compared in the matches);
i. in case roof gardens have the same average rating, the number of similarities established for the match will be used to determine their classification;
j. to ensure active participation in the future, a minimum of 25 executions are recommended for the four ranking lists included and before the main classification list;
k. when a roof garden does not agree to play in a multiple combination series, then the selection process uses the average rating plus the number of runs played during the qualification period;
l. the algorithm repeats this process until reaching the required qualifiers of the series and the location of multiple coincidences for each roof garden and the real possibility of installation.

15.5.1 Mathematical Analysis by the Loads Exerted by the Construction Elements on the Roof

By means of the analysis of variables, the first equation that allows calculating the total weight that will have the roof garden is presented. A second equation is presented later whose improvement is a function of the accumulated precipitation for each cubic meter, where the units of kg/m^2 and that finally the expected result is expressed in kilograms (Guo et al. 2017).

$$Z = \frac{M*C}{D} \qquad (15.2)$$

TABLE 15.2

Value Relation Z (in Kg)

Minimum Z Values	Maximum Z Values
401.9874	871.9874
896.1584	1366.1584
1467.3032	1937.3032
2115.4216	2585.4216
2840.5137	3310.5137
3642.5795	4112.5795
4521.6189	4991.6189
5477.6321	5947.6321
6510.6189	6980.6189
7651.1842	8121.1842

$$Z = \frac{M*C}{D} + \Delta \gamma s \qquad (15.3)$$

Where:

M = Reinforced concrete slab

C = Live load analysis Wp

D = Specific weights of organic layer

γs = Difference of specific weights

Z = Total weight Roof garden.

15.5.2 Experiments

With a simulation of people, the result is presented in Table 15.2.

The above data refer to the possible results that the roof can have, for safety reasons the maximum Z values expressed in kilograms are taken whose last combination n exceeds 8 tons (Figure 15.2).

The results of the Z_{min} values correspond to Equation 15.1, which contemplates the specific weight of the dry organic layer (γ). Consequently, the values of Z_{max} correspond to the equation and the specific weight of the humid organic layer is taken into account, whose equation expresses the addition of the difference of the specific dry and wet weights. The resulting equation ($39.182x^2 + 372.73x + 463.97$) is a function whose result represents the total load, that can be on the top of the building contemplating the roof slab (WD), the average weight of people (WP), and the specific wet weight of the organic layer (Table 15.3).

Green Roof Garden Concept

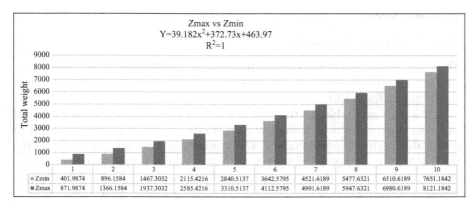

FIGURE 15.2
Minimum and maximum weight (w) comparison.

TABLE 15.3

Analysis of the Maximum Weight (W) According to the Roof Area (m²)

Surface of buildings house room and department		
m²	P	Support W_{max}
100	0.1	19000
200	0.2	38000
300	0.3	57000
400	0.4	76000
500	0.5	95000
600	0.6	114000
700	0.7	133000
800	0.8	152000
900	0.9	171000
1000	1	190000

Notes: According Building regulation of México City, it establishes the following living loads for Buildings:
Apartments and rooms in houses 190 Kg/m²
Meeting places with fixed seats 350 Kg/m²

The experiment results indicate that the main value occurs when the condition of not exceeding 95 tons is complete, for example: in the case of the Habitarea Towers in Juriquilla, Querétaro, which have an architectural design whose roof area is designed by the following dimensions: 35x14 m = 490 m², which works for P = 0.5 (Table 15.3).

15.6 Analysis of Results

The results reflected the information related to the growth of the loads in the roof gardens, the polynomial equation produce an increase in the data results described in Table 15.4.

In the previous table, in the first column (from left to right) the amount of m^3 of organic layer is shown, which is equivalent to the total weight of each value of the second column (Wt) expressed in kilograms (Kg). Similarly, the third column shows the number of users whose equivalences in Kg are expressed in the fourth column. The resulting equation is: 81.9x-1800y = 0, where: x = Number of users, y = m^3 organic layer (Table 15.5).

TABLE 15.4

Array With Final Organic Layer and the Live Load Expressed to Maximum Support

m^3	Wt (Kg)	Number of Users	Wp (Kg)	Maximum weight (Kg)
10.55556	19000	231.99023	19000	38000
21.11111	38000	463.98046	38000	76000
31.66667	57000	695.97070	57000	114000
42.22222	76000	927.96093	76000	152000
52.77778	95000	1159.95116	95000	190000
63.33333	114000	1391.94139	114000	228000
73.88889	133000	1623.93162	133000	266000
84.44444	152000	1855.92186	152000	304000
95	171000	2087.91209	171000	342000
105.55556	190000	2319.90232	190000	380000

TABLE 15.5

Balance of Variables x, y: 50% to 50%

Approximation	Organic Layer	Number of Users	Accumulated
-0.0040815	5.27778	115.995115	19000
0.0008370	10.555555	231.99023	38000
-0.0028350	15.833335	347.98535	57000
0.0020835	21.11111	463.980465	76000
-0.0019980	26.38889	579.97558	95000
0.0029205	31.666665	695.970695	114000
-0.0011610	36.944445	811.96581	133000
0.0041670	42.22222	927.96093	152000
0.0000855	47.5	1043.956045	171000
-0.0039960	52.77778	1159.95116	190000

Green Roof Garden Concept

TABLE 15.6
Equilibrium Coefficient to Find the Optimal Point

Equilibrium Coefficient	Rounding Down	
418.5000000	5	115
918.9000000	10	231
1419.3000000	15	347
119.7000000	21	463
620.1000000	26	579
1120.5000000	31	695
1620.9000000	36	811
321.3000000	42	927
821.7000000	47	1043
1322.1000000	52	1159

In Table 15.6, 50% of both the organic layer and the number of users is obtained with the purpose of achieving a balance between the variables and thereby obtaining the left column of approximations. The results of the left column represent the approximation to 0 that meets the equation $81.9x - 1800y = 0$; however, the kilograms of the organic layer and the number of users must be rounded to the nearest smaller integer for the purposes of real loads.

The equilibrium coefficient is obtained after having rounded the variables x and the nearest integer down. Then, applying the equation $81.9x - 1800y = 0$ corresponding to the number of users and the weight of the organic layer, the aforementioned coefficient is obtained as represented in Figure 15.3.

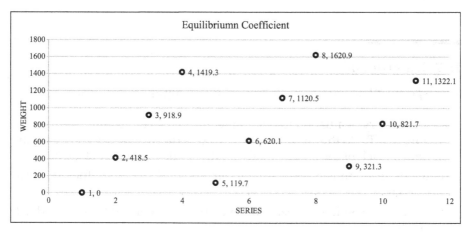

FIGURE 15.3
Dispersion diagram of equilibrium coefficient distribution.

TABLE 15.7

The Orthogonal Array Tests

Roof Garden Increase Index	Commercialization	Service	Recreation and Green Areas	Cost-Benefit	Equipment on the Roof Garden
4	1	2	2	3	4
3	1	2	2	3	3
2	1	3	2	4	4
5	1	3	2	5	2

Finally, in the lower part of the diagram the lowest point is marked with the number 4 which corresponds to the distribution coefficient 119.7 (Table 15.7). This indicates that the number of people that can be on the roof of a building is 463 (Table 15.6). To do this, the garden up to a maximum limit of 21 tons/m^3 (Table 15.7), because the weight of 463 people is 463 × (81.9) = 37,919.7 and 21 × (1800) = 37,800, the sum of the products is 75,719.7 Kg and does not exceed level 4 (Table 15.4). In Table 15.4 the area is in m^2, where 400 m^2 corresponds to 76,000 Kg.

As a result of the research and using Unity software for virtual reality, the prototype was designed according to the data obtained in the previous calculations, which is shown in the Figure 15.4 (the use of the software is for representative purposes only). The previous figure is a proposal of a roof slab with a roof garden and smart farming, whose area = 400 m^2, which can hold up to 463 people, which in essence is the optimal point sought.

15.7 Experiment

The experiment is done according to the following specifications:

a. satisfactory evaluation to obtain the most efficient arrangement of roof gardens. To do this, it is necessary to construct a cluster for keeping the data of each of the representative individuals for each roof garden.
b. the description is made with the purpose of distributing the best form of each.
c. the main experiment needs to have approximately 500 agents and 250 creations.
d. the stop condition is reached after 75 runs; this allowed generating the best selection of each kind and their possible location in a specific model.

Green Roof Garden Concept

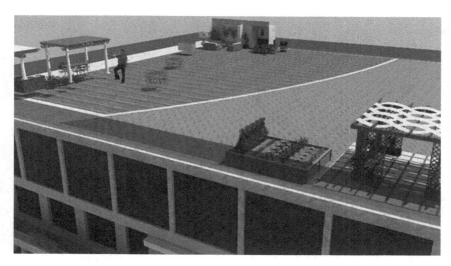

FIGURE 15.4
Roof garden proposed in a smart city including smart farming and organic layer.

Source: own elaboration, design using Unity.

- e. a location is obtained after comparing the different cultural and economic similarities of each roof garden and the evaluation of the multiple matching model (Ustaoglu 2009).
- f. the vector of weights employed for the fitness function, i.e. Wi = [0.6, 0.7, 0.8, 0.5, 0.6, 0.7], which respectively represents the importance of the particular attributes, i.e. roof garden increase index, commercialization, services, recreation and green areas, cost-benefit and equipment on the roof garden.
- g. the algorithm will choose the location of each roof garden based on attribute similarity.
- h. each attribute is represented by a discrete value from 0 to 7, where 0 means absence and 7 the highest value of the attribute.

The experiment design involves a series of steps:

- a. an orthogonal array test with interactions amongst the attribute variables; these variables are studied within a location range (1 to 400) specific to coordinates x and y.
- b. the orthogonal array is L-N (2–5), in other words, 6 times the N executions.
- c. the value of N is defined by the combination of the six possible values of the variables.
- d. the values in the location range.

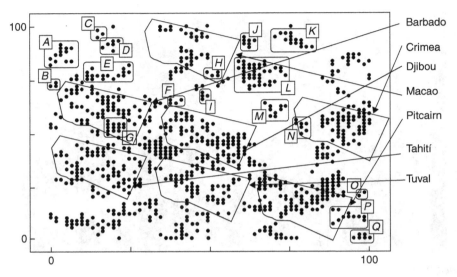

FIGURE 15.5
Location of roof garden installation in urbanization.
Source: own elaboration, design using Expedia.

Table 15.7 demonstrates the list of possible scenarios as the result of combining the values of the attributes and the specific location to represent a specific issue (roof garden).

The results of this table allow the analysis of the effect of the variables in the location, and the selection of all the possible combinations of values. The orthogonal array test and the array aids (Harrell 2001) specify the possibilities for adequate correct solutions (locations) for each roof garden. For the first data to be organized it is necessary to determine the essential attributes.

Different attributes were used to identify the real possibilities of improving a roof garden set in a particular environment and specify the correlations with other roof gardens with similarity necessities (Figure 15.5). The locations were chosen based on the orthogonal test array, because it provides representative (uniformly distributed) coverage of all variable pair combinations.

The clusters B, C, D, F, H, I, M, N, O, and Q are more related with people in families (brothers, sisters, uncles, aunts). The clusters A, E, G, J, K, L and P are clusters more related with social groups in a recreational roof garden. The x axis represents a sustainability model and the y axis represents a facilities fun model to interact with another people. This analysis is done using a dataset of 297 buildings with four levels and the possibility to adapt a roof garden is also analyzed.

The result is related to the behavior of the community with respect to an optimization problem (to culturally select five similar roof gardens). The descriptive information allowed identifying changes related to one or another

Green Roof Garden Concept

FIGURE 15.6
Roof garden final.
Source: own elaboration, design using Unity.

roof garden. The use of cultural algorithm increased the understanding the group behavior in various scenarios and with diverse attributes.

After the experiments it is possible to emphasize the importance of calculating the possible loads that can be allowed on the roof. That is why it is a priority to know the maximum number of people that can be occupied without compromising the structural safety of the building. To provide a better understanding of the model, virtual images can be offered with the proposal, to exemplify a proposal through the application of UNITY, as is shown in Figure 15.6.

In a building of 17 levels, 90 m in height and 20 m x 20 m dimensions, based on final results (400 m^2) it can support 463 people, even with a smart garden.

To prepare a proposal is necessary to calculate and review data related to the building where the roof garden is going to be built and specifically to emphasize the reinforced concrete elements such as beams and columns. As a last recommendation, it is important to review and be sure of the correct distribution of these elements to facilitate the development of the proposal where the live load is balanced with the organic layer (Figure 15.7).

FIGURE 15.7
Proposal of space distribution in a roof garden in Averanda complex, Cuernavaca, Morelos.
Source: photo taken by Ochoa in 2018.

The number of people at the same time must be carefully analyzed to avoid problems both with spacing and recreation, and not to affect the group of plant species in it. Waris and Reynolds (2018) suggested for different models until 27 different combinations of species that which may coexist between them.

Another future work is to collect samples of 77 buildings and analyze those that are under construction, or as it is also known, "projection", since these present characteristics that incorporate the category of intelligent buildings as resistant to earthquakes, fires, and with new loads, such as the installation of solar panels and intelligent control systems (Figure 15.8).

Here, the yellow points represent the buildings constructed, the blue points symbolize the buildings under construction and the red points project the buildings. In the city of Querétaro, there has been an even greater need for corporate offices and housing complexes that have manifested in the current vertical construction boom. There are 28 buildings exceeding 40 m in height and two of the highest are Juriquilla Towers B and Juriquilla Towers A, both with 30 floors, with a height of 116 m and 115 m respectively. There are 30 buildings under construction, where the highest is not strictly the one that has the most floors. The San José Moscati hospital is 130 meters high and 28 levels, while the High Park Corporate 1 is 92 meters high and 29 levels. Finally, there are 19 projected buildings. The Westin Querétaro Hotel is 170 m with a total of 40 floors, this being the tallest building the city has.

For the design of experiments (DOE) there are 77 data files and they will be indicated as A = constructed, B = under construction, C = projected to establish a null hypothesis and an alternative hypothesis. With this, a design

Green Roof Garden Concept 337

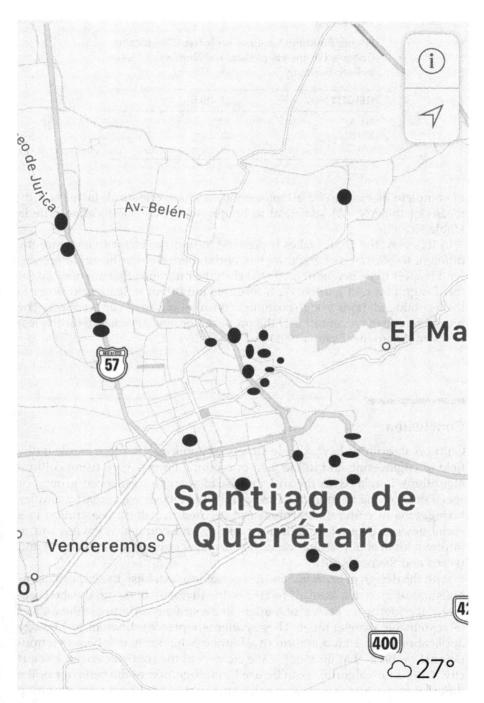

FIGURE 15.8
Map of representative buildings of Querétaro.

TABLE 15.8

Higher Buildings Grouped with Two Classification Criteria: For the Height and the Number of Floors in Each Building

HEIGHT (m)	LEVELS
116 A	30
130 B	28
170 C	40

of complete blocks can be established at random, one block factor and by means of the ANOVA statistical technique with two classification criteria (Table 15.8).

In this way, the DOE makes it possible to find the factor to compare the different levels that each building has, and if there are significant differences with respect to its height in order to select the buildings that are optimal for the design of a roof garden. As a result, is important to learn and listen to the recommendations trough customer consultations to obtain feedback. The results can be the acceptance of the proposed or new alternative for the rest of the roof gardens (Dennis 1991).

Conclusion

Cultural algorithms have a wide variety of applications, for example in the field of engineering and in the field of robotics. Future work using cultural algorithms is related to the distribution of workgroups, social groups or social networking to support diverse problems related with smart manufacturing. Finally, cultural algorithms can be used in pattern recognition in a social database, for example fashion styling and criminal behavior, and to improve models of distribution of goods and services (Guo et al. 2017) and (Waris and Reynolds 2018).

With the design of experiments, it is possible to establish the importance of holistic and systemic analysis related to the diversity of the established economical patterns for each roof garden, like a strategy to support the indices of sustainable development. These patterns represent different technology applications and a unique form of adaptive behavior that solves a computational problem that does not make clusters of the roof gardens in a smart city. The culture algorithm can be used in recognition of the behavior of the different proposals for roof garden design, to obtain the best results of a group of proposals and to find a balance between the variables that are involved in the distribution of the slab, otherwise they would become point loads and

Green Roof Garden Concept   339

FIGURE 15.9
Roof garden in a smart city (based on Google Images 2019).

would bring as consequences fracture points (the latter are analyzed in the diagrams at the time and cutting forces). With this work and the designs of the experiments it is achieved to have better ways of deciding on related proposals of the roof garden and make a support in the urban sustainability indexes in any city where they are applied.

Future Investigation

Further research is necessary to integrate additional data related to the structure of the building and adequate the design to each one – especially for new proposal related to smart farming (Figure 15.9).

Also, creating a group of proposals for roof gardens is recommended in order to allow understanding the similarities shared by the designs, the common characteristics and the diverse methods used to generate more sustainable proposals. Another recommendation is to apply the proposal of Vukčević, Ochoa and Djraguljovič (2005), which indicates that small variations are beyond phenotypic characteristics and are mainly associated with the tastes and related characteristics developed through the time.

Another recommendation has been made for an innovative roof garden in the city of Monterrey, Mexico. This case study presents various forms of use for the roof garden, proposing garden, business, economic and social activities (Figure 15.10).

FIGURE 15.10
Roof garden in Monterrey, NL, Mexico (based on Google Images 2019).

Acknowledgments

Special recognition to de designers, architectures and engineers who contributed for this research, also the designers of the proposal analysis were helpful completing the evaluations and experimentation of the roof gardens test.

References

Adhya, A, Plowright, P. and Stevens, J. 2010. *Defining Sustainable Urbanism: Towards a Responsive Urban Design*, Conference on Technology & Sustainability in the Built Environment. Accessed at: https://cap.ksu.edu.sa/sites/cap.ksu.edu.sa/files/attach/tsbe_1_e_02.pdf.

Brundtland, G. H. 1987. Our Common Future (Brundtland Report). United Nations World Commission on Environment and Development. Retrieved from: www.un-documents.net/our-common-future.pdf.

Dennis, Z. 1991. *Culture and Organizations*. McGraw-Hill: London.

Dourish, P. 2016. Algorithms and their Others: Algorithmic Culture in Context. *Big Data &Society* 3(2). Retrieved from: https://doi.org/10.1177/2053951716665128.

Gallopín, G. 2003. *Ciencia y tecnología para el desarrollo sostenible: una perspectiva latinoamericana y caribeña. Santiago de Chile.* Naciones Unidas, Comisión Económica Para América Latina y el Caribe, División de Desarrollo Sostenible y Asentamientos Humanos, y Taller Regional Latinoamericano y Caribeño sobre Ciencia y Tecnología para el Desarrollo Sostenible, eds. CEPAL, División de Desarrollo Sostenible y Asentamientos Humanos.

Germain, A., Grégoire, B., Hautecoeur, I., Ayalon, R. and Bergeron A. 2008. *Guide to Setting Up Your Own Edible Rooftop Garden. Alternatives and the Rooftop Garden Project.* Retrieved from: http://archives.rooftopgardens.ca/files/howto_EN_FINAL_lowres.pdf.

Google, 2019. *Imagines*. Retrieved from: www.google/images/roofgarden.com.

Guo, Y., Yang, Z., Wang, C. and Gong D. 2017. Cultural Particle Swarm Optimization Algorithms for Uncertain Multi-Objective Problems with Interval Parameters. *Natural Computing* 16(4): 527–548. Retrieved from: https://doi.org/10.1007/s11047-016-9556-3.

Harrell, J. 2001. Orthogonal Array Testing Strategy (OATS) Technique, 9. *Quality Assurance Manager*. Seilevel, Inc. Retrieved from: www.seilevel.com/OATS.html>.

Hernandez, S. 2008. Introducción al urbanismo sustentable o nuevo urbanismo. *Espacios Públicos* 11(23), 298–307. Accessed August 12, 2008: http://ri.uaemex.mx/handle/20.500.11799/39676.

Nolfi, S. and Floreano D. 2000. *Evolutionary Robotics: The Biology, Intelligence, and Technology of Self-Organizing Machines*. MIT Press. Retrieved from: https://doi.org/10.1162/106454601317297031.

Oxford Dictionaries. Algorithm. Definition of Algorithm in English. Oxford Dictionaries English. Accessed January 18, 2019: https://en.oxforddictionaries.com/definition/algorithm.

Ponce, J. and Karahoc A. 2009. Data Mining and Knowledge Discovery in Real Life Applications. I-Tech Education and Publishing. KG, Vienna, Austria. Retrieved from: https://doi.org/10.5772/97.

Reynolds, R. 1994. *Introduction to Cultural Algorithms, in Proceedings of the Third Annual Conference on Evolutionary Programming*, Anthony V. Sebald and Lawrence J. Fogel (eds.) .World Scientific Press: Singapore, 131–139.

Seaver, N. 2017. Algorithms as Culture: Some Tactics for the Ethnography of Algorithmic Systems. *Big Data & Society* 4(2). Retrieved from: https://doi.org/10.1177/2053951717738104.

Striphas, T. 2015. Algorithmic Culture. *European Journal of Cultural Studies* 18(4–5): 395–412. Retrieved from: https://doi.org/10.1177/1367549415577392.

Ustaoglu, Y. 2009. *Simulating the Behavior of a Minority in Turkish Society*. Paper presented at the ASNA 2009, Zurich, Switzerland.

Vukčević, I., Ochoa, A. and DJraguljovič H. 2005. *Similar Cultural Relationships in Montenegro*. WKDD'2007, England.

Waris, F. and Reynolds R. 2018. Optimizing AI Pipelines: A Game-Theoretic Cultural Algorithms Approach. In *018 IEEE Congress on Evolutionary Computation* (CEC), 1–10. Retrieved from: https://doi.org/10.1109/CEC.2018.8477820.

Yan, X., Wu, Q., Zhang, C., Chen, W., Luo, W. and Li W. 2012. An Efficient Function Optimization Algorithm Based on Culture Evolution. *International Journal of Computer Science Issues*, IJCSI 9 (5): 8.

16

School Bus Routing Problems and Solutions for Smart Cities – A Case Study

Carlos Alberto Ochoa and Aida Yarira Reyes Escalante

CONTENTS

16.1 Introduction 343
16.2 School Bus Routing Problem (SBRP)...345
16.3 Particle Swarm Optimization (PSO)..348
16.4 Methodology..349
    16.4.1 Noise Definition 349
    16.4.2 Data Preparation 351
16.5 Results and Discussions...351
    16.5.1 Graph Creation...353
    16.5.2 Operation of the Program..353
Conclusion 354
Suggestions for Future Research...356
References...357

16.1 Introduction

Cities today present important challenges in terms of growth and diversification of problems. One of the most serious problems related to the smart cities is the mobility of its citizens. Here transport can play an important role as it can mobilize large numbers of citizens from one place to another but the times used in the journeys each day are longer and more conflicting.

    The big cities in the world are organizing various programs to encourage new strategies of mobility by increasing the number of vehicles but this increase of vehicles further increases the pollution level. The most common means of transport are private vehicles, private services and public services.

    Smart cities look for people living in urban spaces to be the direct beneficiaries of new technologies and seek to incorporate information and communication technologies (ICT) for their utility. The incorporation of these technologies and tools make the services efficient. Also, with these new technologies and techniques the smart cities can become innovative, competitive, attractive and resilient, thus improving human lives. Technological development is progressing, and alleviating the problems of daily life are a constant task: there are currently an endless number of applications that seek to improve the activities of human beings, and many of them have better information systems, cost reductions, time reductions, better support teams, aspects of health, and so on.

Modern school life has a host of problems that must be addressed to offer a better education. The new education strategies seek to integrate with new technologies, that is, ICT within their classrooms which ultimately provides a better approach with students, improve the relationship between administration, teachers and students.

School life is not limited to the spaces of the classroom, it is constructed from a great variety of areas and moments, for example, relations of classmates; class assignments at school and outside of school; recreational activities; family life; virtual life; travel times and so on. According to population growth, urban areas are farther away, moving from one zone to another is taking longer time. Any trip in the city is via congested avenues due to the increasing number of vehicles and the number of stoplights which cause slow moving traffic and accessibility problems.

Vehicle routing problem (VRP) is an optimization issue that has been used for approximately 50 years for its benefits and applicability, and even today it is still used in the search for better routes of mobility (Lewis et al. 2018). Concerning the problem of transfers within the town, Gakenheimer (1997) indicated that there were disturbing records that accessibility between places had already disappeared. Mobility problems have been studied in more detail since 1998 when the problems caused lost times and important effects in the economic, environmental and social areas, examples of this problem are presented below

- Rio de Janeiro, Brazil, has recorded taking up to 90 minutes to get from one place in the city to another;
- Bogotá, Colombia, where a person can take 60 minutes to move from a place to another;
- Manila, with speeds of seven miles per hour due to vehicular congestion; and
- Bangkok, which, due to the amount of traffic, there were stoppages recorded that, on average, is equivalent to 44 days lost in traffic jams in a year.

Approaches to solving school-related transportation problems are insufficient This is because many of the school districts are the proprietors of the school buses, and one of the big dilemmas is how to manage them efficiently and effectively, which means establishing specific schedules (the time school begins and ends, which can be changed); and optimizing the resources is another determining factor (both the bus and the driver). One of the main problems with school transportation resides in the small location and management of bus stops. This mismanagement of bus routes is reflected in the high maintenance costs of school buses and prolonged time students spend on buses while being taken to school or back home. The primary objective of this chapter is to demonstrate and explain the identification of efficient school bus routes utilizing the particle swarm optimization (PSO) method.

Among the strategies of school service planning, there are two key elements: the route and the scheduling of the service. To resolve this scenario, we combine both elements, as it is important to look for the best alternatives in the complexity of the situations presented, because both are a priority and there needs to be a balance.

It is very important to understand that in 87 societies around the world of different sizes, including Yakutia, the Brazilian state of Minas Gerais and Macau, they want to solve their traffic problems, and the most relevant is to be able to solve the school transport problems.

The following questions were considered for this study:

1. How do you assign a bus stop to a student?
2. How many stops would be required for all students?
3. Are the selected stops relevant?
4. How can the best route be chosen?
5. Are there any alternate routes?
6. How will these decisions affect alternative paths?
7. What will be taken into consideration when deciding on a route ?
8. Does the school bus meet the set time?
9. How do you generate the planning of routes?

Based on the previous problems and questions, the purpose of this chapter is to develop a homogeneous multimodal system to solve the problem of ensuring efficient travel while minimizing waiting times. This plan incorporates trip compatibility, covering both the scheduling problem and the routing problem.

16.2 School Bus Routing Problem (SBRP)

This work originates at the Federal University of Technology Akure / Computer Science Department, Akure (The Federal University of Technology Akure 2019) and different studies related to school bus routing problem

(SBRP) (Shafahi, Wang and Haghani 2017) where they proposed a variation on the vehicle routing problem (VRP). The SBRP is a problem-solving model that involves the route that will follow a limited number of school buses along scheduled routes that are defined in an efficient and optimal way. The SBRP can be divided in five sub-problems (Kang et al. 2015):

a. data preparation: this defines the data required to set up stop allocation, the route definition and the scheduling;
b. selection of stops: this step consists in assigning a stop to each of the students who are users of the buses, which must fulfill the objective of being located closest to where they live;
c. route generation: this covers the decision regarding the order in which the school bus will visit each stop;
d. arrival time: this covers setting the time the student must arrive at school; and
e. schedule of route: this deals with the allocation of school bus routes covering best fit, taking into consideration the time for which students must arrive at school while establishing plans to allow each school bus to transport students from the sector in which they live to their school.

The representation of the problem-solving process based on the SBRP in its five most important moments is presented in Figure 16.1.

The SBRP consists of at least two smaller sub-problems (Díaz-Parra et al., 2012):

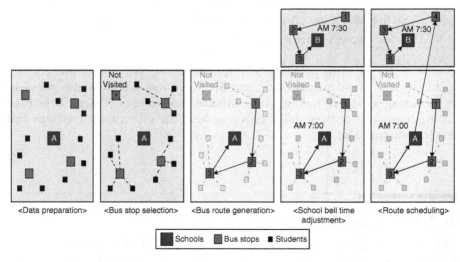

FIGURE 16.1
Representation of problems of school bus algorithm (Kang et al. 2015).

a. the selection of the appropriate bus stop, called "Bus Stop Selection", seeks to select a set of bus stops and assign students to these stops; and
b. "Bus Route Generation" is modeled and implemented to generate the necessary bus routes for a single school.

Furthermore, they established a series of restrictions related to the SBRP (Kang et al. 2015):

a. bus capacity: the maximum number of students that the bus may contain;
b. maximum travel time: the maximum time a student can stay in the bus;
c. time window: the time a bus takes to arrive;
d. homogenous capacity: buses with different capacities;
e. mixed load: different students from different school;
f. knowing the schedules of schools to maximize the number of routes by bus and to reduce the number of buses used for a multi-school system in a school district;
g. the programming of a specific route;
h. the start and end of each route of a bus for a multi-school system.

The results of different studies were indicated by Díaz-Parra (2012) and established seven main characteristics of the SBRP which all models must meet:

a. number of schools (one or multiple)
b. service environment (urban or rural)
c. scope of the problem (morning, afternoon, both)
d. multiple load
e. fleet combination objectives (number of buses used, total distance traveled by bus time taken)
f. total distance, distance of student walks, load balance
g. walk time or distance, earliest pickup time and the minimum number of students to create a route
h. The mathematical model.

In our revision of the literature, we found that the first users of the model of the SBRP date from 1974, when they tried to offer solutions for a problem where the institute had a single school bus and 80 stops. The aim was to propose an algorithm to solve a problem in a district, by the location of each school, the location of each student, the period of use of transportation by the student and the availability of buses (Díaz-Parra 2012). After this

TABLE 16.1

Application of School Bus Routing Problem (SBRP)

Year	Title and References
2018	Using the Metaheuristic Methods for Real-time Optimization of Dynamic School Bus Routing Problem and an Application (Yigit, Unsal and Deperlioglu 2018)
2018	A Heuristic Algorithm for Finding Cost-effective Solutions to Real-world School Bus Routing Problems (Lewis and Smith-Miles 2018)
2018	A Bounded Formulation for the School Bus Scheduling Problem (Zeng, Chopra and Smilowitz 2018)
2018	The School Bus Routing Problem: An Analysis and Algorithm (Lewis, Smith-Miles and Phillips 2018)
2017	IoT-based School Bus Tracking and Arrival Time Prediction (Jisha, Jyothindranath and Kumary 2017)

publication, the applications of the SBRP have become more diverse and they have sought to solve a great diversity of problems. The new contributions to solving problems establishes the state-of-the-art of the investigations carried out using SBRP and indicates the benefits found. A chronological revision of issues related with this research is presented in Table 16.1.

New applications of the algorithm are made for the analysis of various topics, and one of them is to look for the safety of students en route to school, monitor arrival times, improve costs, reduce travel time and increase safety.

16.3 Particle Swarm Optimization (PSO)

PSO was presented for the first time in 1995 in order to create a graphic simulation in which schools are shown a choreography of fish or birds, to show the flows of robust synchronization and coordinated in different aspects (Kennedy and Spears 1998). This application allowed visualization of how the trajectories of the schools and their vectors are seen. The original PSO tests make use of a velocity vector to update the current position of each particle in the swarm. In other population studies it is indicated that the location of each particle is updated based on the social behavior of a population of individuals, the multitude in the case of PSO, which adapts to its environment by returning to promising regions that were previously discovered (Kennedy and Spears 1998). Members of a swarm do not know the behavior of the multitude itself, much less the environment in which they operate. However, the local expression of individuals is what defines the general conduct of the swarm. With this in mind, the PSO algorithm comprises a set of members or particles representing a point in a multidimensional space.

The population studies allow exploring behaviors in defined spaces where it is sought to explain an ideal position, previously defined in the function. Since PSO is oriented to a multidimensional space, it can be adapted to a two-dimensional problem proposed in SBRP, because this case must consider the bus ground displacement (Grosan, Abraham and Chis 2006). Also mentioned, the particles in the swarm explore the problem of space, looking for an optimal position, which is defined by a fitness function. The location of each particle depends on its own experience and that of their peers (other particles), therefore, each particle must have its memory. In addition, this memory saves the best position found, and this position is called personal best or Pbest. The results allow establishing generalized behaviors, such as:

a. homogeneity: the flock moves without a leader, even though temporary leaders seem to appear;
b. locality: the motion of each bird is only influenced by its nearest flock mates. Vision is considered the most important sense for flock organization;
c. collision avoidance;
d. velocity matching: attempt to match velocity with nearby flock mates;
e. flock centering: attempt to stay close to nearby flock mates.

16.4 Methodology

The area of study of our research was in El Paso, Texas. In these cities, the use of transportation is essential for the mobility of students to their schools which ultimately provides benefits to the students.

The city of El Paso, Texas borders with Mexico and adjoins the state of New Mexico. The movements of the school bus are constant because the city is large, and the school trips are by zones. The idea is obviously to transport the maximum number of students with minimum delays and so intelligence to create the best route is key. The transport service is important as it reduces travel cost and volume of vehicles on the road. To minimize the lost time at the school, it is a priority for the bus to arrive on time. In this line, the creation of routing is an appropriate strategy: more students traveling a minimal distance.

16.4.1 Noise Definition

Another factor that must be taken into consideration is "noise". This data set presented extra considerations while defining the functionality of the PSO algorithm in its early stages. Some unexpected data points appeared and were classified as:

a. traffic: defined by common routes that people take to go to their jobs or other places; it should be ideal to identify these routes and try to avoid them.
b. special zones: defined by roads with special restrictions that involve speed limits. For example, hospitals, parks, railroad crossing, markets, places that cannot be accessed by driving or need to be accessed at a reduced speed or frequent stops need to be made.
c. roads with problems: roads being repaired or those that are damaged.

The following assumptions were also made as part of this investigation:

a. school buses are located at the center of a plane and they have to pick up students waiting at pickup points to be brought to school.
b. the number of students waiting at the collection point

$$i \; is \; defined: q_i, (q_i > 0, i = 1, 2, ..., n) \quad (16.1)$$

Where

I = student in a school bus algorithm
Q = bus school in our model

c. The capacity of each bus is limited to a certain number of students.

$$Q \; (q_i \; 0k) \quad (16.2)$$

Where:

Q = capability of school bus
q_i = school bus with students
k = implements associated with student activities and carried by themselves.

The objective function of the school bus problem is composed of two costs:

a. cost incurred for the number of buses used, and
b. driving cost (fuel, maintenance, driver's salary and others), subject to operational restrictions, cost must be minimized.

16.4.2 Data Preparation

The study data is presented in Figures 16.2, 16.3 and 16.4, obtained using Google Maps (2019), covering main streets such as Fort Blvd., Dyer St., Pierce Ave., and Piedras St. The attendance depicted is for Rusk Elementary School.

A graph was created so each street's data could be stored. For the data recollection, the indicated sites where the students took the school bus were visited and the waiting times were measured.

16.5 Results and Discussions

According to the strategies established for answering the questions, the application of the methodology allowed to find the following results:

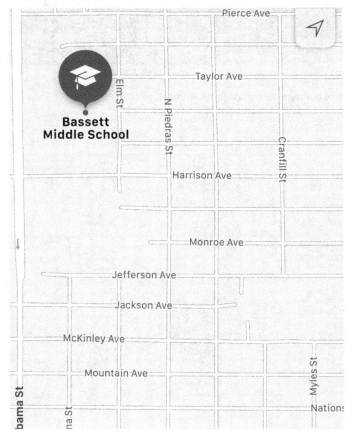

FIGURE 16.2
Area considered for the study (Google Maps 2019).

FIGURE 16.3
Bounded map of the study area (Google Maps 2019).

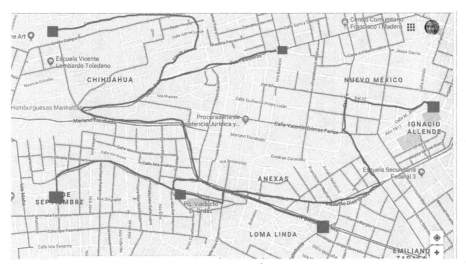

FIGURE 16.4
Area of study in metric system.

16.5.1 Graph Creation

The nodes represent the street and the set of properties are identifications that distinguishes a node from another.

a. initial point: this is the initial position of the node;
b. final point: this is the final position of the node;
c. direction: West/South = 1, East/North = 2, Two ways = 0;
d. speed: the maximum speed in the street (m/h).

For the application of the algorithm, various activities carried out as described below:

a. image preparation: the image is a representation of the map;
b. then the user will click on the map to create a new node;
c. the system will then ask the initial and final points, directions and speed;
d. if the user doesn't finish the creation of all nodes in the map, repeat step b.

16.5.2 Operation of the Program

The proposed program works on an acquired image of Google Maps® (2019) in order to define the information required for the generation of the SBRP route. The first test consisted in the run of several destination points. For reasons of operation and subsequent increase in the complexity of the sectors where the stops were located, it sought to provide greater security. It is possible to define each of the points of the route as to where to have a bus stop based on the number of students that use each of the stops. This is done according to their location closest to their home.

Figure 16.5 shows the points of the bus stop of route 5. Figure 16.6 represents the route created using the bus stops, and Figure 16.7 shows the best route proposal.

The result of the run allows it to create the best route, and this route will enable it to define the time at which the school bus must present at each stop. As a result of watching time, it is determined that the approximate time for each stop was 5 minutes. It is essential to indicate that the number of students is different at each stop and the number of students is also different for each school so the number of buses required differs. A critical point is that the bus must reach its destination with sufficient time for the students to get off the bus.

Another critical aspect for the bus is the travel speed limit, which means that each bus must move at a maximum speed of 30 mph. Each of the routes involved in the study was monitored, and the data was assessed to generate

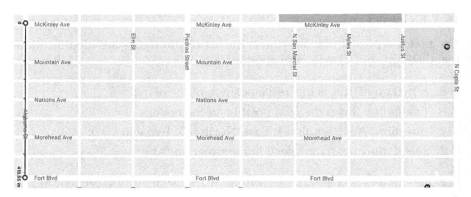

FIGURE 16.5
Creating street graph.

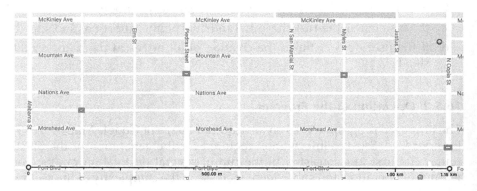

FIGURE 16.6
Particle search.

route maps and waiting times. In Figure 16.8 the image of one of the routes carried out in the study is shown.

Conclusion

With this research, it can be identified that the use of genetic algorithms applied to SBRP can provide a more efficient plan for students to arrive at their school taking an optimal route during a tour based on a set of restrictions. Taking an optimal route implies a shorter path and in the shortest possible time. This type of genetic algorithm can help at people to solve problems associated with the school bus routes. Different types of algorithms were analyzed to see which of them could offer the most

School Bus Routing Problems 355

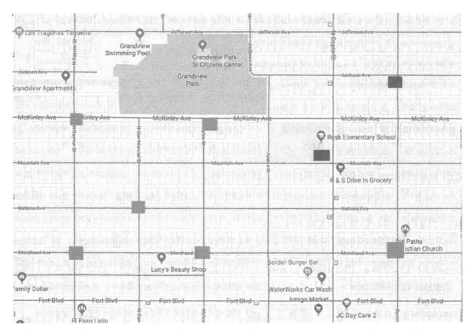

FIGURE 16.7
Route definition.

FIGURE 16.8
Visualization of school bus route in real-time in a school district in El Paso.

optimal solution, so that school districts can make more efficient use of their limited resources, including school routes to be covered by a limited group of buses. It is worth mentioning that in each case study various characteristics of complex problems are presented and a different set of circumstances that affect them are presented, so the type of solutions that can be generated with this algorithm will be different for each case and situation in particular. Each case will require a different fit solution and consideration of the assumptions. This implies that although this is an optimal solution for a specific area, for another it will have to analyse the whole panorama and change all specifications and, probably, use a different type of algorithm. In the application of this type of solution, there is an enormous amount of work to be done in the future. One suggestion would be to analyse the same school bus route with another genetic algorithm and make a comparison of the results.

The use of the school bus algorithm improves the response times associated with transporting children in various societies around the world, regardless of the size of the society, such as Macao, the smallest, and Yakutia, the largest. The importance of these studies is based on the creation of new strategies to produce routes, seeking to improve sections of urban mobility within large cities, mainly those that have high-altitude areas and where large schools are located. These new routes allow internal traffic expense to become less expensive, the number of cars in circulation is reduced, long waiting times are reduced and pollutants are reduced from the number of vehicles circulating.

Suggestions for Future Research

From the results of this study, there are some following suggestions for future research:

a. automate the generation of sectors;
b. the sectors should be made in relation to the density of students per stop;
c. the number of buses available needs to be considered; and
d. the possibility to modify the location of the stops.

A key aspect of future research is to determine adequately when additional equipment must be brought to the school, and this entails the redistribution of spaces within the bus and time considerations, so that everything is sought in real-time, through the use of ubiquitous computing and telemetry.

References

Abraham, A., Grosan, C. and Ramos, V. (eds.) 2006. *Swarm Intelligence in Data Mining. Studies in Computational Intelligence*. Springer-Verlag: Berlin Heidelberg. Retrieved from: www.springer.com/gp/book/9783540349556.

Díaz-Parra, O., Ruiz-Vanoye, J. A., Buenabad-Arias, Á. and Cocón F. 2012. A Vertical Transfer Algorithm for the School Bus Routing Problem. *2012 Fourth World Congress on Nature and Biologically Inspired Computing (NaBIC)*, 66–71. Retrieved from: https://doi.org/10.1109/NaBIC.2012.6402241.

Gakenheimer R. 1999. Urban Mobility in the Developing World. *Transportation Research Part A: Policy and Practice* 33(7–8): 671–689.

Google Maps. 2019. Retrieved from: www.google.com/maps/@31.8091274,-106.4543433,189m/data=!3m1!1e3.

Jisha, R. C., Jyothindranath, A. and Sajitha Kumary L. 2017. IoT-based School Bus Tracking and Arrival Time Prediction. *Proceedings of the IEEE International Conference on Advances in Computing Communications and Informatics (ICACCI)*, 509–514. Retrieved from: https://doi.org/10.1109/icacci.2017.8125890.

Kang, M., Kim, S., Felan, J. T., Rim Choi, H. and Cho M. 2015. Development of a Genetic Algorithm for the School Bus Routing Problem. *International Journal of Software Engineering and Its Applications* 9(5): 107–126.

Kennedy, J. and Spears W. M. 1998. Matching Algorithms to Problems: An Experimental Test of the Particle Swarm and Some Genetic Algorithms on the Multimodal Problem Generator. In *1998 IEEE International Conference on Evolutionary Computation Proceedings. IEEE World Congress on Computational Intelligence (Cat. No.98TH8360)*, 78–83. Anchorage, AK: IEEE. Retrieved from: https://doi.org/10.1109/ICEC.1998.699326.

Lewis, R. and Smith-Miles K. 2018. A Heuristic Algorithm for Finding Cost-effective Solutions to Real-world School Bus Routing Problems. *Journal of Discrete Algorithms*, Combinatorial Algorithms – Special Issue Devoted to Life and Work of Mirka Miller, 52–53: 2–17. Retrieved from: https://doi.org/10.1016/j.jda.2018.11.001.

Lewis, R., Smith-Miles, K. and Phillips K. 2018. The School Bus Routing Problem: An Analysis and Algorithm. In *Combinatorial Algorithms*, edited by L. Brankovic, J. Ryan and W. F. Smyth. Volume 10765. Springer International Publishing: Cham. Retrieved from: https://doi.org/10.1007/978-3-319-78825-8_24.

Shafahi, A., Wang, Z. and Haghani A. 2017. Solving the School Bus Routing Problem by Maximizing Trip Compatibility. *Transportation Research Record* 2667(1): 17–27. Retrieved from: https://doi.org/10.3141/2667-03.

The Federal University of Technology Akure. 2019. *Computer Science*. Retrieved from: http://csc.futa.edu.ng/.

Yigit, T., Unsal, O. and Deperlioglu O. 2018. Using the Metaheuristic Methods for Real-time Optimization of Dynamic School Bus Routing Problem and an Application. *International Journal of Bio-Inspired Computation* 11(2): 123. Retrieved from: https://doi.org/10.1504/IJBIC.2018.091236.

Zeng, L., Chopra, S. and Smilowitz K. 2018. A Bounded Formulation for the School Bus Scheduling Problem. Accessed on March 2019: https://arxiv.org/abs/1803.09040v1.

Index

A
Advanced metering infrastructure, 209
Aggregation, 282
Air quality monitoring system 265
Anaerobic digestion, 252
Anthropogenic, 23, 25, 26
   pressure, 26
   watercourses, 25
Aquatic, 19, 24, 25, 38
   species, 24
   habitats, 25
Automobile, 23

B
Big data, 16, 135, 278
Biogas, 253
Biomedical signal processing, 104
Bootstrap phase, 272
Buddhist, 132
Building envelopes, 30, 51

C
Capillary network, 264
Carbon footprint, 38
Cloud architecture, 11
Cloud computing, 6, 14
Climatic adaptation, 27
Climate smart initiative, 190
Clinical information system, 109
Cloud computing, 135, 137
Cognitive, 285
Comprehensive action plan, 35
Community services layer, 267
Community infrastructure layer, 268
Cool roofs, 54
   effectiveness, 55
   types, 55–56
Cultural algorithms, 323, 324, 338
Cyber- crime, 5
   security, 5
   safety, 5
Cyber physical system, 279

D
Data access mechanism, 283
Data analytics, 135, 138
Data link layer, 280
Data processing and analysis, 282
Demolition, 20
Deployment stages, 263, 272
Development strategies, 29
Design technologies
   active, 65
   passive, 51
Diesel- powered vehicles, 27
Digital democracy, 28
Digital healthcare, 107
Draft action plan, 196

E
E-Books, 135, 141
Eco-
   friendly sidewalks, 31
   logical restoration, 37
Economic
   feasibility, 28
   productivity, 24
Educational data mining, 139
Education methodology, 132
Efficient, 189–190
   mechanical design, 189
   resistance, lubrication, lightweight materials, 189–190
Electronic health record, 111
E-learning management system, 134
Energy
   consumption, 29
   dispersion, 30
Environmental
   challenges, 34
   degradation, 34
Experiments, 328, 335, 336, 339

F
Frankfurt, 26
Flipped classroom, 136, 140, 143
Freiburg, 32

G
Geospatial analytic tools, 25
Green
 smart town planning, 20
 street and transportation, 31
Green communication system
 infrastructure, 10
 sky parks, 11
Green computing, 133, 142, 143
Green healthcare, 91
Green living hospital roof, 96
Green smart building
 principles, 49–51
 standards, 67–68
Green smart city Dubai, 195
Green transportation, 186
Green vehicles, 192–193, 195
 benefits, challenges, 192, 193
 light and heavy vehicles, 193

H
Healthcare, 7
Healthy lifestyle, 7
Heating
 solutions, 30
 systems, 30
Higher building, 23
Horizon 2020, 190
Hydrographic elements, 20
Hyperloop technology, 200

I
Influential factors, 28
In-vivo networking, 45
Instructor-led training, 133
Intelligent transport systems Korea, 196–197
Internet of things, 294, 300–304, 309–310, 315
International standard organizations, 270
Impervious materials, 31
Iron Triangle of green healthcare, 93
Istanbul Electric Tramway and Tunnel Establishments, 307

K
Kardia Band, 119
Knowledge based IDSS, 124

L
Landscape architecture, 28
Legislative and regulatory healthcare waste policies, 99
Low- carbon, 34
Low- energy
 structures, 23
 solutions, 23

M
Mirror ducts, 63
 benefits, 63–64
M-learning, 134
Mobile, 5
 communication, 5
 learning, 6
 network, 264
Mobility, 5
 of device, 6
 of user, 6
MOOC, 132
Multi-criteria decision-making, 314
 AHP, 314
 TOPSIS, 314
Multifunctional centres, 31

N
Natural, 58
 orientation & massing, 59
 stack ventilation, 60–61
Natural resources, 187–188
Network layer, 281
Noise monitoring systems, 265
Non-governmental organizations, 296, 304–305, 309, 311, 313
 living labs, 313
 technoparks, 313
Non-functional opportunities, 263

O
Open governance, 28
Optima MR, 360, 103
Ottawa, 26

P
Particle Swarm Optimization, 345, 348
Patient Safety, 112
Personalized health, 114
Personal area network, 3

Pioneering efforts, 27
Planning
  professionals, 28
  strategies, 25
  structures, 28
Political viability, 28
Pollution causing, 20
Policy Instruments, 28

R
RAPAEL smart glove, 118
Real-time telehealth, 113
Real estate, 7
  modern society, 7
  trade, 7
Renewable
  resources, 23
  energy sources, 36
Remote monitoring, 116
Remote sensing, 24

S
Smart blood sugar, 118
Smart board, 133, 138
Smart city, 4, 5, 294–300, 303–315
  transportation, 5
  technical innovation, 6
Smart living, 299
Smart learner, 136, 137, 139
Smart mobility, 299
Smart sleep Aids, 117
Smart temperature, 118
Smart tourism, 294–296, 299–304, 309, 312–314
Smart web-enabled mechanisms, 297
Sustainable
  ecosystem, 20
  economic development, 24
  public transport system, 32
  solutions, 23, 29
  waste management practices, 36

T
Task scheduling mechanism, 172
Technology, 322
  applications, 338
Telemedicine, 112
Thermal contrasts, 31
Thermal insulation, 23
Topsoil, 85

Transit systems, 36
Transmission power control, 170
Transport networks, 20
Transportation, 7
  emergency vehicles, 8
  multi-layer system, 8
Two factor authentications, 179

U
Ubiquitous, 142
Ultrasound, 102
U-learning, 134
Urbanization, 5
Urbanized regions, 19
Urban
  aesthetics, 26
  development, 19
  planning, 20
  recreational areas, 25
  spaces, 19, 34, 37

V
Vertical Forest, 31
Vigilant monitoring, 175
Vienna
  woods, 26
Vienna Climate Protection Programme (KliP), 26

W
Waste management, 20
Wastewater treatment, 27
Water
  deficiency, 83
  pollution, 83
Waterless urinal system, 249
Wetlands, 35
Wearable technology, 135
World Health Organization (WHO), 76
Wide area network, 3
Wireless technologies, 12
Wings walls, 60

X
X-ray, 101

Z
Zero Energy Buildings, 45
Zemin Istanbul, 311, 313